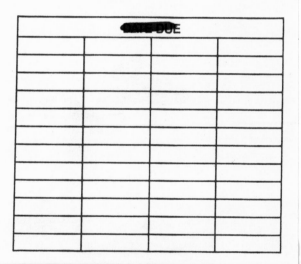

FRONTIERS IN QUANTUM PHYSICS RESEARCH

Frontiers in Quantum Physics Research

Frank Columbus
Volodymyr Krasnoholovets
Editors

Nova Science Publishers, Inc.
New York

Production Coordinator: Donna Dennis
Coordinating Editor: Tatiana Shohov
Senior Production Editors: Susan Boriotti and Donna Dennis
Production Editors: Marius Andronie and Rusudan Razmadze
Office Manager: Annette Hellinger
Graphics: Magdalena Nuñez
Editorial Production: Maya Columbus, Aloysha Klestov, Vladimir Klestov, Matthew Kozlowski and Lorna Loperfido
Circulation: Ave Maria Gonzalez, Vera Popovic, Luis Aviles, Jonathan Boriotti, Alexandra Columbus, Raymond Davis, Cathy DeGregory, Melissa Diaz, Marlene Nuñez, Jeannie Pappas and Frankie Punger
Communications and Acquisitions: Serge P. Shohov

Library of Congress Cataloging-in-Publication Data

p. cm.
Includes index.
ISBN 1-59454-002-0 (hardcover)
1. Quantum theory--Research. I. Krasnoholovets, Volodymyr. II. Columbus, Frank H.

400 Oser Ave, Suite 1600
Hauppauge, New York 11788-3619
Tele. 631-231-7269 Fax 631-231-8175
e-mail: Novascience@earthlink.net
Web Site: http://www.novapublishers.com

Printed in the United States of America

CONTENTS

Preface

Frank Columbus and Volodymyr Krasnoholovets

This new book examines new research in the exploding field of quantum physics. The forefront of contemporary advances in physics lies in the submicroscopic regime, whether it be in atomic, nuclear, condensed-matter, plasma, or particle physics, or in quantum optics, or even in the study of stellar structure. All are based upon quantum theory (i.e., quantum mechanics and quantum field theory) and relativity, which together form the theoretical foundations of modern physics. Many physical quantities whose classical counterparts vary continuously over a range of possible values are in quantum theory constrained to have discontinuous, or discrete, values. The intrinsically deterministic character of classical physics is replaced in quantum theory by intrinsic uncertainty. According to quantum theory, electromagnetic radiation does not always consist of continuous waves; instead it must be viewed under some circumstances as a collection of particle-like photons, the energy and momentum of each being directly proportional to its frequency (or inversely proportional to its wavelength, the photons still possessing some wavelike characteristics).

Dirac's hole theory and quantum field theory are generally thought to be equivalent. In fact field theory can be derived from hole theory through the process of second quantization. However, it can be shown that problems worked in both theories yield different results. The reason for the difference between the two theories will be examined and the effect that this difference has on the way calculations are done in quantum theory will be discussed in the first chapter.

Physical fields and solutions of relativistic wave equations are studied in terms of the functions on the Lorentz group. The principal series of unitary representations of the Lorentz group has been considered in the helicity basis. It is shown that in the helicity basis matrix elements of irreducible representations are expressed via hyperspherical functions. A relation between the hyperspherical functions and other special functions is given. The general Gel'fand-Yaglom relativistically invariant system is defined on a tangent bundle of the Lorentz group manifold. V. V. Varlamov has shown that a 2-dimensional complex sphere is associated with the each point of the group manifold. Solutions of the relativistically invariant system have been found in the form of expansions in the hyperspherical functions defined on the surface of the complex sphere. By way of example, solutions of the Dirac equation are given in terms of the functions on the Lorentz group. Further applications of the proposed approach to the theory of composite particles are discussed.

The Bivacuum model is a consequence of new interpretation of Dirac theory, pointing to equal probability of positive and negative energy. Unified Model (UM) represents the efforts of Alex Kaivarainen for unification of vacuum, matter and fields from few ground postulates. A new concept of Bivacuum is introduced, as a dynamic superfluid matrix of the Universe, composed from non mixing sub-quantum particles of the opposite energies, separated by energy gap. Their collective excitations form mesoscopic vortical structures. These structures, named Bivacuum fermions and antifermions, are presented by infinitive number of double cells-dipoles, each cell containing a pair of correlated rotors and antirotors: V(+) and V(-) of the opposite quantized energy, virtual mass, charge and magnetic moments. The absolute values of rotors and antirotors internal rotational kinetic energy and magnetic moments are postulated to be equal and permanent, in contrast to their virtual mass and internal angle velocities. These dipoles of three generation (e mu, tau), due to very small external zero-point translational momentum, form macroscopic virtual Bose condensate with nonlocal properties. The matter origination in form of sub-elementary fermions or antifermions of three generation is a result of Bivacuum dipoles symmetry shift towards the positive or negative energy, correspondingly.

The asymmetry of velocity rotation of rotor and antirotor of sub-elementary fermion, corresponding to Hidden harmony and standing waves conditions, is responsible for the rest mass and resulting charge of sub-elementary fermion origination. The triplets formed by [sub-elementary fermion + sub-elementary antifermion] with opposite spins, charge and energy and one uncompensated sub-elementary fermion represent electrons. The triplets formed by [sub-elementary fermion + sub-elementary antifermion] and one uncompensated sub-elementary antifermion represent the positrons. The quarks are the result of certain superposition/fusion of (mu) and (tau) electrons and positrons. The bosons, like photons, are composed from the equal number of sub-elementary particles and antiparticles as a system of two [electron + positron] triplets. The [corpuscle (C) - wave (W)] duality is a result of quantum beats between the 'actual' (vortex) and 'complementary' (rotor) states of sub-elementary ferm-ions/antifermions. The [C] phase exists as a mass, electric and magnetic asymmetric dipole with spatial image: [vortex + rotor]. The [W] phase exists in form of Cumulative virtual cloud (CVC) of sub-quantum particles.

The angular momentum of CVC excites the nonlocal massless virtual spin waves (VirSW) in Bivacuum with properties of Nambu - Goldstone modes. It is shown, that the Principle of least action is a consequence of introduced 'Harmonization force (HaF)' of asymmetric Bivacuum. The system [Bivacuum + Matter] has a properties of the active medium, tending to Hidden harmony or Golden mean conditions under HaF influence. The mechanism of quantum entanglement between coherent particles, mediated by HaF and Virtual wave-guides (VirWG) is proposed also. The pace of time for any closed and coherent system of particles is determined by the pace of translational longitudinal and translational contributions to kinetic energy change of this system, related to in-phase changes of electromagnetic and gravitational potentials, correspondingly. Kaivarainen's Unified model has been used also for possible explanation of Bivacuum mediated nonlocal mental interactions: www.emergentmind.org/PDF_files.htm/Kaiv290703.pdf.

Bozhidar Z. Iliev discusses his research in chapters 4-6. In chapter 4, relations between two definitions of the (total) angular momentum operator, as a generator of rotations and in the Lagrangian formalism, are explored in quantum field theory. Generally, these definitions result in different angular momentum operators, which are suitable for different purposes in the theory. From the spin and orbital angular momentum operators (in the Lagrangian formalism) are extracted additive terms which are conserved operators and whose sum is the total angular momentum operator. Chapter 5 focuses on the interrelations between the two definitions of momentum operator, via the canonical energy–momentum tensorial operator and as translation operator (on the operator space) as studied in quantum field theory. These definitions give rise to similar but, generally, different momentum operators, each of them having its own place in the theory. Some speculations on the relations between quantum field theory and quantum mechanics are presented.

Conserved operator quantities in quantum field theory can be defined via the Noether theorem in the Lagrangian formalism and as generators of some transformations. These definitions lead to generally different conserved operators which are suitable for different purposes. Some relations involving conserved operators are analyzed in chapter 6.

The emission of continuous radiation in the $(n;t)-$, $(n;\alpha)-$ and $(n;p)-$nuclear reactions in light atoms and ions is discussed in chapter 7 by Alexei M. Frolov. The angular distributions of the emitted radiation and its spectrum are considered in detail. It is shown that the total radiation $I(t)$ rapidly decreases with time $I(t) \sim t^{-4}$ at large t. It is shown that the radiation emitted during the $(n;t)-$nuclear reactions in the rapidly moving ions and atoms can be used to study the electron-electron position correlations in these systems. The approach developed in our present study can also be applied to describe the emission of bremsstrahlung during the spontaneous and neutron-stimulated nuclear fission in heavy atoms. In particular, the circular motion of the relativistic lithium ions is discussed.

Min Yan, Edward G. Rickey and Yifu Zhu discuss absorptive photon switching by quantum interference in chapter 8. Quantum coherence and interference manifested by electromagnetically induced transparency suppress the linear susceptibilities and enhance the third-order susceptibilities in a four-level atomic system. Under appropriate conditions, the atomic system absorbs two photons, but not one photon. This feature may be used to realize an optical switch in which a laser pulse controls absorption of another laser pulse at different frequencies. In the ideal limit, the switch may operate at single photon levels.

In: Frontiers in Quantum Physics Research
Editor: F. Columbus and V. Krasnoholovets, pp. 1-50
ISBN 1-59454-002-2

The Difference Between Dirac's Hole Theory and Quantum Field Theory

Dan Solomon

Rauland-Borg Corporation
3450 W. Oakton
Skokie, IL 60076

Email: dan.solomon@rauland.com
Phone: 847-324-8337

Abstract

Dirac's hole theory and quantum field theory are generally thought to be equivalent. In fact field theory can be derived from hole theory through the process of second quantization. However, it can be shown that problems worked in both theories yield different results. The reason for the difference between the two theories will be examined and the effect that this difference has on the way calculations are done in quantum theory will be discussed.

1. Introduction

It is generally assumed that Dirac hole theory and quantum field theory are identical. However, it has recently been shown by Coutinho et al. [1][2] that this is not necessarily the case. Coutinho examined the shift in the vacuum energy due to a time independent perturbing potential. It was shown that hole theory and field theory yield different results when this problem was solved using time independent perturbation theory. In this article we will examine hole theory and field theory in order to understand why these differences occur. To avoid mathematical complications we will consider a simple quantum theory where the electrons are non-interacting and respond to an external classical electric potential.

Hole theory was original proposed by Dirac to deal with the problem of negative energy solutions of the Dirac equation. In hole theory the negative energy states are all occupied by a single electron. The electrons in the negative energy "sea" have all the properties of positive energy electrons in that they respond to an electric field and obey the Dirac equation. If one of these negative energy electrons makes a transition to a positive energy state it leaves a "hole" in the negative energy sea which can be identified with a positron. The conceptual difficulty with hole theory is that it includes an infinite number of negative energy electrons which are not directly observable in their unperturbed state. Only changes from the unperturbed state are observable. In field theory the infinite number of electrons that make up the negative energy sea are replaced by the vacuum state vector $|0\rangle$ where $|0\rangle$ is simply the state for which no electrons or positrons are present. Thus in field theory the definition of the vacuum is considerably simplified.

In Section 2 of this discussion we consider the Dirac equation for a single electron in the presence of an external, classical, electric potential. We will show the solutions to the Dirac equation have certain properties. They obey the continuity equation and are gauge invariant. Next, in Section 3, we will consider N-electron theory where the electrons are non-interacting. The N-electron wave function must be anti-symmetric so that the Pauli exclusion principle is obeyed. It will be shown that the N-electron wave function can be considered to be composed of N individual electrons each of which evolves independently in time according to the single particle Dirac equation. The total charge, current, or energy is just the sum of the charge, current, or energy, respectively, of each of the individual electrons. Therefore all the symmetries and properties that are associated with the single particle Dirac equation also apply to N-electron theory. Therefore N-electron theory is gauge invariant and obeys the continuity equation.

In Section 4 and 5 we show how the Schrödinger representation of quantum field theory can be derived from N-electron theory. The first part of this process is essentially equivalent to a change in notation. We go from a cumbersome notation where the N-electron wave function is represented by Slater determinants to a simpler notation using field operators and state vectors. The second part of this process is to treat the "hole" in the infinite electron vacuum sea as a positron. We introduce electron and positron creation operators which act on the vacuum state to produce electrons and positrons, respectively. The concept of a vacuum has, then, been simplified from consisting of an infinite number of negative energy electrons to being the state in which no positive energy electrons or positrons exist. It is shown that the vacuum state is the state for which the free field energy is a minimum. The free field energy is the energy when the electric potential is zero.

The discussion up, to this point, is simply a review of basic quantum mechanics that is generally available in many textbooks. It is assumed that the reader is already somewhat familiar with this material therefore formal proofs are generally not given but references are made to more detailed discussions. The purpose is to present the generally accepted argument that shows that hole theory and the Schrödinger representation of quantum field theory are formally equivalent. The discussion that follows this review contains new and, largely, original material which shows that the two theories are actually not equivalent.

In Sections 6, 7, and 8 we examine the continuity equation and gauge invariance in field theory. Due to the fact that field theory is derived from N-electron theory we expect that gauge invariance and the continuity equation will hold true for field theory since they hold true for N-electron theory. However, it will be shown that there is an inconsistency in field theory. If the continuity equation and gauge invariance are true then there must exist quantum states with less free field energy than the vacuum state. However, if we use the standard definition of the vacuum state, this is not possible. When we examine what is required for the continuity equation and gauge invariance to hold in field theory, it is shown that a quantity called the Schwinger term must be zero. However, as has been shown by Schwinger, this quantity cannot be zero if there are no quantum states with less free field energy than the vacuum state.

Next, in Section 9, we show how this problem with gauge invariance manifests itself in field theory. This problem shows up when the effect of an electromagnetic field on the vacuum is considered. An electromagnetic field perturbs the vacuum and produces a vacuum current which can be calculated using perturbation theory. It is well know that when the vacuum current is calculated the result is not gauge invariant. The non-gauge invariant terms that appear in the calculation of the vacuum current must be removed to obtain a physically acceptable result. It is shown that this problem with the gauge invariance of the vacuum current is directly related to the fact that the Schwinger term is non-zero.

In Section 10 and 11 we take on the question on whether the vacuum state in hole theory is the state with the lowest free field energy. This vacuum state is the physical state where all the negative energy states are occupied by a single electron. These electrons obey the Dirac equation and their energy will change in response to an electric field. If an electromagnetic field is applied and then removed, the resulting perturbed vacuum state will have a different energy from the original state. The change in the energy is just the sum of the change in the energies of each of the individual electrons. It is shown in Section 10 that it is possible to find an electric field which causes the change in the vacuum energy to be negative. That is, energy is extracted from the vacuum due to its interaction with the electric field. This is not possible in field theory because the vacuum is the lowest energy state. This example demonstrates that field theory and hole theory are not equivalent.

The difficulty with this result is that we can no longer assume that the symmetries and properties associated with the single particle Dirac equation are valid for field theory. In particular, it will be shown that that field theory, as currently formulated, is not gauge invariant and the continuity equation is not true. However, as will be shown in Section 12, the vacuum state can be redefined so that quantum field theory can be made equivalent to hole theory. When this is done the principle of gauge invariance and the continuity equation will be valid in quantum field theory. Throughout this discussion we assume "natural units" so that $\hbar = c = 1$.

2. The single particle Dirac Equation

In this section we will review some of the properties of the solutions of the Dirac equation. In particular we are interested in local change conservation, or the continuity equation, and gauge invariance. The single particle Dirac equation is [3][4][5],

$$i\frac{\partial\psi(\vec{x},t)}{\partial t} = H(\vec{x},t)\psi(\vec{x},t) \tag{2.1}$$

where $\psi(\vec{x},t)$ is the wave function and $H(\vec{x},t)$ is the Hamiltonian operator which is defined by,

$$H(\vec{x},t) = H_0(\vec{x}) - q\left(\vec{\alpha}\cdot\vec{A}(\vec{x},t) - A_0(\vec{x},t)\right) \tag{2.2}$$

where $\left(A_0(\vec{x},t),\vec{A}(\vec{x},t)\right)$ is the electric potential, 'q' is the charge on the electron, and $H_0(\vec{x})$ is the free field Hamiltonian, which is the Hamiltonian when the electric potential is zero. Throughout this discussion we assume that the electric potential is an unquantized classical quantity. $H_0(\vec{x})$ is given by,

$$H_0(\vec{x}) = -i\vec{\alpha}\cdot\vec{\nabla} + \beta m \tag{2.3}$$

where m is the mass and $\vec{\alpha}$ and β are the usual 4x4 matrices (see ref. [3]).

If ψ_1 and ψ_2 are two wave functions obeying (2.1), then it can be easily shown that the following relationship holds,

$$\frac{\partial\left(\psi_1^\dagger\psi_2\right)}{\partial t} + \vec{\nabla}\cdot\left(\psi_1^\dagger\vec{\alpha}\psi_2\right) = 0 \tag{2.4}$$

where the '$\dagger$' indicates Hermitian conjugate. Integrate the above over all space to obtain,

$$\frac{\partial}{\partial t}\int\left(\psi_1^\dagger\psi_2\right)d\vec{x} = 0 \tag{2.5}$$

This means that $\int\left(\psi_1^\dagger\psi_2\right)d\vec{x}$ is constant in time. The main result from this is that if ψ_1 and ψ_2 are initially orthogonal, i.e. $\int\left(\psi_1^\dagger\psi_2\right)d\vec{x} = 0$, they will be orthogonal for all time. Also if the wave function ψ is initially normalized (i.e., $\int\left(\psi^\dagger\psi\right)d\vec{x} = 1$) it will be normalized for all time.

Next, define the charge expectation value ρ_e and the current expectation value $\vec{J}_e$ as,

$$\rho_e = q\psi^\dagger\psi \text{ and } \vec{J}_e = q\psi^\dagger\vec{\alpha}\psi \tag{2.6}$$

where ψ is normalized. Use this in (2.4) to show that the quantities ρ_e and $\vec{J}_e$ obey the continuity equation,

$$\frac{\partial\rho_e}{\partial t} + \vec{\nabla}\cdot\vec{J}_e = 0 \tag{2.7}$$

This is the quantum mechanical version of corresponding classical equation for the local conservation of electric charge.

An important property that a physical theory should obey is that of gauge invariance [3]. The electromagnetic field is given in terms of the electric potential by,

$$\vec{E} = -\left(\frac{\partial \vec{A}}{\partial t} + \vec{\nabla} A_0\right) \text{ and } \vec{B} = \vec{\nabla} \times \vec{A} \tag{2.8}$$

A change in the gauge is a change in the electric potential that does not produce a change in the electromagnetic field. Such a change is given by,

$$\vec{A} \to \vec{A}' = \vec{A} - \vec{\nabla}\chi \text{ and } A_0 \to A_0' = A_0 + \frac{\partial \chi}{\partial t} \tag{2.9}$$

where $\chi(\vec{x},t)$ is an arbitrary real valued function. For Dirac field theory to be gauge invariant means that a change in the gauge does not produce a change in any measurable quantities. These include the current and charge expectation values. To show that this is the case substitute (2.9) into (2.2) to obtain,

$$H_g = H + q\left(\vec{\alpha} \cdot \vec{\nabla}\chi + \frac{\partial \chi}{\partial t}\right) \tag{2.10}$$

where H is the original Hamiltonian and H_g is the Hamiltonian after a gauge transformation. Let ψ be the solution of (2.1) and ψ_g be the solution of the gauge transformed equation,

$$i\frac{\partial \psi_g(\vec{x},t)}{\partial t} = H_g \psi_g(\vec{x},t) \tag{2.11}$$

It can be shown that,

$$\psi_g = e^{-iq\chi}\psi \to \psi_g^{\dagger} = \psi^{\dagger} e^{iq\chi} \tag{2.12}$$

When this result is used in (2.6) it is easy to show that the current and charge expectation values are invariant under a gauge transformation.

Now consider the Dirac equation when the electric potential is zero. This may also be referred to as the free field Dirac equation. In this case (2.1) becomes,

$$i\frac{\partial \psi(\vec{x},t)}{\partial t} = H_0(\vec{x})\psi(\vec{x},t) \tag{2.13}$$

The solutions to this equation are of the form,

$$\varphi_n^{(0)}(\vec{x},t)=\varphi_n^{(0)}(\vec{x})e^{-i\lambda_n E_n t};\ \varphi_n^{(0)}(\vec{x})=u_n e^{i\vec{p}_n\cdot\vec{x}} \tag{2.14}$$

where u_n is a 4-spinor (see page 30-32 of [3]). The $\varphi_n^{(0)}(\vec{x})$ satisfy,

$$H_0\varphi_n^{(0)}(\vec{x})=\lambda_n E_n\varphi_n^{(0)}(\vec{x}) \tag{2.15}$$

In the above expression the energy eigenvalue is $\lambda_n E_n$ when E_n is given by $E_n=\sqrt{\vec{p}_n^2+m^2}$, $\vec{p}_n$ is the momentum, $\lambda_n=+1$ for positive energy states, and $\lambda_n=-1$ for negative energy states. The index 'n' is an integer that stands for the triplet $(\lambda_n,\vec{p}_n,s_n)$ where $s=\pm 1/2$ is the spin state.

If the electric potential is zero the triplet $(\lambda_n,\vec{p}_n,s_n)$ does not change as the state $\varphi_n^{(0)}(\vec{x},t)$ evolves forward in time. If an electric potential is applied then the initial state $\varphi_n^{(0)}(\vec{x},t_i)$ will evolve into the final state $\varphi_n(\vec{x},t_f)$. In the case of the final state the index 'n' can no longer be associated with a given triplet $(\lambda_n,\vec{p}_n,s_n)$. In this case the index 'n' means that the final state $\varphi_n(\vec{x},t_f)$ evolves from the initial state $\varphi_n^{(0)}(\vec{x},t_i)$.

The $\varphi_n^{(0)}(\vec{x})$ are normalized and they satisfy the relationships,

$$\int\varphi_n^{(0)\dagger}(\vec{x})\varphi_m^{(0)}(\vec{x})d\vec{x}=\delta_{nm}\ \text{and}\ \sum_n\left(\varphi_n^{(0)\dagger}(\vec{x})\right)_a\left(\varphi_n^{(0)}(\vec{y})\right)_b=\delta_{ab}\delta(\vec{x}-\vec{y}) \tag{2.16}$$

where 'a' and 'b' are spinor indices (see page 107 of [6]). Due to the above relationships an arbitrary 4-spinor wave function can be expressed as a Fourier sum in terms of the basis states $\varphi_n^{(0)}(\vec{x})$.

As can be seen from the above discussion there are both positive and negative energy solutions to the Dirac equation. Therefore there is nothing to prevent a single particle from making a transition to a negative energy state if it is perturbed by some external influence. In order to deal with this problem Dirac made the assumption that all the negative energy states were occupied by a single electron and invoked the Pauli exclusion principle which states that no more then one electron can occupy a given energy state.

The result of this is that we immediately go from a one electron theory to an N-electron theory where $N \to \infty$ (see the discussion in Chapt 4 of [3] and page 274-275 of [7]).

3. N-electron theory

To produce an N-electron theory the N-electron wave function $\Psi^N(\vec{x}_1, \vec{x}_2, ..., \vec{x}_N, t)$ must be defined in such a way that the Pauli principle holds and all N particles are indistinguishable. This is done by defining the wave function so that it is antisymmetric under the exchange of the coordinates. The N particle wave function can be expressed as a Slater determinant [8][9][10],

$$\Psi^N(\vec{x}_1, \vec{x}_2, ..., \vec{x}_N, t) = \frac{1}{\sqrt{N!}} \begin{vmatrix} \psi_1(\vec{x}_1,t) & \psi_1(\vec{x}_2,t) & \cdots & \psi_1(\vec{x}_N,t) \\ \psi_2(\vec{x}_1,t) & \psi_2(\vec{x}_2,t) & \cdots & \psi_2(\vec{x}_N,t) \\ \vdots & \vdots & \vdots & \vdots \\ \psi_N(\vec{x}_1,t) & \psi_N(\vec{x}_2,t) & \cdots & \psi_N(\vec{x}_N,t) \end{vmatrix} \quad (3.1)$$

where the $\psi_n\,(n = 1, 2, \ldots, N)$ are a set of normalized and orthogonal wave functions. A more compact way of writing this is [10],

$$\Psi^N(\vec{x}_1, \vec{x}_2, ..., \vec{x}_N, t) = \frac{1}{\sqrt{N!}} \sum_P (-1)^p P\left(\psi_1(\vec{x}_1,t)\psi_2(\vec{x}_2,t)\cdots\psi_N(\vec{x}_N,t)\right) \quad (3.2)$$

where P is a permutation operator acting on the space coordinates and p is the number of interchanges in P.

This N-electron wave functions obeys the equation,

$$i\frac{\partial}{\partial t}\Psi^N(\vec{x}_1, \vec{x}_2, ..., \vec{x}_N, t) = H^N(\vec{x}_1, \vec{x}_2, ..., \vec{x}_N, t)\Psi^N(\vec{x}_1, \vec{x}_2, ..., \vec{x}_N, t) \quad (3.3)$$

where $H^N(\vec{x}_1, \vec{x}_2, ..., \vec{x}_N, t)$ is the N-electron Hamiltonian. For non-interacting electrons, which is the case throughout this discussion, the N-electron Hamiltonian is given by,

$$H^N(\vec{x}_1, \vec{x}_2, ..., \vec{x}_N, t) = \sum_{n=1}^{N} H(\vec{x}_n, t) \quad (3.4)$$

where $H(\vec{x}_n,t)$ is the single particle Hamiltonian defined by (2.2). The single particle Hamiltonian $H(\vec{x}_n,t)$ acts on the quantities $\psi_j(\vec{x}_n,t)$ that contain the position coordinate $\vec{x}_n$. From the above discussion we can show that each of the functions $\psi_j(\vec{x}_n,t)$ in (3.2) satisfy the single particle Dirac equation (2.1).

The expectation value of a single particle operator $O_{op}(\vec{x})$ is defined as,

$$O_e = \int \psi^{\dagger}(\vec{x}_1,t) O_{op}(\vec{x}_1) \psi(\vec{x}_1,t) d\vec{x}_1 \tag{3.5}$$

where $\psi(\vec{x}_1,t)$ is a normalized single particle wave function. Therefore, from (2.6), we have that the single particle charge and current operators are,

$$\rho_{op} = q\delta(\vec{x}-\vec{x}_1) \text{ and } \vec{J}_{op} = q\vec{\alpha}\delta(\vec{x}-\vec{x}_1) \tag{3.6}$$

The N-electron operator is given by,

$$O_{op}^N = \sum_{n=1}^{N} O_{op}(\vec{x}_n) \tag{3.7}$$

which is just the sum of one particle operators. The expectation value of a normalized N-electron wave function is,

$$O_e^N = \int \Psi^{N\dagger}(\vec{x}_1,\vec{x}_2,...,\vec{x}_N,t) O_{op}^N(\vec{x}_1,\vec{x}_2,...,\vec{x}_N) \Psi^N(\vec{x}_1,\vec{x}_2,...,\vec{x}_N,t) d\vec{x}_1 d\vec{x}_2 ... d\vec{x}_N \tag{3.8}$$

This can be shown to be equal to,

$$O_e^N = \sum_{n=1}^{N} \int \psi_n^{\dagger}(\vec{x},t) O_{op}(\vec{x}) \psi_n(\vec{x},t) d\vec{x} \tag{3.9}$$

That is, the N electron expectation value is just the sum of the single particle expectation values associated with each individual wave function ψ_n. Therefore, the charge and current expectation values for the N electron wave function are,

$$\rho_e^N(\vec{x},t) = \sum_{n=1}^{N} q\psi_n^{\dagger}(\vec{x},t)\psi_n(\vec{x},t) \tag{3.10}$$

and

$$\vec{J}_e^N(\vec{x},t) = \sum_{n=1}^{N} q\psi_n^{\dagger}(\vec{x},t)\vec{\alpha}\psi_n(\vec{x},t) \tag{3.11}$$

In addition, another quantity that we will be interested in later is the free field energy $\xi_f\left(\Psi^N\right)$ of the N-electron state. This is the energy when the electric potential is zero. Based on the above discussion,

$$\xi_f\left(\Psi^N\right)=\sum_{n=1}^{N}\int\psi_n^{\dagger}\left(\vec{x},t\right)H_0\psi_n\left(\vec{x},t\right)d\vec{x} \tag{3.12}$$

The result of the above discussion is that we can view the N-electron system as consisting of N independent orthonormal wave functions $\psi_n\left(\vec{x},t\right)$. Each wave function evolves in time according to the single particle Dirac equation (2.1). The total current and charge expectation value is just the sum of the current and charge expectation value associated with each independent wave function. Similarly, the total free field energy is just the sum of the free field energy of each independent wave function. Therefore, the various symmetries and conservation laws that hold for the one electron wave function hold for the N-electron wave function. In particular, we can easily conclude that the continuity equation and gauge invariance hold for the N-electron wave function.

4. Second Quantization

Consider equation (3.2). Each of the terms $\psi_n\left(\vec{x},t\right)$ can be expanded in terms of the basis states $\varphi_j^{(0)}\left(\vec{x}\right)$, which are defined by equation (2.14). Therefore we can write,

$$\psi_n\left(\vec{x},t\right)=\sum_j c_{nj}\left(t\right)\varphi_j^{(0)}\left(\vec{x}\right) \tag{4.1}$$

where $c_{nj}\left(t\right)$ are the Fourier coefficients. Use this in (3.2) to obtain,

$$\Psi^N=\sum_{j_1,j_2,\ldots,j_N} c_{1,j_1}\left(t\right)c_{2,j_2}\left(t\right)\ldots c_{N,j_N}\left(t\right)\frac{1}{\sqrt{N!}}\sum_P\left(-1\right)^p P\left(\varphi_{j_1}^{(0)}\left(\vec{x}_1\right)\varphi_{j_2}^{(0)}\left(\vec{x}_2\right)\cdots\varphi_{j_N}^{(0)}\left(\vec{x}_N\right)\right) \tag{4.2}$$

where we write Ψ^N instead of $\Psi^N\left(\vec{x}_1,\vec{x}_2,\ldots,\vec{x}_N,t\right)$ to shorten the expression. Note that any of the summation terms in which any of the j_i's are equal is zero. We see, then, that we can expand any arbitrary N electron wave function in terms of N-electron basis states defined by,

$$\Phi^{N}_{j_1,j_2,\ldots,j_N}\left(\vec{x}_1,\vec{x}_2,\ldots,\vec{x}_N\right)=\frac{1}{\sqrt{N!}}\sum_{P}(-1)^{p}\,P\left(\varphi^{(0)}_{j_1}(\vec{x}_1)\varphi^{(0)}_{j_2}(\vec{x}_2)\cdots\varphi^{(0)}_{j_N}(\vec{x}_N)\right) \quad (4.3)$$

Thus for any N-particle state Ψ^N we can write,

$$\Psi^{N}\left(\vec{x}_1,\vec{x}_2,\ldots,\vec{x}_N,t\right)=\sum_{j_1>j_2>,\ldots,>j_N} c_{j_1,j_2,\ldots,j_N}(t)\Phi^{N}_{j_1,j_2,\ldots,j_N}\left(\vec{x}_1,\vec{x}_2,\ldots,\vec{x}_N\right) \quad (4.4)$$

In the above summation none of the j_i's are equal and each ordered set $\left(j_1,j_2,\ldots,j_N\right)$ occurs only once. Next define destruction and creation operators which act on these basis states to the change the number of electrons in each state (see [10]). The destruction operator $\hat{a}_j$ turns the basis state Φ^N into the basis state Φ^{N-1} by eliminating the single particle wave function $\varphi^{(0)}_j$ from Φ^N. If Φ^N does not contain $\varphi^{(0)}_j$ then the action of the operator yields 0. Therefore,

$$\begin{aligned}&\hat{a}_{j_k}\Phi^{N}_{j_1,j_2,\ldots,j_k,\ldots,j_N}\left(\vec{x}_1,\vec{x}_2,\ldots,\vec{x}_N\right)\\ &\quad=(\pm)\Phi^{N-1}_{j_1,j_2,\ldots,j_{k-1},j_{k-1},\ldots,j_N}\left(\vec{x}_1,\vec{x}_2,\ldots,\vec{x}_{N-1}\right)\text{ if }\Phi^N\text{ contains }\varphi^{(0)}_{j_k}\end{aligned} \quad (4.5)$$

where the sign is positive if the number of functions that precede $\varphi^{(0)}_{j_k}$ in Φ^N is odd. Also,

$$\hat{a}_{j_k}\Phi^{N}_{j_1,j_2,\ldots,j_N}\left(\vec{x}_1,\vec{x}_2,\ldots,\vec{x}_N\right)=0\text{ if }\Phi^N\text{ does not contian }\varphi^{(0)}_{j_k} \quad (4.6)$$

The creation operator $\hat{a}^{\dagger}_j$ turns the basis state Φ^N into the basis state Φ^{N+1} by adding the wave function $\varphi^{(0)}_j$ to the state Φ^N that does not already include $\varphi^{(0)}_j$. If Φ^N includes $\varphi^{(0)}_j$ then the result of the action of $\hat{a}^{\dagger}_j$ on Φ^N is zero. Therefore we have,

$$\begin{aligned}&\hat{a}^{\dagger}_{j_k}\Phi^{N}_{j_1,j_2,\ldots,j_{k-1},j_{k+1},\ldots,j_N}\left(\vec{x}_1,\vec{x}_2,\ldots,\vec{x}_N\right)\\ &\quad=(\pm)\Phi^{N+1}_{j_1,j_2,\ldots,j_{k-1},j_k,j_{k+1},\ldots,j_N}\left(\vec{x}_1,\vec{x}_2,\ldots,\vec{x}_{N+1}\right)\text{ if }\Phi^N\text{ does not contain }\varphi^{(0)}_{j_k}\end{aligned} \quad (4.7)$$

where the sign is positive if the number of functions that precede $\varphi_{j_k}^{(0)}$ in Φ^{N+1} is even. Also

$$\hat{a}_{j_k}^{\dagger}\Phi_{j_1,j_2,\ldots,j_k,\ldots,j_N}^{N}\left(\vec{x}_1,\vec{x}_2,\ldots,\vec{x}_N\right)=0 \text{ if } \Phi^N \text{ contains } \varphi_{j_k}^{(0)} \tag{4.8}$$

It can be shown [10] that the operators $\hat{a}_j^{\dagger}$ and $\hat{a}_j$ satisfy the following anti-commutator relationships,

$$\left\{\hat{a}_j,\hat{a}_k\right\}=0;\ \left\{\hat{a}_j^{\dagger},\hat{a}_k^{\dagger}\right\}=0;\ \left\{\hat{a}_j^{\dagger},\hat{a}_k\right\}=\delta_{jk} \tag{4.9}$$

Next define the empty state Φ^0 which contains no particles. This state is defined by,

$$\hat{a}_j\Phi^0=0 \text{ and } \hat{a}_j^{\dagger}\Phi^0=\Phi_j^1=\varphi_j^{(0)}\left(x_1\right) \tag{4.10}$$

Any N-electron basis state Φ^N can be produced by acting on Φ^0 with the creation operators,

$$\Phi_{j_1,j_2,\ldots,j_N}^{N}=\hat{a}_{j_1}^{\dagger}\hat{a}_{j_2}^{\dagger}\ldots\hat{a}_{j_N}^{\dagger}\Phi^0 \tag{4.11}$$

When this result is used in (4.2) along with (4.3) we see that any arbitrary wave function can be expressed in terms of the creation operators. In this case we use bra-ket notation so that we write $\left|\psi^N(t)\right\rangle$ instead of $\Psi^N\left(\vec{x}_1,\vec{x}_2,\ldots,\vec{x}_N,t\right)$. Therefore in place of (4.4) we have,

$$\left|\psi^N(t)\right\rangle=\sum_{j_1>j_2>,\ldots,>j_N} c_{j_1,j_2,\ldots,j_N}(t)\hat{a}_{j_1}^{\dagger}\hat{a}_{j_2}^{\dagger}\ldots\hat{a}_{j_N}^{\dagger}\Phi^0 \tag{4.12}$$

Also, the hermitian conjugate is written as,

$$\left\langle\psi^N(t)\right|=\sum_{j_1>j_2>,\ldots,>j_N}\Phi^0\hat{a}_{j_N}\ldots\hat{a}_{j_2}\hat{a}_{j_1}c_{j_1,j_2,\ldots,j_N}^{*}(t) \tag{4.13}$$

Now, recall the previous discussion defining the expectation values of the N electron operators $O_{op}^N\left(\vec{x}_1,\vec{x}_2,\ldots,\vec{x}_N,t\right)$ that act on the quantities $\psi^N\left(\vec{x}_1,\vec{x}_2,\ldots,\vec{x}_N,t\right)$. We want to define operators $\hat{O}$ that act on the quantities $\left|\psi^N(t)\right\rangle$ and correspond to the operator $O_{op}^{N}(\vec{x}_1,\vec{x}_2,...,\vec{x}_N,t)$ in that the expectation values are the same in both cases. The expectation value associated with some operator $\hat{O}$ for $|\psi^N(t)\rangle$ is defined by,

$$O_e = \langle\psi^N|\hat{O}|\psi^N\rangle \tag{4.14}$$

In order to that this be the same as the expectation for $\psi^N(\vec{x}_1,\vec{x}_2,...,\vec{x}_N,t)$ given by equation (3.9), the operator $\hat{O}$, that corresponds to $O_{op}^{N}(\vec{x}_1,\vec{x}_2,...,\vec{x}_N,t)$, must be defined by,

$$\hat{O} = \sum_{i,j}\hat{a}_i^\dagger\hat{a}_j\int\varphi_i^{(0)\dagger}(\vec{x})O_{op}(\vec{x})\varphi_j^{(0)}(\vec{x})d\vec{x} \tag{4.15}$$

This can be rewritten as,

$$\hat{O} = \int\hat{\psi}^\dagger(\vec{x})O_{op}(\vec{x})\hat{\psi}(\vec{x})d\vec{x} \tag{4.16}$$

where $\hat{\psi}(\vec{x})$ is called the field operator and is defined by,

$$\hat{\psi}(\vec{x}) = \sum_j\hat{a}_j\varphi_j^{(0)}(\vec{x}) \text{ and } \hat{\psi}^\dagger(\vec{x}) = \sum_j\hat{a}_j^\dagger\varphi_j^{(0)\dagger}(\vec{x}) \tag{4.17}$$

The current and charge operators then become,

$$\hat{\vec{J}}(\vec{x}) = q\hat{\psi}^\dagger(\vec{x})\vec{\alpha}\hat{\psi}(\vec{x}) \text{ and } \hat{\rho}(\vec{x}) = q\hat{\psi}^\dagger(\vec{x})\hat{\psi}(\vec{x}) \tag{4.18}$$

The Hamiltonian operator is given by,

$$\hat{H} = \int\hat{\psi}^\dagger(\vec{x})H(\vec{x},t)\hat{\psi}(\vec{x})d\vec{x} \tag{4.19}$$

The Dirac equation (3.3) becomes,

$$i\frac{\partial|\Omega(t)\rangle}{\partial t} = \hat{H}|\Omega(t)\rangle \text{ and } -i\frac{\partial\langle\Omega(t)|}{\partial t} = \langle\Omega(t)|\hat{H} \tag{4.20}$$

where $|\Omega(t)\rangle$, instead of $|\psi^N\rangle$, is used to designate the state vector.

Use (2.2) in (4.19) to obtain,

$$\hat{H} = \int\hat{\psi}^\dagger(\vec{x})\left(H_0(\vec{x}) - q\left(\vec{\alpha}\cdot\vec{A}(\vec{x},t) - A_0(\vec{x},t)\right)\right)\hat{\psi}(\vec{x})d\vec{x} \tag{4.21}$$

This can be written as,

$$\hat{H} = \hat{H}_0 - \int\hat{\vec{J}}(\vec{x})\cdot\vec{A}(\vec{x},t)d\vec{x} + \int\hat{\rho}(\vec{x})\cdot A_0(\vec{x},t)d\vec{x} \tag{4.22}$$

where we have used (4.18) and define,

$$\hat{H}_0 = \int \hat{\psi}^{\dagger}(\vec{x}) H_0(\vec{x}) \hat{\psi}(\vec{x}) d\vec{x} \tag{4.23}$$

Use (4.17) along with (2.15) and (2.16) in the above to obtain,

$$\hat{H}_0 = \sum_n \lambda_n E_n \hat{a}_n^{\dagger} \hat{a}_n \tag{4.24}$$

The above relationships ((4.16) through (4.24)) represent quantum field theory in the Schrödinger representation. The basic elements of the theory are the state vectors $|\Omega(t)\rangle$ and the field operator $\hat{\psi}(\vec{x})$. The state vectors $|\Omega(t)\rangle$ evolve in time according to equation (4.20) while the field operators are constant in time. Operators associated with observables are defined in terms of the field operators (e.g. equation (4.18)). According to the derivation given here quantum field theory is equivalent to N-electron theory. Essentially the cumbersome notation of N-electron theory is replaced by the more compact notion of field theory.

Note that there is an alternative version of field theory that is commonly used called the Heisenberg representation. In this case the state vector is constant in time and the time dependence of the theory is associated with the field operators. It can be shown that both representations have the same expectation values and are, therefore, equivalent.

5. The vacuum state

The final step in the formulation of quantum field theory is the definition of the vacuum state $|0\rangle$. We start with the hole theory concept that the vacuum state is the state where all negative energy states are occupied and the positive energy states are empty. Order the index 'n' associated with the basis wave function $\varphi_n^{(0)}(\vec{x})$ so that $n > 0$ implies a positive energy state and $n < 0$ implies a negative energy state. Further order the states so that the magnitude of 'n' increases with the magnitude of the energy. Following Greiner[3] the vacuum state in quantum field theory is defined by,

$$|0\rangle = \left(\hat{a}_{-1}^{\dagger} \hat{a}_{-2}^{\dagger} \hat{a}_{-3}^{\dagger} \ldots\right) \Phi^0 = \prod_{n=1}^{\infty} \hat{a}_{-n}^{\dagger} \Phi^0 \tag{5.1}$$

Thus all negative energy states are occupied and all positive energy states are unoccupied. From (4.24) it is seen that $|0\rangle$ is an eigenstate of the free field Hamiltonian operator $\hat{H}_0$,

$$\hat{H}_0 |0\rangle = \xi_{vac} |0\rangle \tag{5.2}$$

where the energy eigenvalue ξ_{vac} is given by,

$$\xi_{vac} = \sum_{n=1}^{\infty} \lambda_{-n} E_{-n} = -\sum_{n=1}^{\infty} E_{-n} \tag{5.3}$$

Note that ξ_{vac} is an infinite negative number. This does not concern us, at this point, because we are interested in changes of energy from some initial state and not the actual value. Now additional eigenstates $|k\rangle$ can be formed by acting on the vacuum state with the creation operators $\hat{a}_j^{\dagger}$ or the destruction operators $\hat{a}_{-j}$ where 'j' is a positive nonzero integer. The effect of operating with $\hat{a}_j^{\dagger}$ is to create an electron with positive energy $\lambda_j E_j$. The effect of operating with $\hat{a}_{-j}$ is to destroy an electron with negative energy $\lambda_{-j} E_{-j}$. In either case the energy of the state is increased.

The final step in the process of going from hole theory to field theory is to replace the operator $\hat{a}_{-j}$, that destroys a negative energy electron, by the operator $\hat{d}_j^{\dagger}$, that creates a positive energy positron. Thus we make the following change in notation,

$$\hat{b}_j = \hat{a}_j \rightarrow \hat{b}_j^{\dagger} = \hat{a}_j^{\dagger} \text{ and } \hat{d}_j = \hat{a}_{-j}^{\dagger} \rightarrow \hat{d}_j^{\dagger} = \hat{a}_{-j} \text{ where } j > 0 \tag{5.4}$$

In the above expressions $\hat{d}_j$ and $\hat{d}_j^{\dagger}$ are positron destruction and creation operators, respectively. The $\hat{b}_j$ and $\hat{b}_j^{\dagger}$ are the electron destruction and creation operators, respectively, that act on electrons in positive energy states. Use this notation in (4.17) to obtain,

$$\hat{\psi}(\vec{x}) = \sum_{j=1}^{\infty} \left(\hat{b}_j \varphi_j(\vec{x}) + \hat{d}_j^{\dagger} \varphi_{-j}(\vec{x}) \right) \tag{5.5}$$

and,

$$\hat{\psi}^{\dagger}(\vec{x}) = \sum_{j=1}^{\infty}\left(\hat{b}_j^{\dagger}\varphi_j^{\dagger}(\vec{x}) + \hat{d}_j\varphi_{-j}^{\dagger}(\vec{x})\right) \tag{5.6}$$

The free field Hamiltonian operator becomes,

$$\hat{H}_0 = \sum_{n=1}^{\infty}\left(\lambda_n E_n \hat{b}_n^{\dagger}\hat{b}_n + \lambda_{-n}E_{-n}\hat{d}_n\hat{d}_n^{\dagger}\right) = \sum_{n=1}^{\infty}\left(E_n\hat{b}_n^{\dagger}\hat{b}_n - E_n\hat{d}_n\hat{d}_n^{\dagger}\right) \tag{5.7}$$

where we have used $E_n = E_{-n}$, $\lambda_n = 1$, and $\lambda_{-n} = -1$ for $n > 0$. The destruction and creation operators satisfy,

$$\left\{\hat{d}_j, \hat{d}_k^{\dagger}\right\} = \delta_{jk}\,;\ \left\{\hat{b}_j, \hat{b}_k^{\dagger}\right\} = \delta_{jk}\,;\ \text{all other anti-commutators are zero} \tag{5.8}$$

The vacuum state $|0\rangle$ is now defined by,

$$\hat{d}_j|0\rangle = \hat{b}_j|0\rangle = 0 \text{ and } \langle 0|\hat{d}_j^{\dagger} = \langle 0|\hat{b}_j^{\dagger} = 0 \tag{5.9}$$

Use (5.8) in (5.7) to obtain,

$$\hat{H}_0 = \sum_{n=1}^{\infty}E_n\left(\hat{b}_n^{\dagger}\hat{b}_n - \left(1 - \hat{d}_n^{\dagger}\hat{d}_n\right)\right) = \sum_{n=1}^{\infty}\left(E_n\hat{b}_n^{\dagger}\hat{b}_n + E_n\hat{d}_n^{\dagger}\hat{d}_n\right) - \xi_{vac} \tag{5.10}$$

At this point redefine $\hat{H}_0$ by adding the constant ξ_{vac} to obtain,

$$\hat{H}_0 = \sum_{n=1}^{\infty}\left(E_n\hat{b}_n^{\dagger}\hat{b}_n + E_n\hat{d}_n^{\dagger}\hat{d}_n\right) \tag{5.11}$$

This last step does not affect any results and simply corresponds to a shift in energy, making the energy of the vacuum state equal to zero. This will simplify some of the mathematical analysis but does not change any of the results of the following discussion.

We can now think of the vacuum state $|0\rangle$ as the state which contains no electrons, no positrons, and has zero energy. New eigenstates states $|k\rangle$ are created by acting on $|0\rangle$ with electron and positron creation operators, e.g,

$$|k\rangle = \hat{b}_{j_1}^{\dagger}\hat{b}_{j_2}^{\dagger}\ldots\hat{b}_{j_s}^{\dagger}\hat{d}_{v_1}^{\dagger}\hat{d}_{v_2}^{\dagger}\ldots\hat{d}_{v_r}^{\dagger}|0\rangle \tag{5.12}$$

From this definition and (5.11) we have,

$$\hat{H}_0|k\rangle = \xi_{|k\rangle}|k\rangle \tag{5.13}$$

where $\xi_{|k\rangle}$ is the energy eigenvalue of the eigenstate $|k\rangle$ and is given by,

$$\xi_{|k\rangle} = \left(E_{j_1} + E_{j_2} + \ldots + E_{j_s}\right) + \left(E_{v_1} + E_{v_2} + \ldots + E_{v_r}\right) \tag{5.14}$$

Since all the quantities E_j and E_v in the above equation are greater than zero we have that,

$$\xi_{|k\rangle} > \xi_{|0\rangle} = 0 \text{ for all } |k\rangle \neq |0\rangle \tag{5.15}$$

The eigenstates $|k\rangle$ form an orthonormal set in fock space and satisfy [11],

$$\langle k|q\rangle = \delta_{kq} \text{ and } \sum_{|k\rangle} |k\rangle\langle k| = 1 \tag{5.16}$$

where the summation is over all eigenstates $|k\rangle$. Any arbitrary state vector $|\Omega\rangle$ can be expressed as an expansion in terms of the eigenstates $|k\rangle$ as,

$$|\Omega\rangle = \sum_{|k\rangle} c_{|k\rangle} |k\rangle \tag{5.17}$$

The free field energy of a given normalized state $|\Omega\rangle$ is defined by,

$$\xi_f\left(|\Omega\rangle\right) = \langle\Omega|\hat{H}_0|\Omega\rangle \tag{5.18}$$

From the above discussion we have that,

$$\xi_f\left(|\Omega\rangle\right) = \langle\Omega|\hat{H}_0|\Omega\rangle = \sum_{|k\rangle} \left|c_{|k\rangle}\right|^2 \xi_{|k\rangle} \tag{5.19}$$

Since the $\xi_{|k\rangle}$ are all positive we have that,

$$\xi_f\left(|\Omega\rangle\right) \geq \xi_f\left(|0\rangle\right) = 0 \text{ for all } |\Omega\rangle \tag{5.20}$$

Therefore the vacuum state $|0\rangle$ represents a lower bound to the free field energy.

6. Gauge Invariance and Field Theory

The discussion up to this point has essentially been a review of various elements of quantum theory. These elements, in one form or another, are found in many basic works on quantum mechanics. The purpose of this review was to trace the path from the single particle Dirac equation to N-electron theory and then to the Schrödinger representation of quantum field theory in which it shown that N-electron theory and field theory are equivalent. Hole theory can be considered as N-electron theory with $N \to \infty$ and the vacuum defined as the state in which all negative energy states are occupied. N-electron theory can be shown to obey the continuity equation and to be gauge invariant. Therefore we expect that quantum field theory should also obey the continuity equation and be gauge invariant. In the remainder of this section we will show that this is not, necessarily, the case (see also ref. [12]).

Consider a normalized state vector $|\Omega(t)\rangle$. Assume that at some initial time, say $t = t_i$, $|\Omega(t_i)\rangle$ satisfies (5.20), i.e., $\xi_f(|\Omega(t_i)\rangle) \geq \xi_f(|0\rangle)$. Now let $|\Omega(t)\rangle$ evolve according to equation (4.20). If the electric potential is non-zero then, in general, the free field energy $\xi_f(|\Omega(t)\rangle)$ will change in time. It will be shown that if we assume that field theory is gauge invariant and the continuity equation holds then it is possible to specify an electric potential so that (5.20) is not true at some final time $t_f > t_i$. Therefore if (5.20) is true then the principle of gauge invariance or the continuity equation or both is not valid for field theory.

Start by taking the time derivative of $\xi_f(|\Omega(t)\rangle)$ and use (4.20) and (5.18) to obtain,

$$\frac{\partial \xi_f(|\Omega(t)\rangle)}{\partial t} = i\langle\Omega(t)|\left[\hat{H}, \hat{H}_0\right]|\Omega(t)\rangle \tag{6.1}$$

Next use (4.22) and substitute for $\hat{H}_0$ in the above to obtain,

$$\frac{\partial \xi_f(|\Omega(t)\rangle)}{\partial t} = i\langle\Omega(t)|\left[\hat{H}, \left(\hat{H} + \int \hat{\vec{J}}(\vec{x}) \cdot \vec{A}(\vec{x},t)d\vec{x} - \int \hat{\rho}(\vec{x}) A_0(\vec{x},t)d\vec{x}\right)\right]|\Omega(t)\rangle \tag{6.2}$$

Rearrange terms to yield,

$$\frac{\partial \xi_f(|\Omega(t)\rangle)}{\partial t} = i\begin{pmatrix} \int \langle\Omega(t)|\left[\hat{H}, \hat{\vec{J}}(\vec{x})\right]|\Omega(t)\rangle \cdot \vec{A}(\vec{x},t)d\vec{x} \\ -\int \langle\Omega(t)|\left[\hat{H}, \hat{\rho}(\vec{x})\right]|\Omega(t)\rangle A_0(\vec{x},t)d\vec{x} \end{pmatrix} \tag{6.3}$$

The current and charge expectation values are defined by,

$$\vec{J}_e(\vec{x},t) = \langle\Omega(t)|\hat{\vec{J}}(\vec{x})|\Omega(t)\rangle \text{ and } \rho_e(\vec{x},t) = \langle\Omega(t)|\hat{\rho}(\vec{x})|\Omega(t)\rangle \tag{6.4}$$

Use this along with (4.20) to obtain,

$$\frac{\partial \vec{J}_e(\vec{x},t)}{\partial t} = i\langle\Omega(t)|\left[\hat{H},\hat{\vec{J}}(\vec{x})\right]|\Omega(t)\rangle \text{ and } \frac{\partial \rho_e(\vec{x},t)}{\partial t} = i\langle\Omega(t)|\left[\hat{H},\hat{\rho}(\vec{x})\right]|\Omega(t)\rangle \quad (6.5)$$

Use this result in (6.3) to obtain,

$$\frac{\partial \xi_f(|\Omega(t)\rangle)}{\partial t} = \left(\int \frac{\partial \vec{J}_e(\vec{x},t)}{\partial t}\cdot\vec{A}(\vec{x},t)d\vec{x} - \int \frac{\partial \rho_e(\vec{x},t)}{\partial t}A_0(\vec{x},t)d\vec{x}\right) \quad (6.6)$$

Next, since we assume that the continuity equation is true for quantum field theory, refer to (2.7) and substitute for $\partial\rho_e/\partial t$ to obtain,

$$\frac{\partial \xi_f(|\Omega(t)\rangle)}{\partial t} = \left(\int \frac{\partial \vec{J}_e(\vec{x},t)}{\partial t}\cdot\vec{A}(\vec{x},t)d\vec{x} + \int \vec{\nabla}\cdot\vec{J}_e(\vec{x},t)A_0(\vec{x},t)d\vec{x}\right) \quad (6.7)$$

Now let the electric potential be given by,

$$\left(A_0(\vec{x},t),\vec{A}(\vec{x},t)\right) = \left(\frac{\partial\chi(\vec{x},t)}{\partial t}, -\vec{\nabla}\chi(\vec{x},t)\right) \quad (6.8)$$

where $\chi(\vec{x},t)$ is an arbitrary real valued function. Use this in (6.7) to obtain,

$$\frac{\partial \xi_f(|\Omega(t)\rangle)}{\partial t} = \left(-\int \frac{\partial \vec{J}_e(\vec{x},t)}{\partial t}\cdot\vec{\nabla}\chi(\vec{x},t)d\vec{x} + \int \vec{\nabla}\cdot\vec{J}_e(\vec{x},t)\frac{\partial\chi(\vec{x},t)}{\partial t}d\vec{x}\right) \quad (6.9)$$

Integrate the first integral by parts and assume reasonable boundary conditions to obtain,

$$\frac{\partial \xi_f(|\Omega(t)\rangle)}{\partial t} = \left(\int \frac{\partial \vec{\nabla}\cdot\vec{J}_e(\vec{x},t)}{\partial t}\chi(\vec{x},t)d\vec{x} + \int \vec{\nabla}\cdot\vec{J}_e(\vec{x},t)\frac{\partial\chi(\vec{x},t)}{\partial t}d\vec{x}\right) \quad (6.10)$$

This becomes,

$$\frac{\partial \xi_f(|\Omega(t)\rangle)}{\partial t} = \frac{\partial}{\partial t}\int \chi(\vec{x},t)\vec{\nabla}\cdot\vec{J}_e(\vec{x},t)d\vec{x} \quad (6.11)$$

Integrate the above from the initial time $t = t_i$ to some final time $t_f > t_i$. This yields,

$$\xi_f(|\Omega(t_f)\rangle) - \xi_f(|\Omega(t_i)\rangle) = \int \chi(\vec{x},t_f)\vec{\nabla}\cdot\vec{J}_e(\vec{x},t_f)d\vec{x} - \int \chi(\vec{x},t_i)\vec{\nabla}\cdot\vec{J}_e(\vec{x},t_i)d\vec{x} \quad (6.12)$$

Next invoke the principle of gauge invariance. When the electric potential of (6.8) is substituted into (2.8) it can be seen that the electromagnetic field is zero for all functions

$\chi(\vec{x},t)$. Therefore, from the assumption of gauge invariance, the current expectation value $\vec{J}_e(\vec{x},t)$ is independent of $\chi(\vec{x},t)$. This means that $\chi(\vec{x},t)$ can be varied in an arbitrary manner without affecting $\vec{J}_e(\vec{x},t)$. Therefore, if $\vec{\nabla}\cdot\vec{J}_e(\vec{x},t_f)$ is non-zero then it is always possible to find a $\chi(\vec{x},t)$ which makes the final free field energy, $\xi_f(|\Omega(t_f)\rangle)$, a negative number with an arbitrarily large magnitude. For example, let $\chi(\vec{x},t_i)=0$ and $\chi(\vec{x},t_f)=-f\vec{\nabla}\cdot\vec{J}_e(\vec{x},t_f)$ where f is a constant. Use this in (6.12) to obtain,

$$\xi_f(|\Omega(t_f)\rangle)=\xi_f(|\Omega(t_i)\rangle)-f\int\left(\vec{\nabla}\cdot\vec{J}_e(\vec{x},t_f)\right)^2 d\vec{x} \tag{6.13}$$

If $\vec{\nabla}\cdot\vec{J}_e(\vec{x},t_f)$ is non-zero, the integral on the right is always positive so that as $f\to\infty$, $\xi_f(|\Omega(t_f)\rangle)\to-\infty$. Therefore there is no lower bound for the free field energy and the relationship given in (5.20) cannot be true for all state vectors $|\Omega(t)\rangle$.

Now the above result depends on the assumption that the divergence of the current expectation value is non-zero. How do we know that this will be the case? If we assume that quantum theory is a correct model of the real world then we can always find a quantum state for which $\vec{\nabla}\cdot\vec{J}_e(\vec{x},t_f)$ is non-zero because there are many examples in the real world where this is, indeed, the case.

In the discussion leading up to (6.13) we have used equation (4.20) for the evolution of the state vector $|\Omega(t)\rangle$, assumed that the continuity equation is true, and assumed that the theory is gauge invariant. We find that these conditions are not consistent with the relationship given by (5.20). Therefore, one or more of these relationships must not be valid.

7. The Schwinger Term

In the previous section we assumed that the continuity equation was true. The basis of this assumption was that quantum field theory is derived from N-electron theory and that the continuity equation is certainly true for N-electron theory. In this section will we will discuss the requirements for continuity equation to be true in the Schrödinger representation of quantum field theory. The expression for the continuity equation is given by (2.7). Use this and (6.4) to obtain,

$$\frac{\partial\langle\Omega(t)|\hat{\rho}(\vec{x})|\Omega(t)\rangle}{\partial t}=-\langle\Omega(t)|\vec{\nabla}\cdot\hat{\vec{J}}(\vec{x})|\Omega(t)\rangle \tag{7.1}$$

Next use (4.20) in the above to obtain,

$$i\langle\Omega(t)|\left[\hat{H},\hat{\rho}(\vec{x})\right]|\Omega(t)\rangle=-\langle\Omega(t)|\vec{\nabla}\cdot\hat{\vec{J}}(\vec{x})|\Omega(t)\rangle \tag{7.2}$$

Use (4.22) in the above to yield,

$$i\langle\Omega(t)|\begin{pmatrix}\left[\hat{H}_0,\hat{\rho}(\vec{x})\right]-\int\left[\hat{\vec{J}}(\vec{y}),\hat{\rho}(\vec{x})\right]\cdot\vec{A}(\vec{y},t)d\vec{y}\\+\int\left[\hat{\rho}(\vec{y}),\hat{\rho}(\vec{x})\right]A_0(\vec{y},t)d\vec{y}\end{pmatrix}|\Omega(t)\rangle=-\langle\Omega(t)|\vec{\nabla}\cdot\hat{\vec{J}}(\vec{x})|\Omega(t)\rangle \tag{7.3}$$

For the above equation to be true for arbitrary values of the state vector $|\Omega(t)\rangle$ and the electric potential $\left(A_0(\vec{x},t),\vec{A}(\vec{x},t)\right)$ the following relationships must hold,

$$\left[\hat{\vec{J}}(\vec{y}),\hat{\rho}(\vec{x})\right]=0 \tag{7.4}$$

$$i\left[\hat{H}_0,\hat{\rho}(\vec{x})\right]=-\vec{\nabla}\cdot\hat{\vec{J}}(\vec{x}) \tag{7.5}$$

$$\left[\hat{\rho}(\vec{y}),\hat{\rho}(\vec{x})\right]=0 \tag{7.6}$$

These relationships are a sufficient and necessary condition for the continuity equation to be true in Dirac field theory. However it has been show by Schwinger [13] that if (5.15) and (7.5) are true then (7.4) cannot be true. Define the Schwinger term by

$$ST(\vec{y},\vec{x})=\left[\hat{\rho}(\vec{y}),\hat{\vec{J}}(\vec{x})\right] \tag{7.7}$$

According to (7.4) $ST(\vec{y},\vec{x})$ must be zero for the continuity equation to be valid.

Take the divergence of the Schwinger term $\left[\hat{\rho}(\vec{y}),\hat{\vec{J}}(\vec{x})\right]$ and use (7.5) to obtain,

$$\vec{\nabla}_{\vec{x}}\cdot\left[\hat{\rho}(\vec{y}),\hat{\vec{J}}(\vec{x})\right]=\left[\hat{\rho}(\vec{y}),\vec{\nabla}\cdot\hat{\vec{J}}(\vec{x})\right]=-i\left[\hat{\rho}(\vec{y}),\left[\hat{H}_0,\hat{\rho}(\vec{x})\right]\right] \tag{7.8}$$

Next expand the commutator to yield,

$$i\vec{\nabla}_{\vec{x}} \cdot \left[\hat{\rho}(\vec{y}), \hat{\vec{J}}(\vec{x}) \right] = -\hat{H}_0 \hat{\rho}(\vec{x})\hat{\rho}(\vec{y}) + \hat{\rho}(\vec{x})\hat{H}_0\hat{\rho}(\vec{y}) + \hat{\rho}(\vec{y})\hat{H}_0\hat{\rho}(\vec{x}) - \hat{\rho}(\vec{y})\hat{\rho}(\vec{x})\hat{H}_0 \tag{7.9}$$

Sandwich the above expression between the state vector $\langle 0|$ and its dual $|0\rangle$ and use $\hat{H}_0|0\rangle = 0$ and $\langle 0|\hat{H}_0 = 0$ to obtain,

$$i\vec{\nabla}_{\vec{x}} \cdot \langle 0|\left[\hat{\rho}(\vec{y}), \hat{\vec{J}}(\vec{x}) \right]|0\rangle = \langle 0|\hat{\rho}(\vec{x})\hat{H}_0\hat{\rho}(\vec{y})|0\rangle + \langle 0|\hat{\rho}(\vec{y})\hat{H}_0\hat{\rho}(\vec{x})|0\rangle \tag{7.10}$$

Next set $\vec{y} = \vec{x}$ to obtain,

$$i\vec{\nabla}_{\vec{x}} \cdot \langle 0|\left[\hat{\rho}(\vec{y}), \hat{\vec{J}}(\vec{x}) \right]|0\rangle \underset{\vec{y}=\vec{x}}{=} 2\langle 0|\hat{\rho}(\vec{x})\hat{H}_0\hat{\rho}(\vec{x})|0\rangle \tag{7.11}$$

Use (5.16) in the above to obtain,

$$i\vec{\nabla}_{\vec{x}} \cdot \langle 0|\left[\hat{\rho}(\vec{y}), \hat{\vec{J}}(\vec{x}) \right]|0\rangle \underset{\vec{y}=\vec{x}}{=} 2\sum_{|n\rangle,|m\rangle} \langle 0|\hat{\rho}(\vec{x})|n\rangle\langle n|\hat{H}_0|m\rangle\langle m|\hat{\rho}(\vec{x})|0\rangle \tag{7.12}$$

Next use (5.16) and (5.13) to obtain,

$$i\vec{\nabla}_{\vec{x}} \cdot \langle 0|\left[\hat{\rho}(\vec{y}), \hat{\vec{J}}(\vec{x}) \right]|0\rangle \underset{\vec{y}=\vec{x}}{=} 2\sum_{|n\rangle} \xi_{|n\rangle} \langle 0|\hat{\rho}(\vec{x})|n\rangle\langle n|\hat{\rho}(\vec{x})|0\rangle = 2\sum_{|n\rangle} \xi_{|n\rangle} \left|\langle 0|\hat{\rho}(\vec{x})|n\rangle\right|^2 \tag{7.13}$$

Now, in general, the quantity $\langle 0|\hat{\rho}(\vec{x})|n\rangle$ is not zero [13] and since $\xi_{|n\rangle} > 0$ for all $|n\rangle \neq |0\rangle$ the above expression is non-zero and positive. Therefore the Schwinger term must be non-zero. This is, of course, in direct contradiction to (7.4).

As a result of the above discussion we see that the various elements of quantum field theory do not produce consistent results. On one hand if the continuity equation is true then the Schwinger term must be zero. On the other hand if the vacuum state $|0\rangle$ is a lower bound to the free field energy (per Eqs. (5.15) or (5.20)) then the Schwinger term cannot be zero.

8. Evaluating the Schwinger term

To confirm the results of the previous section we will use the material of section V to show that the Schwinger term is, indeed, nonzero. Define,

$$\vec{I}(\vec{y},\vec{x}) = \langle 0|\left[\hat{\rho}(\vec{y}),\hat{\vec{J}}(\vec{x})\right]|0\rangle = \langle 0|ST(\vec{y},\vec{x})|0\rangle \tag{8.1}$$

If $ST(\vec{y},\vec{x})$ is zero then $\vec{I}(\vec{y},\vec{x})$ must be zero. Use (4.18) in the above to yield,

$$\vec{I}(\vec{y},\vec{x}) = q^2\langle 0|\left[\hat{\psi}^\dagger(\vec{y})\hat{\psi}(\vec{y}),\hat{\psi}^\dagger(\vec{x})\vec{\alpha}\hat{\psi}(\vec{x})\right]|0\rangle \tag{8.2}$$

Use (5.5) and (5.6) in the above to obtain,

$$\vec{I}(\vec{y},\vec{x}) = q^2\sum_{n,m,r,s=1}^{\infty}\langle 0|\begin{bmatrix}\begin{pmatrix}\hat{b}_n^\dagger\varphi_n^{(0)\dagger}(\vec{y})\\+\hat{d}_n\varphi_{-n}^{(0)\dagger}(\vec{y})\end{pmatrix}\begin{pmatrix}\hat{b}_m\varphi_m^{(0)}(\vec{y})\\+\hat{d}_m^\dagger\varphi_{-m}^{(0)}(\vec{y})\end{pmatrix},\\ \begin{pmatrix}\hat{b}_r^\dagger\varphi_r^{(0)\dagger}(\vec{x})\\+\hat{d}_r\varphi_{-r}^{(0)\dagger}(\vec{x})\end{pmatrix}\vec{\alpha}\begin{pmatrix}\hat{b}_s\varphi_s^{(0)}(\vec{x})\\+\hat{d}_s^\dagger\varphi_{-s}^{(0)}(\vec{x})\end{pmatrix}\end{bmatrix}|0\rangle \tag{8.3}$$

Use (5.8) and (5.9) in the above to yield,

$$\vec{I}(\vec{y},\vec{x}) = q^2\sum_{\substack{n,m,\\r,s=1}}^{\infty}\begin{pmatrix}\langle 0|\hat{b}_n^\dagger\hat{d}_m^\dagger,\hat{d}_r\hat{b}_s|0\rangle\left(\varphi_n^{(0)\dagger}(\vec{y})\varphi_{-m}^{(0)}(\vec{y})\right)\left(\varphi_{-r}^{(0)\dagger}(\vec{x})\vec{\alpha}\varphi_s^{(0)}(\vec{x})\right)\\+\langle 0|\hat{d}_n\hat{b}_m,\hat{b}_r^\dagger\hat{d}_s^\dagger|0\rangle\left(\varphi_{-n}^{(0)\dagger}(\vec{y})\varphi_m^{(0)}(\vec{y})\right)\left(\varphi_r^{(0)\dagger}(\vec{x})\vec{\alpha}\varphi_s^{(0)}(\vec{x})\right)\\+\langle 0|\hat{d}_n\hat{d}_m^\dagger,\hat{d}_r\hat{d}_s^\dagger|0\rangle\left(\varphi_{-n}^{(0)\dagger}(\vec{y})\varphi_{-m}^{(0)}(\vec{y})\right)\left(\varphi_{-r}^{(0)\dagger}(\vec{x})\vec{\alpha}\varphi_{-s}^{(0)}(\vec{x})\right)\end{pmatrix} \tag{8.4}$$

From (5.8) and (5.9) we have the following relationships,

$$\langle 0|\hat{b}_n^\dagger\hat{d}_m^\dagger,\hat{d}_r\hat{b}_s|0\rangle = \langle 0|\hat{b}_n^\dagger\hat{d}_m^\dagger\hat{d}_r\hat{b}_s|0\rangle - \langle 0|\hat{d}_r\hat{b}_s\hat{b}_n^\dagger\hat{d}_m^\dagger|0\rangle = -\delta_{mr}\delta_{ns}$$

$$\langle 0|\hat{d}_n\hat{b}_m,\hat{b}_r^\dagger\hat{d}_s^\dagger|0\rangle = \langle 0|\hat{d}_n\hat{b}_m\hat{b}_r^\dagger\hat{d}_s^\dagger|0\rangle - \langle 0|\hat{b}_r^\dagger\hat{d}_s^\dagger\hat{d}_n\hat{b}_m|0\rangle = \delta_{sn}\delta_{rm}$$

$$\langle 0|\hat{d}_n\hat{d}_m^\dagger,\hat{d}_r\hat{d}_s^\dagger|0\rangle = \langle 0|\hat{d}_n\hat{d}_m^\dagger\hat{d}_r\hat{d}_s^\dagger|0\rangle - \langle 0|\hat{d}_r\hat{d}_s^\dagger\hat{d}_n\hat{d}_m^\dagger|0\rangle = \delta_{sr}\delta_{mn} - \delta_{sr}\delta_{mn} = 0$$

Use these in (8.4) to obtain,

$$\vec{I}(\vec{y},\vec{x}) = q^2\sum_{n,m=1}^{\infty}\begin{pmatrix}\left(\varphi_{-n}^{(0)\dagger}(\vec{y})\varphi_m^{(0)}(\vec{y})\right)\left(\varphi_m^{(0)\dagger}(\vec{x})\vec{\alpha}\varphi_{-n}^{(0)}(\vec{x})\right)\\-\left(\varphi_n^{(0)\dagger}(\vec{y})\varphi_{-m}^{(0)}(\vec{y})\right)\left(\varphi_{-m}^{(0)\dagger}(\vec{x})\vec{\alpha}\varphi_n^{(0)}(\vec{x})\right)\end{pmatrix} \tag{8.5}$$

By relabeling some of the dummy indices this can be rewritten as,

$$\vec{I}(\vec{y},\vec{x}) = \left\{q^2\sum_{n,m=1}^{\infty}\left(\varphi_{-n}^{(0)\dagger}(\vec{y})\varphi_m^{(0)}(\vec{y})\right)\left(\varphi_m^{(0)\dagger}(\vec{x})\vec{\alpha}\varphi_{-n}^{(0)}(\vec{x})\right)\right\} - (\text{c.c.}) \tag{8.6}$$

where the expression c.c. means to take the complex conjugate of the previous term. Next take the divergence of the above expression with respect to $\vec{x}$ to obtain,

$$\vec{\nabla}_{\vec{x}} \cdot \vec{I}(\vec{y},\vec{x}) = \left\{ q^2 \sum_{n,m=1}^{\infty} \left(\varphi_{-n}^{(0)\dagger}(\vec{y}) \varphi_m^{(0)}(\vec{y}) \right) \vec{\nabla} \cdot \left(\varphi_m^{(0)\dagger}(\vec{x}) \vec{\alpha} \varphi_{-n}^{(0)}(\vec{x}) \right) \right\} - (\text{c.c.}) \quad (8.7)$$

To evaluate this expression use the following,

$$\vec{\nabla} \cdot \left(\varphi_m^{(0)\dagger}(\vec{x}) \vec{\alpha} \varphi_{-n}^{(0)}(\vec{x}) \right) = \left(\left(\vec{\alpha} \cdot \vec{\nabla} \varphi_m^{(0)}(\vec{x}) \right)^{\dagger} \varphi_{-n}^{(0)}(\vec{x}) \right) + \left(\varphi_m^{(0)\dagger}(\vec{x}) \vec{\alpha} \cdot \vec{\nabla} \varphi_{-n}^{(0)}(\vec{x}) \right) (8.8)$$

Next use (2.3) in the above to obtain,

$$\vec{\nabla} \cdot \left(\varphi_m^{(0)\dagger}(\vec{x}) \vec{\alpha} \varphi_{-n}^{(0)}(\vec{x}) \right) = \left(\left(iH_0 \varphi_m^{(0)}(\vec{x}) \right)^{\dagger} \varphi_{-n}^{(0)}(\vec{x}) \right) + \left(\varphi_m^{(0)\dagger}(\vec{x}) iH_0 \varphi_{-n}^{(0)}(\vec{x}) \right) (8.9)$$

Next use (2.15) to obtain,

$$\vec{\nabla} \cdot \left(\varphi_m^{(0)\dagger}(\vec{x}) \vec{\alpha} \varphi_{-n}^{(0)}(\vec{x}) \right) = i \varphi_m^{(0)\dagger}(\vec{x}) \left(\lambda_{-n} E_{-n} - \lambda_m E_m \right) \varphi_{-n}^{(0)}(\vec{x}) \quad (8.10)$$

Use this in (8.7) to yield,

$$\vec{\nabla}_{\vec{x}} \cdot \vec{I}(\vec{y},\vec{x}) = \left\{ iq^2 \sum_{n,m=1}^{\infty} \left(\varphi_{-n}^{(0)\dagger}(\vec{y}) \varphi_m^{(0)}(\vec{y}) \right) \left(\varphi_m^{(0)\dagger}(\vec{x}) \varphi_{-n}^{(0)}(\vec{x}) \right) \begin{pmatrix} \lambda_{-n} E_{-n} \\ -\lambda_m E_m \end{pmatrix} \right\} - (\text{c.c.})$$

(8.11)

Use the fact that, in the above expression, $\lambda_{-n} = -1$, $\lambda_m = 1$, $E_{-n} = E_n$ and set $\vec{y} = \vec{x}$ to obtain,

$$\vec{\nabla}_{\vec{x}} \cdot \vec{I}(\vec{y},\vec{x}) \underset{\vec{y}=\vec{x}}{=} -2 \left\{ iq^2 \sum_{n,m=1}^{\infty} \left(\varphi_{-n}^{(0)\dagger}(\vec{x}) \varphi_m^{(0)}(\vec{x}) \right) \left(\varphi_m^{(0)\dagger}(\vec{x}) \varphi_{-n}^{(0)}(\vec{x}) \right) \left(E_n + E_m \right) \right\} \quad (8.12)$$

This yields,

$$\vec{\nabla}_{\vec{x}} \cdot \vec{I}(\vec{y},\vec{x}) \underset{\vec{y}=\vec{x}}{=} -2 \left\{ iq^2 \sum_{n,m=1}^{\infty} \left| \left(\varphi_m^{(0)\dagger}(\vec{x}) \varphi_{-n}^{(0)}(\vec{x}) \right) \right|^2 \left(E_n + E_m \right) \right\} \quad (8.13)$$

Each term in the sum is positive so that the above expression is not equal to zero. This confirms that the Schwinger term $ST(\vec{y},\vec{x})$ is non-zero. This result is expected because, as previously discussed, the Schwinger term must be non-zero if (5.15) is true.

9. The Schwinger term and Gauge invariance

In Section 7 it was shown that the Schwinger term must be zero for the continuity equation to be true for arbitrary values of the electric potential. In this section we will also show that the Schwinger term must be zero for the theory to be gauge invariant. To demonstrate this, take the time derivative of the current expectation value (6.4) to yield,

$$\frac{\partial \vec{J}_e(\vec{x},t)}{\partial t} = i\left\langle \Omega(t)\right|\left[\hat{H}(t),\hat{\vec{J}}(\vec{x})\right]\left|\Omega(t)\right\rangle \tag{9.1}$$

Use (4.22) in the above to obtain,

$$\frac{\partial \vec{J}_e(\vec{x},t)}{\partial t} = i\left\langle \Omega(t)\right|\left(\begin{array}{l}\left[\hat{H}_0,\hat{\vec{J}}(\vec{x})\right]-\int\left[\hat{\vec{J}}(\vec{y})\cdot\vec{A}(\vec{y},t),\hat{\vec{J}}(\vec{x})\right]d\vec{y}\\ +\int\left[\hat{\rho}(\vec{y}),\hat{\vec{J}}(\vec{x})\right]A_0(\vec{y},t)d\vec{y}\end{array}\right)\left|\Omega(t)\right\rangle \tag{9.2}$$

Next perform the gauge transformation (2.9) to obtain,

$$\frac{\partial \vec{J}_e(\vec{x},t)}{\partial t} = i\left\langle \Omega(t)\right|\left(\begin{array}{l}\left[\hat{H}_0,\hat{\vec{J}}(\vec{x})\right]-\int\left[\hat{\vec{J}}(\vec{y})\cdot\left(\vec{A}(\vec{y},t)-\vec{\nabla}\chi(\vec{y},t)\right),\hat{\vec{J}}(\vec{x})\right]d\vec{y}\\ +\int\left[\hat{\rho}(\vec{y}),\hat{\vec{J}}(\vec{x})\right]\left(A_0(\vec{y},t)+\frac{\partial\chi(\vec{y},t)}{\partial t}\right)d\vec{y}\end{array}\right)\left|\Omega(t)\right\rangle \tag{9.3}$$

The quantity $\partial\vec{J}_e/\partial t$ is a physical observable and therefore, if the theory is gauge invariant, must not depend on the quantities χ or $\partial\chi/\partial t$. Now, at a particular instant of time $\partial\chi/\partial t$ can be varied in an arbitrary manner without changing the values, at that instant of time, of any of the other quantities on the right hand side of the equals sign in the above equation. Therefore for $\partial\vec{J}_e/\partial t$ to be independent of $\partial\chi/\partial t$ we must have that $ST(\vec{y},\vec{x})=\left[\hat{\rho}(\vec{y}),\hat{\vec{J}}(\vec{x})\right]=0$. However, we have just shown that the Schwinger term is nonzero therefore we expect that there should be a problem with the gauge invariance of Dirac field theory.

This problem shows up when the vacuum polarization tensor $\pi^{\mu\nu}$ is calculated. The first order change in the vacuum current due to an applied electric potential can be shown to be,

$$J^{\mu}_{vac}\left(\vec{x},t\right)=\int\pi^{\mu\nu}\left(\vec{x}-\vec{x}',t-t',\right)A_{\nu}\left(\vec{x}',t'\right)d\vec{x}'dt' \tag{9.4}$$

It is well known that when the vacuum polarization tensor, $\pi^{\mu\nu}$, is calculated, using perturbation theory, the result is not gauge invariant (see Chapter 14 of [3], Sect. 22 of [6], Chapter 5 of [14], and [15]). The non-gauge invariant terms must be removed from the results of the calculation in order to obtain a physically correct result. This may involve some form of regularization, where other functions are introduced that happen to have the correct behavior so that the non-gauge invariant terms are cancelled. However there is no physical explanation for introducing these functions [15]. They are merely mathematical devices used to force the desired gauge invariant result.

Consider, for example, a calculation of the vacuum polarization tensor by Heitler (see page 322 of ref. 6). Heitler's solution for the Fourier transform of the vacuum polarization tensor is given by

$$\pi^{uv}\left(k^{\alpha}\right)=\pi^{uv}_{G}\left(k^{\alpha}\right)+\pi^{uv}_{NG}\left(k^{\alpha}\right) \tag{9.5}$$

where k^{α} is the 4 momentum of the electromagnetic field. The first term on the right hand side is given by,

$$\pi^{uv}_{G}\left(k^{\alpha}\right)=\left(\frac{2q^2}{3\pi}\right)\left(k^{\mu}k^{\nu}-g^{\mu\nu}k^2\right)\int_{2m}^{\infty}dz\frac{\left(z^2+2m^2\right)\sqrt{\left(z^2-4m^2\right)}}{z^2\left(z^2-k^2\right)} \tag{9.6}$$

This term is gauge invariant because $k_{\nu}\pi^{uv}_{G}=0$.

The second term on the right of (9.5) is

$$\pi^{uv}_{NG}\left(k^{\alpha}\right)=\left(\frac{2q^2}{3\pi}\right)g^{\mu}_{\nu}\left(1-g^{\mu 0}\right)\int_{2m}^{\infty}dz\frac{\left(z^2+2m^2\right)\sqrt{\left(z^2-4m^2\right)}}{z^2} \tag{9.6}$$

where there is no summation over the two μ superscripts that appear on the right. Note that π^{uv}_{NG} is not gauge invariant because $k_{\nu}\pi^{uv}_{NG}\neq 0$. Therefore to get a physically correct result it is necessary to "correct" equation (9.5) by dropping π_{NG}^{uv} from the solution.

The problem that we are addressing is why does this extra non-gauge invariant term appear in a theory that is supposed to be gauge invariant? In order to understand the source of the problem we shall use perturbation theory to calculate the vacuum current and show that the lack of gauge invariance of the result is directly related to the fact the Schwinger term is non-zero. This problem was originally addressed in [12]. The following analysis closely follows this reference.

The first order change in the vacuum current due to an applied electric potential is given by the following expression which is derived in Appendix A (see also [12] and Eq. 8.3 of [16]).

$$\vec{J}_{vac}(\vec{x},t) = i\langle 0|\left[\hat{\vec{J}}_I(\vec{x},t), \int d\vec{y}\int_{-\infty}^{t} dt'\left(-\hat{\vec{J}}_I(\vec{y},t')\cdot\vec{A}(\vec{y},t') + \hat{\rho}_I(\vec{y},t')A_0(\vec{y},t')\right)\right]|0\rangle \quad (9.7)$$

where the operators $\hat{\vec{J}}_I(\vec{x},t)$ and $\hat{\rho}_I(\vec{x},t)$ are the current and charge operators, respectively, in the interaction representation. They are defined by,

$$\hat{\vec{J}}_I(\vec{x},t) = e^{i\hat{H}_0 t}\hat{\vec{J}}(\vec{x})e^{-i\hat{H}_0 t} \text{ and } \hat{\rho}_I(\vec{x},t) = e^{i\hat{H}_0 t}\hat{\rho}(\vec{x})e^{-i\hat{H}_0 t} \quad (9.8)$$

They can be shown to satisfy the continuity equation (see Eq. 3.11 of [16]),

$$\frac{\partial\hat{\rho}_I(\vec{x},t)}{\partial t} = -\vec{\nabla}\cdot\hat{\vec{J}}_I(\vec{x},t) \quad (9.9)$$

The change in the vacuum current $\delta_g\vec{J}_{vac}(\vec{x},t)$ due to a gauge transformation is obtained by using (2.9) in (9.7) to yield

$$\delta_g\vec{J}_{vac}(\vec{x},t) = i\langle 0|\left[\hat{\vec{J}}_I(\vec{x},t), \int d\vec{y}\int_{-\infty}^{t} dt'\left(\hat{\vec{J}}_I(\vec{y},t')\cdot\vec{\nabla}\chi(\vec{y},t') + \hat{\rho}_I(\vec{y},t')\frac{\partial\chi(\vec{y},t')}{\partial t'}\right)\right]|0\rangle$$

(9.10)

If quantum field theory is gauge invariant then a gauge transformation should produce no change in any observable quantity. Therefore $\delta_g\vec{J}_{vac}(\vec{x},t)$ should be zero. To see if this is the case we will solve the above equation as follows. First consider the following relationship,

$$\int_{-\infty}^{t} dt' \hat{\rho}_I(\vec{y},t') \frac{\partial \chi(\vec{y},t')}{\partial t'} = \Big|_{-\infty}^{t} \hat{\rho}_I(\vec{y},t')\chi(\vec{y},t') - \int_{-\infty}^{t} dt' \chi(\vec{y},t') \frac{\partial \hat{\rho}_I(\vec{y},t')}{\partial t'} \quad (9.11)$$

Assume that $\chi(\vec{y},t) = 0$ at $t \to -\infty$. Use this and (9.9) in the above expression to obtain

$$\int_{-\infty}^{t} dt' \hat{\rho}_I(\vec{y},t') \frac{\partial \chi(\vec{y},t')}{\partial t'} = \hat{\rho}_I(\vec{y},t)\chi(\vec{y},t) + \int_{-\infty}^{t} dt' \chi(\vec{y},t') \vec{\nabla} \cdot \hat{\vec{J}}_I(\vec{y},t') \quad (9.12)$$

Substitute this into (9.10) to obtain

$$\begin{aligned} \delta_g \vec{J}_{vac}(\vec{x},t) = i\langle 0 | \left[\hat{\vec{J}}_I(\vec{x},t), \int d\vec{y} \int_{-\infty}^{t} dt' \left(\hat{\vec{J}}_I(\vec{y},t') \cdot \vec{\nabla}\chi(\vec{y},t') + \chi(\vec{y},t')\vec{\nabla} \cdot \hat{\vec{J}}_I(\vec{y},t') \right) \right] | 0 \rangle \\ + i\langle 0 | \left[\hat{\vec{J}}_I(\vec{x},t), \int \hat{\rho}_I(\vec{y},t)\chi(\vec{y},t) d\vec{y} \right] | 0 \rangle \end{aligned} \quad (9.13)$$

Rearrange terms to obtain

$$\begin{aligned} \delta_g \vec{J}_{vac}(\vec{x},t) = i\langle 0 | \left[\hat{\vec{J}}_I(\vec{x},t), \int_{-\infty}^{t} dt' \int d\vec{y} \vec{\nabla} \cdot \left(\hat{\vec{J}}_I(\vec{y},t')\chi(\vec{y},t') \right) \right] | 0 \rangle \\ + i\langle 0 | \left[\hat{\vec{J}}_I(\vec{x},t), \int \hat{\rho}_I(\vec{y},t)\chi(\vec{y},t) d\vec{y} \right] | 0 \rangle \end{aligned} \quad (9.14)$$

Assume reasonable boundary conditions at $|\vec{y}| \to \infty$ so that

$$\int d\vec{y} \vec{\nabla} \cdot \left(\hat{\vec{J}}_I(\vec{y},t')\chi(\vec{y},t') \right) = 0 \quad (9.15)$$

Use this to obtain

$$\begin{aligned} \delta_g \vec{J}_{vac}(\vec{x},t) &= i\langle 0 | \left[\hat{\vec{J}}_I(\vec{x},t), \int \hat{\rho}_I(\vec{y},t)\chi(\vec{y},t) d\vec{y} \right] | 0 \rangle \\ &= i \int \langle 0 | \left[\hat{\vec{J}}_I(\vec{x},t), \hat{\rho}_I(\vec{y},t) \right] | 0 \rangle \chi(\vec{y},t) d\vec{y} \end{aligned} \quad (9.16)$$

Use (9.8) and the fact that $H_0 |0\rangle = \langle 0| H_0 = 0$ to show that,

$$\langle 0 | \left[\hat{\vec{J}}_I(\vec{x},t), \hat{\rho}_I(\vec{y},t) \right] | 0 \rangle = \langle 0 | \left[\hat{\vec{J}}(\vec{x}), \hat{\rho}(\vec{y}) \right] | 0 \rangle = -\langle 0 | ST(\vec{y},\vec{x}) | 0 \rangle \quad (9.17)$$

Use this in (9.16) to obtain,

$$\delta_g \vec{J}_{vac}(\vec{x},t) = -i \int \langle 0 | ST(\vec{y},\vec{x}) | 0 \rangle \chi(\vec{y},t) d\vec{y} \quad (9.18)$$

Therefore for $\delta_g \vec{J}_{vac}(\vec{x},t)$ to be zero, for arbitrary $\chi(\vec{y},t)$, the quantity $\langle 0|ST(\vec{x},\vec{y})|0\rangle$ must be zero. However, we have shown in Section VIII that this quantity is not zero. Therefore perturbation theory does not produce a gauge invariant result for the vacuum current. This is why regulation is required. It is needed to remove the non-gauge invariant terms that occur due to the fact that the Schwinger term is nonzero.

10. Removing energy from the vacuum state in hole theory

In the last several sections we have examined the problems that quantum field theory has with gauge invariance and the continuity equation. We have shown that these problems are related to the fact that in quantum field theory the vacuum state $|0\rangle$ is the state with the lowest free field energy. In this section we will examine the vacuum state in hole theory to determine if it is the state with the lowest free field energy. It will be shown that this is not the case. That is, in hole theory there exist states with less free field energy than the vacuum state.

Recall that in hole theory the vacuum state corresponds to the state in which every negative energy state is filled up with a single electron and all the positive energy states are unoccupied. Each of these negative energy electrons obeys the single particle Dirac equation. If an external electric field is applied for a finite amount of time then the energy of each negative energy electron will change in response to the electric field. The total change in the energy of the vacuum state is the sum of the change in the energy of each individual electron. It will be shown that it is possible to find an electric field where this change is negative. That is, energy is extracted from the initial vacuum state due to its interaction with the electric field.

To show this we will work the following problem. Assume that at some initial time $t = t_i$ the system is in the unperturbed vacuum state. This is the state where all the negative energy states $\varphi_{-j}^{(0)}(\vec{x},t_i)$ (where j is positive) are occupied by a single electron. (Note that we are using the ordering convention introduced in the paragraph leading up to equation (5.1)). Using the notation of Section III the vacuum state can be written as,

$$\psi^N(\vec{x}_1,\vec{x}_2,...,\vec{x}_N,t_i) \underset{N\to\infty}{=} \frac{1}{\sqrt{N!}}\sum_P(-1)^p P\left(\varphi_{-1}^{(0)}(\vec{x}_1,t_i)\varphi_{-2}^{(0)}(\vec{x}_2,t_i)\cdots\varphi_{-N}^{(0)}(\vec{x}_N,t_i)\right) \quad (10.1)$$

Next, at $t = t_i$, apply an electric potential which is removed at some later time $t = t_f$. Under the action of the electric potential each initial state $\varphi_{-j}^{(0)}(\vec{x},t_i)$ evolves into the final state $\varphi_{-j}(\vec{x},t_f)$. Therefore the perturbed vacuum state, at time $t = t_f$, is,

$$\psi^N(\vec{x}_1,\vec{x}_2,...,\vec{x}_N,t_f) \underset{N\to\infty}{=} \frac{1}{\sqrt{N!}}\sum_P(-1)^p P\left(\varphi_{-1}(\vec{x}_1,t_f)\varphi_{-2}(\vec{x}_2,t_f)\cdots\varphi_{-N}(\vec{x}_N,t_f)\right) \quad (10.2)$$

To simplify notation define,

$$\left\langle f(\vec{x})\right\rangle = \int f(\vec{x})d\vec{x} \quad (10.3)$$

The total change in the free field energy of the vacuum state, $\Delta\xi_{f,vac}(t_i \to t_f)$, is the sum of changes of the free field energy of each individual vacuum electron. Therefore,

$$\Delta\xi_{f,vac}(t_i \to t_f) \underset{N\to\infty}{=} \sum_{j=1}^{N}\Delta\xi_{f,-j}(t_i \to t_f) \quad (10.4)$$

where $\Delta\xi_{f,n}(t_i \to t_f)$ is the change in the free field energy of the electron in state 'n'. Since the initial state for this electron is $\varphi_n^{(0)}(\vec{x},t_i)$ and the final state is $\varphi_n(\vec{x},t_f)$,

$$\Delta\xi_{f,n}(t_i \to t_f) = \left\langle\varphi_n^\dagger(\vec{x},t_f)H_0\varphi_n(\vec{x},t_f)\right\rangle - \left\langle\varphi_n^{(0)\dagger}(\vec{x},t_i)H_0\varphi_n^{(0)}(\vec{x},t_i)\right\rangle \quad (10.5)$$

Now we want to determine $\Delta\xi_{f,vac}(t_i \to t_f)$ for a given electric potential applied during the time interval between t_i and t_f. To do this it is necessary to evaluate $\varphi_n(\vec{x},t)$. The change in $\varphi_n(\vec{x},t)$, due to an applied electric potential, must be determined using perturbation theory since it is, in general, not possible to find exact solutions to these kind of problems. This problem has already been considered for the case of field theory in Appendix A. Exactly the same analysis can be applied for the single particle wave function. The result is that $\varphi_n(\vec{x},t)$ can be expanded as a series in the charge 'q' i.e.,

$$\varphi_n(\vec{x},t)=\varphi_n^{(0)}(\vec{x},t)+q\varphi_n^{(1)}(\vec{x},t)+q^2\varphi_n^{(2)}(\vec{x},t)+O(q^3) \tag{10.6}$$

where $O(q^3)$ means terms to the third order in the perturbation or higher. Use this in (10.5) to obtain $\Delta\xi_{f,n}(t_i\to t_f)$. It is shown in Appendix B that

$$\Delta\xi_{f,n}(t_i\to t_f)=\xi_f(\varphi_n(\vec{x},t_f))-\lambda_n E_n=q^2\left(\left\langle\varphi_n^{(1)\dagger}H_0\varphi_n^{(1)}\right\rangle-\lambda_n E_n\left\langle\varphi_n^{(1)\dagger}\varphi_n^{(1)}\right\rangle\right)+O(q^3) \tag{10.7}$$

where,

$$\varphi_n^{(1)}(\vec{x},t)=-i\int_{-\infty}^{t}dt'e^{-iH_0(t-t')}V(\vec{x},t')\varphi_n^{(0)}(\vec{x},t') \tag{10.8}$$

and $V=(-\vec{\alpha}\cdot\vec{A}+A_0)$. This result can be used in (10.4) to obtain $\Delta\xi_{f,vac}(t_i\to t_f)$ to the second order in q.

Now, in order to avoid unnecessary mathematical details, this problem will be worked in 1-1D space-time where we will take the space axis in the z-direction. In 1-1D space-time the Dirac equation can be written as,

$$i\frac{\partial\psi}{\partial t}=\left(\sigma_x\left(-i\frac{\partial}{\partial z}-qA_z\right)+m\sigma_z+qA_0\right)\psi \tag{10.9}$$

where σ_x and σ_z are the Pauli spin matrices,

$$\sigma_x=\begin{pmatrix}0&1\\1&0\end{pmatrix}\quad\text{and }\sigma_z=\begin{pmatrix}1&0\\0&-1\end{pmatrix} \tag{10.10}$$

The orthonormal free field solutions (the electric potential is zero) of (10.9) are given by,

$$\varphi_{\lambda,p}^{(0)}(z,t)=u_{\lambda,p}e^{-i(\lambda Et-pz)} \tag{10.11}$$

where,

$$u_{\lambda,p}=N_{\lambda,p}\begin{pmatrix}1\\ p/(\lambda E+m)\end{pmatrix};\ N_{\lambda,p}=\sqrt{\frac{\lambda E+m}{2L\lambda E}};\quad E=\sqrt{p^2+m^2} \tag{10.12}$$

In the above expressions $L\to\infty$ is the 1 dimensional integration volume, p is the momentum, $\lambda=1$ for positive energy states, and $\lambda=-1$ for negative energy states. The function $\varphi^{(0)}_{\lambda,p}(z,t)$ shall satisfy periodic boundary conditions $\varphi^{(0)}_{\lambda,p}(z,t)=\varphi^{(0)}_{\lambda,p}(z+L,t)$. This condition is realized if the momentum is given by,

$$p_r=\frac{2\pi r}{L}\ \text{where r is an integer.} \tag{10.13}$$

Let the electric potential be,

$$A_z=0 \text{ and } A_0=4\cos(kz)\left(\frac{\sin(mt)}{t}\right)=\left(e^{ikz}+e^{-ikz}\right)\int_{-m}^{+m}e^{iqt}dq \tag{10.14}$$

where m is the mass of the electron and $k<m$. It is obvious from the above expression that $A_0\to 0$ at $t\to\pm\infty$. Under the action of this electric potential each initial wave function $\varphi^{(0)}_{\lambda,p}(z,t_i)$, where $t_i\to-\infty$, evolves into the final wave function $\varphi_{\lambda,p}(z,t_f)$ where $t_f\to+\infty$. $\varphi_{\lambda,p}(z,t_f)$ can be expanded as a series in orders of 'q' as in (10.6). Using perturbation theory it is shown in Appendix C that,

$$\varphi^{(1)}_{\lambda,p_r}(z,t_f)=-i\left\{\left(c_{\lambda,p_r,k}\varphi^{(0)}_{\lambda,p_r+k}(z,t_f)\right)+\left(c_{\lambda,p_r,-k}\varphi^{(0)}_{\lambda,p_r-k}(z,t_f)\right)\right\} \tag{10.15}$$

where,

$$c_{\lambda,p_r,k}=2\pi L u^{\dagger}_{\lambda,p_r+k}u_{\lambda,p_r} \tag{10.16}$$

Use Eq. (10.15) and the fact that,

$$H_0\varphi^{(0)}_{\lambda',p_r}=\lambda' E_{p_r}\varphi^{(0)}_{\lambda',p_r} \text{ and } \left\langle\varphi^{(0)\dagger}_{\lambda,p_s}\varphi^{(0)}_{\lambda',p_r}\right\rangle=\delta_{sr}\delta_{\lambda'\lambda} \tag{10.17}$$

to obtain,

$$\left\langle\varphi^{(1)\dagger}_{\lambda,p_r}H_0\varphi^{(1)}_{\lambda,p_r}\right\rangle=\left|c_{\lambda,p_r,k}\right|^2\lambda E_{p_r+k}+\left|c_{\lambda,p_r,-k}\right|^2\lambda E_{p_r-k} \tag{10.18}$$

and,

$$\left\langle\varphi^{(1)\dagger}_{\lambda,p_r}\varphi^{(1)}_{\lambda,p_r}\right\rangle=\left|c_{\lambda,p_r,k}\right|^2+\left|c_{\lambda,p_r,-k}\right|^2 \tag{10.19}$$

Use the above in (10.7) to obtain,

$$\Delta\xi_f(\lambda,p_r;t_i\to t_f)=q^2\begin{pmatrix}\left|c_{\lambda,p_r,k}\right|^2\left(\lambda E_{p_r+k}-\lambda E_{p_r}\right)\\+\left|c_{\lambda,p_r,-k}\right|^2\left(\lambda E_{p_r-k}-\lambda E_{p_r}\right)\end{pmatrix}+O\left(q^3\right) \tag{10.20}$$

where $\Delta\xi_{f}(\lambda, p_{r}; t_{i} \to t_{f})$ is the change in free field energy of the initial state $\varphi_{\lambda,p_r}^{(0)}(z, t_i)$ when it evolves into the final state $\varphi_{\lambda,p_r}(z, t_f)$ under the action of the electric potential. Use (10.16) and (10.12) the above to obtain,

$$\Delta\xi_{f}(\lambda, p_{r}; t_{i} \to t_{f}) = 2\pi^{2}q^{2}\lambda k\left(\frac{(p_{r}+k)}{E_{p_{r}+k}} - \frac{(p_{r}-k)}{E_{p_{r}-k}}\right) + O\left(q^{3}\right) \tag{10.21}$$

The details of the derivation of this equation are provided in Appendix D.

Now we are interested in the change of the free field energy of the electrons in negative energy states in which case $\lambda = -1$. Therefore, for negative energy states,

$$\Delta\xi_{f}(\lambda = -1, p_{r}; t_{i} \to t_{f}) = -2\pi^{2}q^{2}k\left(\frac{(p_{r}+k)}{E_{p_{r}+k}} - \frac{(p_{r}-k)}{E_{p_{r}-k}}\right) + O\left(q^{3}\right) \tag{10.22}$$

In the limit that $q \to 0$ the $O\left(q^{3}\right)$ term can be dropped. Therefore,

$$\Delta\xi_{f}(\lambda = -1, p_{r}; t_{i} \to t_{f}) \underset{q\to 0}{=} -2\pi^{2}q^{2}k\left(\frac{(p_{r}+k)}{E_{p_{r}+k}} - \frac{(p_{r}-k)}{E_{p_{r}-k}}\right) \tag{10.23}$$

It is shown in Appendix E that this quantity is negative for all p_r. Therefore the change in the free field energy of each vacuum electron is negative. Therefore the total change in the free field energy of the vacuum is negative. This means that, in hole theory, there exist quantum states with less energy than the vacuum state.

11. Discussion

We have just shown that in hole theory energy can be extracted from the vacuum state due to its interaction with an electric field. Now suppose we had worked the same problem in quantum field theory. From Appendix A the perturbed vacuum state $\left|0_{p}\right\rangle$, due to an interaction with an electromagnetic field, is given by,

$$\left|0_{p}\right\rangle = \left(1 - iqe^{-iH_{0}t_{f}}\int_{t_{i}}^{t_{f}}\hat{V}_{I}(t)\,dt + \ldots\right)\left|0\right\rangle \tag{11.1}$$

Therefore, for an arbitrary perturbation, the final state $\left|0_p\right\rangle$ is the sum of the vacuum state $|0\rangle$ and various states of the form $\hat{b}^\dagger_{j_1}\hat{d}^\dagger_{j_2}|0\rangle$, $\hat{b}^\dagger_{j_1}\hat{b}^\dagger_{j_2}\hat{d}^\dagger_{j_3}\hat{d}^\dagger_{j_4}|0\rangle$, etc. As discussed previously all these states are eigenstates of the free field Hamiltonian $\hat{H}_0$ with positive energy eigenvalues. Therefore the free field energy of $\left|0_p\right\rangle$ is positive. In field theory the effect of an interaction with the electric field can never result in the decrease in the free field energy of the vacuum state.

To understand why these differences between hole theory and field theory occur consider the expression for the vacuum state in hole theory which is given by equation (10.1). We will rewrite this using our notation for the 1-1D wave function where $\varphi_n^{(0)} \to \varphi_{\lambda,p_r}^{(0)}$,

$$\Psi^{2N+1}(t_i) \underset{\substack{N\to\infty \\ \lambda=-1}}{=} \sum_P \frac{(-1)^p}{\sqrt{(2N+1)!}} P\begin{pmatrix} \varphi_{\lambda,p_0}^{(0)}(z_1,t_i)\varphi_{\lambda,p_1}^{(0)}(z_2,t_i)\varphi_{\lambda,p_{-1}}^{(0)}(z_3,t_i) \\ \cdots\varphi_{\lambda,p_N}^{(0)}(z_{2N},t_i)\varphi_{\lambda,p_{-N}}^{(0)}(z_{2N+1},t_i) \end{pmatrix} \tag{11.2}$$

In this expression the wave functions are ordered in terms of increasing magnitude of momentum. Note that because of the way we have done the ordering the number of electrons is 2N+1. Now as we have shown each of the initial unperturbed wave functions $\varphi_{\lambda,p_r}^{(0)}(z,t_i)$ has evolved under the action of the electric potential into the final wave function $\varphi_{\lambda,p_r}(z,t_f)$. Therefore $\Psi^{2N+1}(t_i)$ evolves into,

$$\Psi^{2N+1}(t_f) \underset{\substack{N\to\infty \\ \lambda=-1}}{=} \sum_P \frac{(-1)^p}{\sqrt{(2N+1)!}} P\begin{pmatrix} \varphi_{\lambda,p_0}(z_1,t_f)\varphi_{\lambda,p_1}(z_2,t_f)\varphi_{\lambda,p_{-1}}(z_3,t_f) \\ \cdots\varphi_{\lambda,p_N}(z_{2N},t_f)\varphi_{\lambda,p_{-N}}(z_{2N+1},t_f) \end{pmatrix} \tag{11.3}$$

From the results of the previous section we have that to the lowest order,

$$\varphi_{\lambda,p_r}(z,t_f) \underset{q\to 0}{=} \varphi_{\lambda,p_r}^{(0)}(z,t_f) - iq\left\{\left(c_{\lambda,p_r,k}\phi_{\lambda,p_r+k}^{(0)}(z,t_f)\right) + \left(c_{\lambda,p_r,-k}\phi_{\lambda,p_r-k}^{(0)}(z,t_f)\right)\right\} \tag{11.4}$$

Next assume that $k = 2\pi s/L$ where 's' is a positive integer. We shall write k_s instead of k to reflect this fact. In this case $p_r \pm k_s = p_{r\pm s}$ so that the above equation becomes,

$$\varphi_{\lambda,p_r}\left(z,t_f\right) \underset{q\to 0}{=} \varphi^{(0)}_{\lambda,p_r}\left(z,t_f\right) - iq\left\{\left(c_{\lambda,p_r,k_s}\varphi^{(0)}_{\lambda,p_{r+s}}\left(z,t_f\right)\right) + \left(c_{\lambda,p_r,k_{-s}}\varphi^{(0)}_{\lambda,p_{r-s}}\left(z,t_f\right)\right)\right\} \quad (11.5)$$

Use this in (11.3) to obtain,

$$\Psi^{2N+1}\left(t_f\right) \underset{\substack{N\to\infty \\ \lambda=-1 \\ q\to 0}}{=} \sum_P \frac{(-1)^p}{\sqrt{(2N+1)!}} P \left(\begin{array}{l} \left(\varphi^{(0)}_{\lambda,p_0}\left(z_1,t_f\right) - iq \left\{ \begin{array}{l} \left(c_{\lambda,p_0,k_s}\phi^{(0)}_{\lambda,p_{0+s}}\left(z_1,t_f\right)\right) \\ +\left(c_{\lambda,p_0,k_{-s}}\phi^{(0)}_{\lambda,p_{0-s}}\left(z_1,t_f\right)\right) \end{array} \right\} \right) \\ \cdots \\ \left(\varphi^{(0)}_{\lambda,p_N}\left(z_{2N},t_f\right) - iq \left\{ \begin{array}{l} \left(c_{\lambda,p_N,k_s}\phi^{(0)}_{\lambda,p_{N+s}}\left(z_{2N},t_f\right)\right) \\ +\left(c_{\lambda,p_N,k_{-s}}\phi^{(0)}_{\lambda,p_{N-s}}\left(z_{2N},t_f\right)\right) \end{array} \right\} \right) \\ \left(\varphi^{(0)}_{\lambda,p_{-N}}\left(z_{2N+1},t_f\right) - iq \left\{ \begin{array}{l} \left(c_{\lambda,p_{-N},k_s}\phi^{(0)}_{\lambda,p_{-N+s}}\left(z_{2N+1},t_f\right)\right) \\ +\left(c_{\lambda,p_{-N},k_{-s}}\phi^{(0)}_{\lambda,p_{-N-s}}\left(z_{2N+1},t_f\right)\right) \end{array} \right\} \right) \end{array} \right)$$

(11.6)

Now in the above expression we will only retain terms to the first order in q and write the result using the language of creation operators per section IV to obtain,

$$\left|\Psi^{2N+1}\left(t_f\right)\right\rangle \underset{\substack{N\to\infty \\ \lambda=-1 \\ q\to 0}}{=} \left(1 - iq\sum_{r=-N}^{N} \left(\begin{array}{l} e^{-i\lambda\left(E_{p_{r+s}}-E_{p_r}\right)t_f} c_{\lambda,p_r,k_s}\hat{a}^{\dagger}_{\lambda,p_{r+s}}\hat{a}_{\lambda,p_r} \\ +e^{-i\lambda\left(E_{p_{r-s}}-E_{p_r}\right)t_f} c_{\lambda,p_r,k_{-s}}\hat{a}^{\dagger}_{\lambda,p_{r-s}}\hat{a}_{\lambda,p_r} \end{array} \right) \right) \left(e^{-iE_{vac}t_f}\left|\Psi^{2N+1}_{vac}\right\rangle \right)$$

(11.7)

where,

$$\left|\Psi^{2N+1}_{vac}\right\rangle \underset{\substack{N\to\infty \\ \lambda=-1}}{=} \left(\hat{a}^{\dagger}_{\lambda,p_0}\hat{a}^{\dagger}_{\lambda,p_1}\hat{a}^{\dagger}_{\lambda,p_{-1}}\hat{a}^{\dagger}_{\lambda,p_2}\hat{a}^{\dagger}_{\lambda,p_{-2}}\cdots\hat{a}^{\dagger}_{\lambda,p_N}\hat{a}^{\dagger}_{\lambda,p_{-N}}\right)\Phi^0 \quad (11.8)$$

and

$$E_{vac} = -\left(E_{p_0} + E_{p_1} + E_{p_{-1}} + \ldots + E_{p_N} + E_{p_{-N}}\right) \quad (11.9)$$

and where $\hat{a}^{\dagger}_{\lambda,p_r}$ ($\hat{a}_{\lambda,p_r}$) creates (destroys) the state $\varphi^{(0)}_{\lambda,p_r}(z)$. The initial vacuum state (11.2) is given by,

$$\left|\Psi^{2N+1}\left(t_i\right)\right\rangle \underset{N\to\infty}{=} e^{-iE_{vac}t_i}\left|\Psi^{2N+1}_{vac}\right\rangle \quad (11.10)$$

According to (11.8) $\left|\Psi_{\text{vac}}^{2N+1}\right\rangle$ is composed of the product of raising operators $\hat{a}_{-1,p_r}^{\dagger}$ where $-N \le r \le N$. The term $\hat{a}_{-1,p_{r+s}}^{\dagger}\hat{a}_{-1,p_r}$, in (11.7), acts on the state $\left|\Psi_{\text{vac}}^{2N+1}\right\rangle$ by removing the operator $\hat{a}_{-1,p_r}^{\dagger}$ from $\left|\Psi_{\text{vac}}^{2N+1}\right\rangle$ and replacing it with $\hat{a}_{-1,p_{r+s}}^{\dagger}$. The result will be zero if $\hat{a}_{-1,p_{r+s}}^{\dagger}$ is already one of the operators in the product of operators that make up $\left|\Psi_{\text{vac}}^{2N+1}\right\rangle$. This will be the case if $-N \le r+s \le N$. Therefore, since we assume 's' is positive, the term $\hat{a}_{-1,p_{r+s}}^{\dagger}\hat{a}_{-1,p_r}$ only produces a nonzero contribution when $r > N-s$. Similarly the term $\hat{a}_{-1,p_{r-s}}^{\dagger}\hat{a}_{-1,p_r}$ will only produce a nonzero contribution if $s-N > r$. Therefore we can write (11.7) as,

$$\left|\Psi^{2N+1}(t_f)\right\rangle_{q\to 0} = e^{-iE_{\text{vac}}t_f}\left(\Psi_{\text{vac}}^{2N+1} - iq\Psi_A^{2N+1} - iq\Psi_B^{2N+1}\right) \tag{11.11}$$

where,

$$\left|\Psi_A^{2N+1}\right\rangle_{\substack{N\to\infty \\ \lambda=-1}} = \left(\sum_{r=N-s+1}^{N} e^{-i\lambda\left(E_{p_{r+s}}-E_{p_r}\right)t_f} c_{\lambda,p_r,k_s}\left(\hat{a}_{\lambda,p_{r+s}}^{\dagger}\hat{a}_{\lambda,p_r}\Psi_{\text{vac}}^{2N+1}\right)\right) \tag{11.12}$$

and

$$\left|\Psi_B^{2N+1}\right\rangle_{\substack{N\to\infty \\ \lambda=-1}} = \left(\sum_{r=-N}^{r=s-N-1} e^{-i\lambda\left(E_{p_{r-s}}-E_{p_r}\right)t_f} c_{\lambda,p_r,k_{-s}}\left(\hat{a}_{\lambda,p_{r-s}}^{\dagger}\hat{a}_{\lambda,p_r}\Psi_{\text{vac}}^{2N+1}\right)\right) \tag{11.13}$$

Now consider the terms $\hat{a}_{-1,p_{r+s}}^{\dagger}\hat{a}_{-1,p_r}\left|\Psi_{\text{vac}}^{2N+1}\right\rangle$ that appear in (11.12). The action of $\hat{a}_{-1,p_{r+s}}^{\dagger}\hat{a}_{-1,p_r}$ on $\left|\Psi_{\text{vac}}^{2N+1}\right\rangle$ is to destroy the state $\varphi_{-1,p_r}^{(0)}(z)$ with energy $-E_{p_r}$ and create the state $\varphi_{-1,p_{r+s}}^{(0)}(z)$ with energy $-E_{p_{r+s}}$ where $r+s > N$. Therefore the free field energy of the state $\hat{a}_{-1,p_{r+s}}^{\dagger}\hat{a}_{-1,p_r}\left|\Psi_{\text{vac}}^{2N+1}\right\rangle$ is given by,

$$\xi_f\left(\hat{a}_{-1,p_{r+s}}^{\dagger}\hat{a}_{-1,p_r}\left|\Psi_{\text{vac}}^{2N+1}\right\rangle\right) = E_{\text{vac}} - \left(E_{p_{r+s}} - E_{p_r}\right) < E_{\text{vac}} \text{ where } r+s > N \tag{11.14}$$

The term $\hat{a}^{\dagger}_{-1,p_{r+s}} \hat{a}_{-1,p_r} \left|\Psi^{2N+1}_{vac}\right\rangle$ is an eigenfunction of the free field Hamiltonian operator with an eigenvalue $E_{vac} - \left(E_{p_{r+s}} - E_{p_r}\right)$ which is less than E_{vac}, the energy of the vacuum state. Similarly the energy eigenvalue of the terms $\hat{a}^{\dagger}_{-1,p_{r-s}} \hat{a}_{-1,p_r} \left|\Psi^{2N+1}_{vac}\right\rangle$ that appear in (11.13) is less than E_{vac}. Therefore the energy of the perturbed vacuum state $\left|\Psi^{2N+1}(t_f)\right\rangle$ consists of a sum of states which include the original unperturbed vacuum state $\left|\Psi^{2N+1}_{vac}\right\rangle$ and states of the form $\hat{a}^{\dagger}_{-1,p_{r+s}} \hat{a}_{-1,p_r} \left|\Psi^{2N+1}_{vac}\right\rangle$ and $\hat{a}^{\dagger}_{-1,p_{r-s}} \hat{a}_{-1,p_r} \left|\Psi^{2N+1}_{vac}\right\rangle$ that have less energy then the vacuum. The result is that the energy of the perturbed state $\left|\Psi^{2N+1}(t_f)\right\rangle$ is less than that of the initial vacuum state $\left|\Psi^{2N+1}(t_i)\right\rangle$.

The reason for this result is that, in hole theory, we have defined the vacuum $\left|\Psi^{2N+1}_{vac}\right\rangle$ using a limiting procedure. As defined in equations (11.2), (11.10), and (11.8) the vacuum is the quantum state which consists of the product of terms of the form $\hat{a}^{\dagger}_{-1,p_r}$ where $|r| \leq N$ and $N \to \infty$. One can think of the vacuum as consisting of the quantum state in which the band of negative energy states between $-m$ and $-E_{p_N}$ are occupied, where $N \to \infty$. All positive energy states are unoccupied and all states with energy less than $-E_{p_N}$ are unoccupied. The effect of the operator $\hat{a}^{\dagger}_{-1,p_{r+s}} \hat{a}_{-1,p_r}$, where $r + s > N$ and $r < N$, on $\left|\Psi^{2N+1}_{vac}\right\rangle$ is to destroy an electron in the occupied negative energy band and to create an electron in one of the unoccupied negative energy states underneath the occupied band. This new state has less energy than the initial vacuum state. Therefore hole theory includes eigenstates with less energy than the vacuum state.

12. Redefining the Vacuum state

As can be seen from the prior discussion hole theory allows for existence of quantum states with less energy than the vacuum and field theory does not. Therefore field theory is not equivalent to hole theory. The difficulty with this result is that hole theory included all the conservation laws and symmetries associated with the single particle Dirac equation. In particular these include gauge invariance and the continuity equation. Therefore if hole theory and field theory are not equivalent we can no longer assume that field theory is gauge invariant or obeys the continuity equation. And, as has been shown this is indeed the case. It has been shown that the fact that there are no states with less free field energy than the vacuum state means that field theory is not gauge invariant and does not obey the continuity equation. This result is born out when calculations are made using perturbation theory. The offending non-gauge invariant terms that appear in these results must be removed to make the result physically correct. It should be stressed that these physically incorrect solutions are *mathematically* correct. Since the underlying theory is not gauge invariant the results of mathematical calculations should produce non-gauge invariant results [12].

What will be shown next is that it is possible to define the vacuum in field theory so that the equivalence between hole theory and field theory is restored. Refer to the definition of the vacuum state vector $|0\rangle$ given by (5.1). According to this definition $|0\rangle$ is the state in which all negative energy states are occupied by a single electron and all positive energy states are unoccupied. The top of the negative energy band has an energy of $-m$. Therefore for $|0\rangle$ all energy states with energy less than $-m$ are occupied. Define the state vector $|0_c\rangle$ as the quantum state in which all positive energy states are unoccupied, each negative energy state in the band of states between $-m$ and $-E_c$ is occupied, and negative energy states with energy less than $-E_c$ are unoccupied where $E_c \to \infty$. The state $|0_c\rangle$ is, then, defined by,

$$|0_c\rangle = \prod_{n \in \text{band}} \hat{a}_n^\dagger |0, \text{bare}\rangle \tag{12.1}$$

where the notation $n \in \text{band}$ means the product is taken over the band of states whose energy is between $-m$ and $-E_c$ where $E_c \to \infty$. This can also be expressed as,

$$\begin{aligned} \hat{a}_n \left|0_c\right\rangle &= 0 \text{ for } \lambda_n E_{\vec{p}_n} > m \\ \hat{a}_n^\dagger \left|0_c\right\rangle &= 0 \text{ for } -m \geq \lambda_n E_{\vec{p}_n} \geq -E_c \\ \hat{a}_n \left|0_c\right\rangle &= 0 \text{ for } -E_c > \lambda_n E_{\vec{p}_n} \end{aligned} \tag{12.2}$$

Now let us take $\left|0_c\right\rangle$ as the vacuum state instead of $\left|0\right\rangle$. Compare the definition of $\left|0_c\right\rangle$ (equation (12.1)) with $\left|0\right\rangle$ (equation (5.1)). Note the state $\left|0_c\right\rangle$ is almost identical to $\left|0\right\rangle$ with the exception that for $\left|0_c\right\rangle$ the bottom of the negative energy band is handled by a limiting process. The cutoff energy $-E_c$ is assumed to finite and is taken to negative infinity at the end of a calculation. States with less energy then $-E_c$ are unoccupied. If one of these states becomes occupied then the new state will have less free field energy then $\left|0_c\right\rangle$. Therefore $\left|0_c\right\rangle$ is no longer the lower bound to the free field energy. In fact there is no lower bound to the free field energy. Therefore using $\left|0_c\right\rangle$ as the vacuum state meets the requirement of Section V where it was shown that, in order for field theory to be gauge invariant and obey the continuity equation, there must be no lower bound to the free field energy. Also, as we will show below, if $\left|0_c\right\rangle$ is used as the vacuum state then the Schwinger term is zero.

Recall the discussion in Section VIII where we have shown that quantity $I(\vec{y},\vec{x}) = \left\langle 0\right| ST(\vec{y},\vec{x}) \left|0\right\rangle$ was non-zero which meant that the Schwinger term $ST(\vec{y},\vec{x})$ was non-zero. Here we evaluate the quantity defined by

$$I_c(\vec{y},\vec{x}) = \left\langle 0_c\right| ST(\vec{y},\vec{x}) \left|0_c\right\rangle = \left\langle 0_c\right| \left[\hat{\rho}(\vec{y}), \hat{\vec{J}}(\vec{x})\right] \left|0_c\right\rangle \tag{12.3}$$

Use (4.17) and (4.18) in the above to yield,

$$I_c(\vec{y},\vec{x}) = q^2 \sum_{nmrs} \left\langle 0_c\right| \left[\hat{a}_n^\dagger \hat{a}_m, \hat{a}_r^\dagger \hat{a}_s\right] \left|0_c\right\rangle \left(\varphi_n^{(0)\dagger}(\vec{y}) \varphi_m^{(0)}(\vec{y})\right) \left(\varphi_r^{(0)\dagger}(\vec{x}) \vec{\alpha} \varphi_s^{(0)}(\vec{x})\right) \tag{12.4}$$

Use (4.9) in the above to obtain,

$$I_c(\vec{y},\vec{x}) = q^2 \sum_{nmrs} \langle 0_c | \begin{pmatrix} \delta_{mr}\hat{a}_n^\dagger \hat{a}_s \\ -\delta_{ns}\hat{a}_r^\dagger \hat{a}_m \end{pmatrix} | 0_c \rangle \left(\varphi_n^{(0)\dagger}(\vec{y})\varphi_m^{(0)}(\vec{y})\right)\left(\varphi_r^{(0)\dagger}(\vec{x})\vec{\alpha}\varphi_s^{(0)}(\vec{x})\right) \quad (12.5)$$

Use (12.2) in the above and redefine some of the dummy variables to yield,

$$I_c(\vec{y},\vec{x}) = q^2 \sum_{s\in \text{band}} \sum_r \begin{Bmatrix} \left(\varphi_s^{(0)\dagger}(\vec{y})\varphi_r^{(0)}(\vec{y})\right)\left(\varphi_r^{(0)\dagger}(\vec{x})\vec{\alpha}\varphi_s^{(0)}(\vec{x})\right) \\ -\left(\varphi_s^{(0)\dagger}(\vec{x})\vec{\alpha}\varphi_r^{(0)}(\vec{x})\right)\left(\varphi_r^{(0)\dagger}(\vec{y})\varphi_s^{(0)}(\vec{y})\right) \end{Bmatrix} \quad (12.6)$$

The notation $s \in \text{band}$ means the index 's' is summed over the states whose energy is in the band from $-m$ to $-E_c$. Note that the summation over 'r' is over all states.

Take the summation over 'r' and use (2.16) in the above to obtain,

$$I_c(\vec{y},\vec{x}) = q^2 \sum_{s\in \text{band}} \begin{Bmatrix} \left(\varphi_s^{(0)\dagger}(\vec{y})\vec{\alpha}\varphi_s^{(0)}(\vec{x})\right)\delta^{(3)}(\vec{x}-\vec{y}) \\ -\left(\varphi_s^{(0)\dagger}(\vec{x})\vec{\alpha}\varphi_s^{(0)}(\vec{y})\right)\delta^{(3)}(\vec{x}-\vec{y}) \end{Bmatrix} \quad (12.7)$$

Next use the relationship,

$$f(\vec{y})\delta^{(3)}(\vec{x}-\vec{y}) = f(\vec{x}) \quad (12.8)$$

to obtain,

$$I_c(\vec{y},\vec{x}) = q^2 \sum_{s\in \text{band}} \delta^{(3)}(\vec{x}-\vec{y}) \begin{Bmatrix} \left(\varphi_s^{(0)\dagger}(\vec{x})\vec{\alpha}\varphi_s^{(0)}(\vec{x})\right) \\ -\left(\varphi_s^{(0)\dagger}(\vec{x})\vec{\alpha}\varphi_s^{(0)}(\vec{x})\right) \end{Bmatrix} = 0 \quad (12.9)$$

Therefore the quantity $\langle 0_c | ST(\vec{y},\vec{x}) | 0_c \rangle$ is zero. This allows for the possibility that the Schwinger term is zero. Now the Schwinger term is zero if all quantities of the form $\langle k' | ST(\vec{y},\vec{x}) | k \rangle$ are zero where $|k\rangle$ and $|k'\rangle$ are energy eigenstates. To show that this is the case evaluate,

$$\langle k' | ST(\vec{y},\vec{x}) | k \rangle = q^2 \sum_{nmrs} \langle k' | \left[\hat{a}_n^\dagger \hat{a}_m, \hat{a}_r^\dagger \hat{a}_s\right] | k \rangle \left(\varphi_n^{(0)\dagger}(\vec{y})\varphi_m^{(0)}(\vec{y})\right)\left(\varphi_r^{(0)\dagger}(\vec{x})\vec{\alpha}\varphi_s^{(0)}(\vec{x})\right)$$

(12.10)

Next use (4.9) in the above to yield,

$$\langle k' | ST(\vec{y},\vec{x}) | k \rangle = q^2 \sum_{nmrs} \langle k' | \begin{pmatrix} \delta_{mr}\hat{a}_n^\dagger \hat{a}_s \\ -\delta_{ns}\hat{a}_r^\dagger \hat{a}_m \end{pmatrix} | k \rangle \left(\varphi_n^{(0)\dagger}(\vec{y})\varphi_m^{(0)}(\vec{y})\right)\left(\varphi_r^{(0)\dagger}(\vec{x})\vec{\alpha}\varphi_s^{(0)}(\vec{x})\right)$$

(12.11)

This becomes,

$$\langle k'|ST(\vec{y},\vec{x})|k\rangle = q^2\begin{pmatrix} \sum\limits_{nms} \langle k'|\left(\hat{a}_n^\dagger \hat{a}_s\right)|k\rangle\left(\varphi_n^{(0)\dagger}(\vec{y})\varphi_m^{(0)}(\vec{y})\right)\left(\varphi_m^{(0)\dagger}(\vec{x})\vec{\alpha}\varphi_s^{(0)}(\vec{x})\right) \\ -\sum\limits_{nmr} \langle k'|\left(\hat{a}_r^\dagger \hat{a}_m\right)|k\rangle\left(\varphi_r^{(0)\dagger}(\vec{x})\vec{\alpha}\varphi_n^{(0)}(\vec{x})\right)\left(\varphi_n^{(0)\dagger}(\vec{y})\varphi_m^{(0)}(\vec{y})\right) \end{pmatrix} \tag{12.12}$$

Next use (2.16) and (12.8) redefine some dummy variables to obtain,

$$\langle k'|ST(\vec{y},\vec{x})|k\rangle = q^2\delta(\vec{x}-\vec{y})\begin{pmatrix} \sum\limits_{ns} \langle k'|\left(\hat{a}_n^\dagger \hat{a}_s\right)|k\rangle\left(\varphi_n^{(0)\dagger}(\vec{y})\vec{\alpha}\varphi_s^{(0)}(\vec{x})\right) \\ -\sum\limits_{ns} \langle k'|\left(\hat{a}_n^\dagger \hat{a}_s\right)|k\rangle\left(\varphi_n^{(0)\dagger}(\vec{x})\vec{\alpha}\varphi_s^{(0)}(\vec{y})\right) \end{pmatrix} \tag{12.13}$$

$$= q^2\delta(\vec{x}-\vec{y})\sum_{ns}\begin{pmatrix} \langle k'|\left(\hat{a}_n^\dagger \hat{a}_s\right)|k\rangle\left(\varphi_n^{(0)\dagger}(\vec{x})\vec{\alpha}\varphi_s^{(0)}(\vec{x})\right) \\ -\langle k'|\left(\hat{a}_n^\dagger \hat{a}_s\right)|k\rangle\left(\varphi_n^{(0)\dagger}(\vec{x})\vec{\alpha}\varphi_s^{(0)}(\vec{x})\right) \end{pmatrix} = 0$$

Therefore if $|0_c\rangle$ is used as the vacuum state then the Schwinger term $ST(\vec{y},\vec{x}) = 0$ because all quantities $\langle k'|ST(\vec{y},\vec{x})|k\rangle$ are zero. And, as has been shown, this is a necessary condition for quantum field theory to be gauge invariant and for the continuity equation to be true.

13. Conclusion

It has been shown that hole theory and the Schrödinger representation of quantum field theory are not equivalent. This is due to the fact that, in field theory, there can be no states with less free field energy then the vacuum state $|0\rangle$, however, as was shown in Section X, this is not the case for hole theory. In hole theory it is possible to extract energy from the vacuum state through interaction with an appropriately applied electric field.

As a result of this lack of equivalence between the two theories we must examine whether the conservation laws and symmetries that are associated with the single particle Dirac equation are valid in field theory. It is seen that they are not. Field theory is not gauge invariant and the continuity equation does not hold. This is due to the fact that the

Schwinger term is not zero. And the Schwinger term is not zero because field theory does not allow for the existence of states with less free field energy than the vacuum state $|0\rangle$.

In order to restore the equivalence between hole theory and field theory it is necessary to modify the definition of the vacuum state. Instead of the using the state vector $|0\rangle$ for the vacuum the state vector $|0_c\rangle$ is used instead. When $|0_c\rangle$ is used there can exist quantum states with less free field energy than the vacuum which is a necessary requirement for the Schwinger term to be zero. It is shown that the Schwinger term is zero when $|0_c\rangle$ is used as the vacuum state. In this case field theory will be gauge invariance and the continuity equation will be valid.

References

1 F.A.B. Coutinho, D. Kaing, Y. Nagami, and L. Tomio, *Can. J. of Phys.*, **80**, 837 (2002). (see also *quant-ph/0010039*).

2 F.A.B. Coutinho, Y. Nagami, and L. Tomio, *Phy. Rev.* A, **59**, 2624 (1999).

3 W. Greiner, B. Muller, and T. Rafelski, "*Quantum Electrodynamics of Strong Fields*", Springer-Verlag, Berlin (1985).

4 M. Guidry, "*Gauge Field Theories*", John Wiley & Sons, Inc., New York (1991).

5 W. Griener, "*Relativistic Quantum Mechanics*", Springer-Verlag, Berlin (1990).

6 W. Heitler. *The quantum theory of radiation.* Dover Publications, Inc., New York (1954).

7 P.A.M. Dirac, "*The Principles of Quantum Mechanics*", Claredon Press, Oxford (1958).

8 E.K.U. Gross, E. Runge, O. Heinonen, "*Many Particle Theory*", Adam Hilger, Bristol (1991).

9 P. Roman, "*Advanced Quantum Theory*", Addison-Wesley Publishing Co., Inc., Reading, Massachussetts, (1965).

10 S. Raines, "*Many-Electron Theory*", North Holland Publishing Co., Amsterdam (1972).

11 A. Messiah, "*Quantum Mechanics Vol 1.*" North Holland Publishing Company, Amsterdam, (1961).

12 D. Solomon, *Can. J. Phys.* **76**, 111 (1998). (see also quant-ph/9905021).

13 J. Schwinger, *Phys. Rev. Lett.*, **3**, 296 (1959).

14 K. Nishijima. "*Fields and Particles: Field theory and Dispersion Relations.*" W.A. Benjamin, New York, (1969).

15 W. Pauli and F. Villers, *Rev. Mod. Phys.* **21,** 434 (1949).

16 W. Pauli, "*Pauli lectures on physics Vol. 6. Selected topics in field quantization*", MIT Press, Cambridge, Mass. 1973.

17 J.J. Sakurai, "*Advanced Quantum Mechanics*", Addison-Wesley Publishing Company, Inc., Redwood City, California, (1967).

Appendix A

In order to use perturbation theory we must convert from the Schrödinger picture to the interaction picture (see Chapter 4-2 of [17]). Refer back to equation (4.20). Write the Hamiltonian as,

$$\hat{H} = \hat{H}_0 + \hat{V} \tag{A.1}$$

where $\hat{H}_0$ is the unperturbed free field Hamiltonian and $\hat{V}$ is the perturbation. From (4.22),

$$\hat{V} = -\int \hat{J}(\vec{x}) \cdot \vec{A}(\vec{x},t)d\vec{x} + \int \rho(\vec{x}) \cdot A_0(\vec{x},t)d\vec{x} \tag{A.2}$$

Define the interaction state vector by,

$$|\Omega_I\rangle = e^{i\hat{H}_0 t}|\Omega\rangle \tag{A.3}$$

Interaction operators $\hat{O}_I$ are defined in terms of Schrödinger operators $\hat{O}$ according to,

$$\hat{O}_I = e^{i\hat{H}_0 t}\hat{O}e^{-i\hat{H}_0 t} \tag{A.4}$$

The expectation values of operators have the same value in both representations, i.e., $O_e = \langle\Omega_I|\hat{O}_I|\Omega_I\rangle = \langle\Omega|\hat{O}|\Omega\rangle$. Use the above expressions in (4.20) to obtain,

$$i\frac{\partial|\Omega_I\rangle}{\partial t} = \hat{V}_I|\Omega_I\rangle \tag{A.5}$$

A formal solution of above is given by,

$$|\Omega_I(t)\rangle = \left(1 - i\int_{t_0}^{t}\hat{V}_I(t_1)dt_1 + (-i)^2\int_{t_0}^{t}\hat{V}_I(t_1)dt_1\int_{t_0}^{t_1}\hat{V}_I(t_2)dt_2\right)|\Omega_I(t_0)\rangle \tag{A.6}$$

The current expectation value is then given by,

$$\vec{J}_e(\vec{x},t) = \langle\Omega_I(t)|\hat{\vec{J}}_I(\vec{x},t)|\Omega_I(t)\rangle$$
$$= \langle\Omega_I(t_0)|\hat{\vec{J}}_I(\vec{x},t)|\Omega_I(t_0)\rangle - i\langle\Omega_I(t_0)|\left[\hat{\vec{J}}_I(\vec{x},t), \int_{t_0}^{t}\hat{V}_I(t_1)dt_1\right]|\Omega_I(t_0)\rangle + O(V^2) \tag{A.7}$$

Let the initial state, at time $t = t_0$, be the vacuum state $|0\rangle$. The first order vacuum current at time t is then,

$$\vec{J}_{vac}(\vec{x},t) = \langle 0_I|\hat{\vec{J}}_I(\vec{x},t)|0_I\rangle - i\langle 0_I|\left[\hat{\vec{J}}_I(\vec{x},t), \int_{t_0}^{t}\hat{V}_I(t_1)dt_1\right]|0_I\rangle \tag{A.8}$$

where $|0_I\rangle$ is the interaction vacuum state and is given by,

$$|0_I\rangle = e^{i\hat{H}_0 t}|0\rangle = |0\rangle \tag{A.9}$$

It is easy to show that $\langle 0_I|\hat{\vec{J}}_I(\vec{x},t)|0_I\rangle = 0$. Use this, along with (A.9) and let the initial time $t_0 = -\infty$ to obtain equation (9.7) in the text.

Appendix B

Expand $\varphi_n(\vec{x},t)$ as a series in the charge 'q',

$$\varphi_n(\vec{x},t) = \varphi_n^{(0)}(\vec{x},t) + q\varphi_n^{(1)}(\vec{x},t) + q^2\varphi_n^{(2)}(\vec{x},t) + O(q^3) \tag{B.1}$$

where $O(q^3)$ means terms to the third order in the perturbation or higher. The free field energy of the state $\varphi_n(\vec{x},t)$ is,

$$\xi_f(\varphi_n) \equiv \langle\varphi_n^\dagger H_0\varphi_n\rangle = \left\langle\left(\varphi_n^{(0)\dagger} + q\varphi_n^{(1)\dagger} + q^2\varphi_n^{(2)\dagger}\right)H_0\left(\varphi_n^{(0)} + q\varphi_n^{(1)} + q^2\varphi_n^{(2)}\right)\right\rangle + O(q^3) \tag{B.2}$$

Rearrange terms to obtain,

$$\xi_f(\varphi_n) = \langle\varphi_n^{(0)\dagger}H_0\varphi_n^{(0)}\rangle + q\left(\langle\varphi_n^{(0)\dagger}H_0\varphi_n^{(1)}\rangle + \langle\varphi_n^{(1)\dagger}H_0\varphi_n^{(0)}\rangle\right)$$
$$+ q^2\left(\langle\varphi_n^{(1)\dagger}H_0\varphi_n^{(1)}\rangle + \langle\varphi_n^{(0)\dagger}H_0\varphi_n^{(2)}\rangle + \langle\varphi_n^{(2)\dagger}H_0\varphi_n^{(0)}\rangle\right) + O(q^3) \tag{B.3}$$

Use $H_0\varphi_n^{(0)} = \lambda_n E_n\varphi_n^{(0)}$ in the above to obtain,

$$\xi_f(\varphi_n) = q^2 \left\langle \varphi_n^{(1)\dagger} H_0 \varphi_n^{(1)} \right\rangle + \lambda_n E_n \begin{pmatrix} \left\langle \varphi_n^{(0)\dagger}\varphi_n^{(0)} \right\rangle + q\left(\left\langle \varphi_n^{(0)\dagger}\varphi_n^{(1)} \right\rangle + \left\langle \varphi_n^{(1)\dagger}\varphi_n^{(0)} \right\rangle \right) \\ +q^2\left(\left\langle \varphi_n^{(0)\dagger}\varphi_n^{(2)} \right\rangle + \left\langle \varphi_n^{(2)\dagger}\varphi_n^{(0)} \right\rangle \right) \end{pmatrix} + O\left(q^3\right) \tag{B.4}$$

The Dirac equation does not affect the normalization condition therefore,

$$\left\langle \varphi_n^\dagger \varphi_n \right\rangle = 1 \tag{B.5}$$

Use (B.1) in the above to obtain,

$$1 = \left\langle \varphi_n^{(0)\dagger}\varphi_n^{(0)} \right\rangle + q\begin{pmatrix} \left\langle \varphi_n^{(1)\dagger}\varphi_n^{(0)} \right\rangle \\ +\left\langle \varphi_n^{(0)\dagger}\varphi_n^{(1)} \right\rangle \end{pmatrix} + q^2 \begin{pmatrix} \left\langle \varphi_n^{(1)\dagger}\varphi_n^{(1)} \right\rangle + \left\langle \varphi_n^{(0)\dagger}\varphi_n^{(2)} \right\rangle \\ +\left\langle \varphi_n^{(2)\dagger}\varphi_n^{(0)} \right\rangle \end{pmatrix} + O\left(q^3\right) \tag{B.6}$$

Rearrange terms to yield,

$$1 - q^2\left\langle \varphi_n^{(1)\dagger}\varphi_n^{(1)} \right\rangle = \left\langle \varphi_n^{(0)\dagger}\varphi_n^{(0)} \right\rangle + q\begin{pmatrix} \left\langle \varphi_n^{(1)\dagger}\varphi_n^{(0)} \right\rangle \\ +\left\langle \varphi_n^{(0)\dagger}\varphi_n^{(1)} \right\rangle \end{pmatrix} + q^2 \begin{pmatrix} \left\langle \varphi_n^{(0)\dagger}\varphi_n^{(2)} \right\rangle \\ +\left\langle \varphi_n^{(2)\dagger}\varphi_n^{(0)} \right\rangle \end{pmatrix} + O\left(q^3\right) \tag{B.7}$$

Use this in (B.4) to obtain,

$$\xi_f(\varphi_n) = q^2 \left\langle \varphi_n^{(1)\dagger} H_0 \varphi_n^{(1)} \right\rangle + \lambda_n E_n \left(1 - q^2 \left\langle \varphi_n^{(1)\dagger}\varphi_n^{(1)} \right\rangle \right) + O\left(q^3\right) \tag{B.8}$$

Rearrange terms to obtain,

$$\xi_f(\varphi_n) = \lambda_n E_n + q^2 \left(\left\langle \varphi_n^{(1)\dagger} H_0 \varphi_n^{(1)} \right\rangle - \lambda_n E_n \left\langle \varphi_n^{(1)\dagger}\varphi_n^{(1)} \right\rangle \right) + O\left(q^3\right) \tag{B.9}$$

Therefore the change in the free field energy of state φ_n is,

$$\Delta\xi_{f,n}(t_i \to t_f) = \xi_f(\varphi_n) - \lambda_n E_n = q^2 \left(\left\langle \varphi_n^{(1)\dagger} H_0 \varphi_n^{(1)} \right\rangle - \lambda_n E_n \left\langle \varphi_n^{(1)\dagger}\varphi_n^{(1)} \right\rangle \right) + O\left(q^3\right) \tag{B.10}$$

which is (10.7) in the text.

To evaluate the above we have to obtain $\varphi_n^{(1)}$. Based on the discussion in Appendix A this is given by,

$$\varphi_n^{(1)}(\vec{x},t) = -iqe^{-iH_0 t} \int_{t_i}^{t} dt' V_I(\vec{x},t')\varphi_n^{(0)}(\vec{x},0) \tag{B.11}$$

In the above expression V_I is given by $V_I = e^{-iH_0 t} V e^{+iH_0 t}$ where $V = \left(-\vec{\alpha}\cdot\vec{A} + A_0\right)$. Use this in the above to obtain,

$$\varphi_n^{(1)}(\vec{x},t) = -i\int_{-\infty}^{t} dt' e^{-iH_0(t-t')} V(\vec{x},t')\varphi_n^{(0)}(\vec{x},t') \tag{B.12}$$

Next expand the quantity $V(\vec{x},t')\varphi_n^{(0)}(\vec{x},t')$ in terms of the basis states $\varphi_s^{(0)}(\vec{x},t')$ to obtain,

$$\begin{aligned} & e^{iH_0(t-t')} V(\vec{x},t')\varphi_n^{(0)}(\vec{x},t') \\ &= \sum_s \left(e^{-iH_0(t-t')}\varphi_s^{(0)}(\vec{x},t')\right)\int \varphi_s^{(0)\dagger}(\vec{x}',t')\left(V(\vec{x}',t')\varphi_n^{(0)}(\vec{x}',t')\right)d\vec{x}' \end{aligned} \tag{B.13}$$

Use this and $e^{-iH_0(t-t')}\varphi_s^{(0)}(\vec{x},t') = \varphi_s^{(0)}(\vec{x},t)$ to obtain,

$$\varphi_n^{(1)}(\vec{x},t) = -i\int_{-\infty}^{t} dt' \sum_s \varphi_s^{(0)}(\vec{x},t)\int \varphi_s^{(0)\dagger}(\vec{x}',t') V(\vec{x}',t')\varphi_n^{(0)}(\vec{x}',t')\,d\vec{x}' \tag{B.14}$$

Appendix C

From (B.14) and the discussion of the 1-1D Dirac equation we obtain,

$$\varphi_{\lambda,p_r}^{(1)}(z,t=+\infty) = -i\sum_{\lambda'=\pm 1}\sum_{p_s}\varphi_{\lambda',p_s}^{(0)}(z,t)\int_{-\infty}^{+\infty} dt' \int dz' \varphi_{\lambda',p_s}^{(0)\dagger}(z',t') A_0(z',t')\varphi_{\lambda,p_r}^{(0)}(z',t') \tag{C.1}$$

for the case where $A_z = 0$. Next use (10.11) in the above to obtain,

$$\varphi_{\lambda,p_r}^{(1)}(z,t) = -i\sum_{\lambda'=\pm 1}\sum_{p_s}\varphi_{\lambda',p_s}^{(0)}(z,t)\int_{-\infty}^{t} dt' \int dz' A_0(z',t') u_{\lambda',p_s}^{\dagger} u_{\lambda,p_r} e^{-i\left(\lambda E_{p_r} - \lambda' E_{p_s}\right)t'} e^{i(p_r - p_s)z'} \tag{C.2}$$

Use (10.14) in the above to yield,

$$\varphi_{\lambda,p_r}^{(1)}(z,t=\infty) = -i\sum_{\lambda'=\pm 1}\sum_{p_s}\varphi_{\lambda',p_s}^{(0)}(z,t) u_{\lambda',p_s}^{\dagger} u_{\lambda,p_r}\int_{-\infty}^{+\infty} dt' \int dz' \begin{pmatrix} e^{-i\left(\lambda E_{p_r} - \lambda' E_{p_s}\right)t'} e^{i(p_r - p_s)z'} \\ \times\left(e^{ikz'} + e^{-ikz'}\right)\int_{-m}^{+m} e^{iqt'} dq \end{pmatrix} \tag{C.3}$$

Integrate over z' and t' to obtain,

$$\varphi^{(1)}_{\lambda,p_r}(z,t=\infty) = -4\pi^2 i \sum_{\lambda'=\pm1}\sum_{p_s}\left\{\begin{array}{l}\varphi^{(0)}_{\lambda',p_s}(z,t)u^{\dagger}_{\lambda',p_s}u_{\lambda,p_r} \\ \times\begin{pmatrix}\delta(k+p_r-p_s) \\ +\delta(-k+p_r-p_s)\end{pmatrix}\int_{-m}^{+m}\delta\begin{pmatrix}-\lambda E_{p_r} \\ +\lambda' E_{p_s}+q\end{pmatrix}dq\end{array}\right\} \tag{C.4}$$

Make the substitution,

$$\sum_{p_s} \rightarrow \int_{-\infty}^{+\infty}\frac{Ldp_s}{2\pi} \tag{C.5}$$

in the above to obtain,

$$\varphi^{(1)}_{\lambda,p_r}(z,t=\infty) = -2\pi L i \sum_{\lambda'=\pm1}\int dp_s\left\{\begin{array}{l}\varphi^{(0)}_{\lambda',p_s}(z,t)u^{\dagger}_{\lambda',p_s}u_{\lambda,p_r} \\ \times\begin{pmatrix}\delta(k+p_r-p_s) \\ +\delta(-k+p_r-p_s)\end{pmatrix}\int_{-m}^{+m}\delta\begin{pmatrix}q-\lambda E_{p_r} \\ +\lambda' E_{p_s}\end{pmatrix}dq\end{array}\right\} \tag{C.6}$$

Next integrate with respect to momentum p_s to obtain,

$$\varphi^{(1)}_{\lambda,p_r}(z,t=\infty) = -2\pi L i \sum_{\lambda'=\pm1}\left\{\begin{array}{l}\varphi^{(0)}_{\lambda',p_r+k}(z,t)u^{\dagger}_{\lambda',p_r+k}u_{\lambda,p_r}\int_{-m}^{+m}\delta\begin{pmatrix}q-\lambda E_{p_r} \\ +\lambda' E_{p_r+k}\end{pmatrix}dq \\ +(k\rightarrow -k)\end{array}\right\} \tag{C.7}$$

Now for sufficiently small k ($k < m$) we have that,

$$\int_{-m}^{+m}\delta\left(q-\lambda E_{p_r}+\lambda' E_{p_r+k}\right)dq = \delta(\lambda-\lambda') \tag{C.8}$$

Use this in (C.7) to obtain,

$$\varphi^{(1)}_{\lambda,p_r}(z,t=\infty) = -2\pi L i\left\{\left(\varphi^{(0)}_{\lambda,p_r+k}(z,t)u^{\dagger}_{\lambda,p_r+k}u_{\lambda,p_r}\right)+(k\rightarrow -k)\right\} \tag{C.9}$$

Define,

$$c_{\lambda,p_r,k} = 2\pi L u^{\dagger}_{\lambda,p_r+k}u_{\lambda,p_r} \tag{C.10}$$

Use this in (C.9) to obtain,

$$\varphi^{(1)}_{\lambda,p_r} = -i\left\{c_{\lambda,p_r,k}\varphi^{(0)}_{\lambda,p_r+k}+c_{\lambda,p_r,-k}\varphi^{(0)}_{\lambda,p_r-k}\right\} \tag{C.11}$$

Appendix D

We want to evaluate equation (10.20) and show that it results in equation (10.21). Use (10.16) and (10.12) to obtain,

$$c_{\lambda,p_r,k} = 2\pi\sqrt{\frac{\lambda E_{p_r+k}+m}{2\lambda E_{p_r+k}}}\sqrt{\frac{\lambda E_{p_r}+m}{2\lambda E_{p_r}}}\left(1+\frac{p_r(p_r+k)}{(\lambda E_{p_r+k}+m)(\lambda E_{p_r}+m)}\right) \tag{D.1}$$

Use $(\lambda E_p)^2 - m^2 = p^2$ in the above to obtain,

$$c_{\lambda,p_r,k} = 2\pi\sqrt{\frac{\lambda E_{p_r+k}+m}{2\lambda E_{p_r+k}}}\sqrt{\frac{\lambda E_{p_r}+m}{2\lambda E_{p_r}}}\left(1+\frac{(\lambda E_{p_r+k}-m)(\lambda E_{p_r}-m)}{p_r(p_r+k)}\right) \tag{D.2}$$

Use this result to yield,

$$\left|c_{\lambda,p_r,k}\right|^2 = 4\pi^2\left(\frac{(\lambda E_{p_r+k})(\lambda E_{p_r})+p_r(p_r+k)+m^2}{2(\lambda E_{p_r+k})(\lambda E_{p_r})}\right) \tag{D.3}$$

Use this in (10.20) along with the fact that $\lambda^2 = 1$ to obtain

$$\Delta\xi_f(\lambda,p_r;t_i\to t_f) = 2\pi^2q^2\lambda\begin{pmatrix}\left(1+\frac{p_r(p_r+k)+m^2}{E_{p_r+k}E_{p_r}}\right)(E_{p_r+k}-E_{p_r})\\ +\left(1+\frac{p_r(p_r-k)+m^2}{E_{p_r-k}E_{p_r}}\right)(E_{p_r-k}-E_{p_r})\end{pmatrix}+O(q^3) \tag{D.4}$$

This yields,

$$\Delta\xi_f(\lambda,p_r;t_i\to t_f) = 2\pi^2q^2\lambda\begin{pmatrix}\left(1+\frac{E_{p_r+k}}{E_{p_r}}-\frac{k(p_r+k)}{E_{p_r+k}E_{p_r}}\right)(E_{p_r+k}-E_{p_r})\\ +(k\to -k)\end{pmatrix}+O(q^3) \tag{D.5}$$

Some additional algebraic manipulation yields,

$$\Delta\xi_f\left(\lambda, p_r; t_i \to t_f\right) = 2\pi^2 q^2 \lambda \left(\begin{pmatrix} -E_{p_r} + \dfrac{E^2_{p_r+k}}{E_{p_r}} - \dfrac{k(p_r+k)}{E_{p_r}} + \dfrac{k(p_r+k)}{E_{p_r+k}} \end{pmatrix} \\ +(k \to -k) \right) + O\left(q^3\right)$$

q (D.6)

Use some simple algebra to obtain,

$$\Delta\xi_f\left(\lambda, p_r; t_i \to t_f\right) = 2\pi^2 q^2 \lambda \left(\left(\frac{p_r k}{E_{p_r}} + \frac{k(p_r+k)}{E_{p_r+k}} \right) + \left(\frac{-p_r k}{E_{p_r}} - \frac{k(p_r-k)}{E_{p_r-k}} \right) \right) + O\left(q^3\right) \tag{D.7}$$

Use this result to yield (10.21).

Appendix E

Assume k is positive. Then it can be shown that,

$$\frac{(p_r+k)}{E_{p_r+k}} > \frac{(p_r-k)}{E_{p_r-k}} \text{ for all } p_r \tag{E.1}$$

First consider the case where p_r is positive. The relationship is obviously true for $k > p_r$. Now let $p_r > k$. In this case both sides of (E.1) are positive therefore we can square both sides to obtain,

$$(p_r+k)^2 E^2_{p_r-k} > (p_r-k)^2 E^2_{p_r+k} \tag{E.2}$$

From this we obtain,

$$(p_r+k)^2\left((p_r-k)^2+m^2\right) > (p_r-k)^2\left((p_r+k)^2+m^2\right) \tag{E.3}$$

This yields,

$$(p_r+k)^2 > (p_r-k)^2 \tag{E.4}$$

which is true for positive k and $p_r > k$. If p_r is negative then (E.1) becomes,

$$\frac{\left(-|p_r|+k\right)}{E_{|p_r|-k}} > \frac{\left(-|p_r|-k\right)}{E_{|p_r|+k}} \tag{E.5}$$

This yields,

$$\frac{\left(|p_r|+k\right)}{E_{|p_r|+k}} > \frac{\left(|p_r|-k\right)}{E_{|p_r|-k}} \tag{E.6}$$

which is obviously true from the previous discussion.

In: Frontiers in Quantum Physics Research
Editor: F. Columbus and V. Krasnoholovets, pp. 51-82
ISBN 1-59454-002-2

Fields on the Lorentz Group: Helicity Basis and Relativistic Wave Equations

V. V. Varlamov
Department of Mathematics, Siberia State University of Industry,
Kirova 42, Novokuznetsk 654007, Russia.

Abstract

Physical fields and solutions of relativistic wave equations are studied in terms of the functions on the Lorentz group. The principal series of unitary representations of the Lorentz group has been considered in the helicity basis. It is shown that in the helicity basis matrix elements of irreducible representations are expressed via hyperspherical functions. A relation between the hyperspherical functions and other special functions is given. The general Gel'fand-Yaglom relativistically invariant system is defined on a tangent bundle of the Lorentz group manifold. It is shown that a 2-dimensional complex sphere is associated with the each point of the group manifold. Solutions of the relativistically invariant system have been found in the form of expansions in the hyperspherical functions defined on the surface of the complex sphere. By way of example, solutions of the Dirac equation are given in terms of the functions on the Lorentz group. Further applications of the proposed approach to the theory of composite particles are discussed.

PACS numbers: **03.65.Pm,02.30.Gp,02.10.Tq**

1 Introduction

Traditionally in quantum field theory particles of arbitrary spin are described within finite-dimensional spin-tensor representations of the Lorentz group. It follows that a mathematical design of the physical fields depends on the structure of the Lorentz group. In the present paper we will attempt to consider this dependence in detail. As is known [22], a root subgroup of a semisimple Lie group O_4 (a rotation group of the 4-dimensional space) is a normal divisor of O_4. For that reason the 6-parameter group O_4 is semisimple, and is represented by a direct product of the two 3-parameter unimodular groups. By analogy with the group O_4, a double covering $SL(2,\mathbb{C})$ of the proper orthochronous Lorentz group $\mathfrak{G}_+$ (a rotation group of the 4-dimensional spacetime continuum) is semisimple, and is represented by a direct product of the two 3-parameter special unimodular groups, $SL(2,\mathbb{C}) \simeq SU(2)\otimes SU(2)$. An explicit form of this isomorphism can be obtained by means of a complexification of the group $SU(2)$, that is, $SL(2,\mathbb{C}) \simeq \mathsf{complex}(SU(2)) \simeq SU(2)\otimes SU(2)$ [48]. Moreover, in the works [2, 17], inspired by Ryder book [38], the Lorentz group is represented by a product $SU_R(2) \otimes SU_L(2)$, and spinors $\psi(p^\mu) = (\phi_R(p^\mu), \phi_L(p^\mu))^T$ are transformed within $(j_1, j_2) \oplus (j_2, j_1)$ representation space. The components $\phi_R(p^\mu)$ and $\phi_L(p^\mu)$ correspond to different helicity states (right- and left-handed spinors).

On the other hand, all the physical fields of arbitrary spin are solutions of relativistic wave equations. In turn, higher-spin relativistic wave equations are formulated within the system of irreducible (both finite- and infinite-dimensional) representations of the Lorentz group. The first construction of a relativistically invariant theory of arbitrary half integer or integer spin particles was given by Majorana in 1932 [31]. Over a period of a century many authors (among them Dirac [12], Fierz and Pauli [18], Rarita and Schwinger [35], de Broglie [11], Bhabha [9], Bargmann and Wigner [7], Gel'fand and Yaglom [19], Joos and Weinberg [25, 53]) investigated the higher-spin equations. The thousands papers devoted to this problem. Despite considerable work and progress, there remain fundamental diffculties with each of the various theoretical approaches which have so far been proposed.

As is known, equations of motion play a basic role in physics. Relativistic wave equations (RWE) play the same role in quantum field theory. The wave function (main object of quantum theory) is a solution of RWE. For that reason all textbooks on quantum field theory began with a brief introduction to RWE. As a rule, solutions of the Dirac equation are represented by plane–wave solutions (see for example [39], and also many other textbooks). However, the plane–wave solutions are strongly degenerate. These solutions do not contain parameters of the Lorentz group and weakly reflect a relativistic nature of the wave function described by the Dirac equation.

In the present paper solutions of relativistically invariant equations are found in the form of expansions in generalized hyperspherical functions. In turn, matrix elements of irreducible representations of the Lorentz group are expressed via the hyperspherical functions (for more details see recent paper [48]). By this reason the wave function directly depends on the parameters of the Lorentz group. Moreover, it allows us to consider solutions of RWE as the functions on the group $\mathfrak{G}_+$ and further to apply methods of harmonic analysis to the product $SU(2) \otimes SU(2)$. A starting point of the research is an isomorphism $SL(2,\mathbb{C}) \sim \mathsf{complex}(SU(2))$. The well–known Van der Waerden representation for the Lorentz group [52], which gives rise to a helicity basis, is a direct consequence of this isomorphism. The following important point is a definition of the Gel'fand–Yaglom equations in the helicity basis (it is equivalent to a definition of these equations on the tangent bundle of the Lorentz group manifold $\mathfrak{L}$). It is shown that a 2-dimensional complex sphere is associated with the each point of the group manifold $\mathfrak{L}$ (this sphere was first considered by Smorodinsky and Huszar at the study of helicity states [40]). The separation of variables in the Gel'fand-Yaglom relativistically invariant system is realized via the definition of hyperspherical functions on the surface of the 2-dimensional complex sphere. At this point, all the variables are parameters of the Lorentz group. Group theoretical description of RWE allows us to present all the physical fields on an equal footing. Physical fields of any spin (both massive and massless) are the functions on the group $\mathfrak{G}_+$. By way of example, we consider a Dirac field (electron-positron field). In conclusion we point out a relationship between the proposed approach and a quantum field theory on the Poincaré group, and also we discuss some further applications of the approach to the study of composite particles in terms of indecomposable RWE.

2 Helicity Basis

Let $\mathfrak{g} \to T_{\mathfrak{g}}$ be an arbitrary linear representation of the proper orthochronous Lorentz group $\mathfrak{G}_+$ and let $\mathsf{A}_i(t) = T_{a_i(t)}$ be an infinitesimal operator corresponding the rotation $a_i(t) \in \mathfrak{G}_+$. Analogously, we have $\mathsf{B}_i(t) = T_{b_i(t)}$, where $b_i(t) \in \mathfrak{G}_+$ is a hyperbolic rotation. The operators A_i and B_i satisfy the following commutation relations[1]:

$$\left.\begin{array}{lll}
[\mathsf{A}_1, \mathsf{A}_2] = \mathsf{A}_3, & [\mathsf{A}_2, \mathsf{A}_3] = \mathsf{A}_1, & [\mathsf{A}_3, \mathsf{A}_1] = \mathsf{A}_2, \\
[\mathsf{B}_1, \mathsf{B}_2] = -\mathsf{A}_3, & [\mathsf{B}_2, \mathsf{B}_3] = -\mathsf{A}_1, & [\mathsf{B}_3, \mathsf{B}_1] = -\mathsf{A}_2, \\
[\mathsf{A}_1, \mathsf{B}_1] = 0, & [\mathsf{A}_2, \mathsf{B}_2] = 0, & [\mathsf{A}_3, \mathsf{B}_3] = 0, \\
[\mathsf{A}_1, \mathsf{B}_2] = \mathsf{B}_3, & [\mathsf{A}_1, \mathsf{B}_3] = -\mathsf{B}_2, & \\
[\mathsf{A}_2, \mathsf{B}_3] = \mathsf{B}_1, & [\mathsf{A}_2, \mathsf{B}_1] = -\mathsf{B}_3, & \\
[\mathsf{A}_3, \mathsf{B}_1] = \mathsf{B}_2, & [\mathsf{A}_3, \mathsf{B}_2] = -\mathsf{B}_1. &
\end{array}\right\} \quad (1)$$

Let us consider the operators

$$\mathsf{X}_l = \frac{1}{2}i(\mathsf{A}_l + i\mathsf{B}_l), \quad \mathsf{Y}_l = \frac{1}{2}i(\mathsf{A}_l - i\mathsf{B}_l), \quad (l = 1, 2, 3). \quad (2)$$

Using the relations (1), we find that

$$[\mathsf{X}_k, \mathsf{X}_l] = i\varepsilon_{klm}\mathsf{X}_m, \quad [\mathsf{Y}_l, \mathsf{Y}_m] = i\varepsilon_{lmn}\mathsf{Y}_n, \quad [\mathsf{X}_l, \mathsf{Y}_m] = 0. \quad (3)$$

Further, introducing generators of the form

$$\left.\begin{array}{ll}
\mathsf{X}_+ = \mathsf{X}_1 + i\mathsf{X}_2, & \mathsf{X}_- = \mathsf{X}_1 - i\mathsf{X}_2, \\
\mathsf{Y}_+ = \mathsf{Y}_1 + i\mathsf{Y}_2, & \mathsf{Y}_- = \mathsf{Y}_1 - i\mathsf{Y}_2,
\end{array}\right\} \quad (4)$$

[1]Denoting $\mathsf{I}^{23} = \mathsf{A}_1$, $\mathsf{I}^{31} = \mathsf{A}_2$, $\mathsf{I}^{12} = \mathsf{A}_3$, and $\mathsf{I}^{01} = \mathsf{B}_1$, $\mathsf{I}^{02} = \mathsf{B}_2$, $\mathsf{I}^{03} = \mathsf{B}_3$ we can write the relations (1) in a more compact form:

$$\left[\mathsf{I}^{\mu\nu}, \mathsf{I}^{\lambda\rho}\right] = \delta_{\mu\rho}\mathsf{I}^{\lambda\nu} + \delta_{\nu\lambda}\mathsf{I}^{\mu\rho} - \delta_{\nu\rho}\mathsf{I}^{\mu\lambda} - \delta_{\mu\lambda}\mathsf{I}^{\nu\rho}.$$

we see that in virtue of commutativity of the relations (3) a space of an irreducible finite–dimensional representation of the group $\mathfrak{G}_+$ can be spanned on the totality of $(2l+1)(2\dot{l}+1)$ basis vectors $\mid l, m; \dot{l}, \dot{m}\rangle$, where $l, m, \dot{l}, \dot{m}$ are integer or half–integer numbers, $-l \leq m \leq l$, $-\dot{l} \leq \dot{m} \leq \dot{l}$. Therefore,

$$
\begin{aligned}
&\mathsf{X}_- \mid l, m; \dot{l}, \dot{m}\rangle = \sqrt{(l+m)(l-m+1)} \mid l, m-1, \dot{l}, \dot{m}\rangle \quad (m > -l),\\
&\mathsf{X}_+ \mid l, m; \dot{l}, \dot{m}\rangle = \sqrt{(l-m)(l+m+1)} \mid l, m+1; \dot{l}, \dot{m}\rangle \quad (m < l),\\
&\mathsf{X}_3 \mid l, m; \dot{l}, \dot{m}\rangle = m \mid l, m; \dot{l}, \dot{m}\rangle,\\
&\mathsf{Y}_- \mid l, m; \dot{l}, \dot{m}\rangle = \sqrt{(\dot{l}+\dot{m})(\dot{l}-\dot{m}+1)} \mid l, m; \dot{l}, \dot{m}-1\rangle \quad (\dot{m} > -\dot{l}),\\
&\mathsf{Y}_+ \mid l, m; \dot{l}, \dot{m}\rangle = \sqrt{(\dot{l}-\dot{m})(\dot{l}+\dot{m}+1)} \mid l, m; \dot{l}, \dot{m}+1\rangle \quad (\dot{m} < \dot{l}),\\
&\mathsf{Y}_3 \mid l, m; \dot{l}, \dot{m}\rangle = \dot{m} \mid l, m; \dot{l}, \dot{m}\rangle.
\end{aligned} \tag{5}
$$

From the relations (3) it follows that each of the sets of infinitisimal operators X and Y generates the group $SU(2)$ and these two groups commute with each other. Thus, from the relations (3) and (5) it follows that the group $\mathfrak{G}_+$, in essence, is equivalent to the group $SU(2) \otimes SU(2)$. In contrast to the Gel'fand–Naimark representation for the Lorentz group [20, 32], which does not find a broad application in physics, a representation (5) is a most useful in theoretical physics (see, for example, [1, 39, 37, 38]). This representation for the Lorentz group was first given by Van der Waerden in his brilliant book [52]. It should be noted here that the helicity basis, defined by the formulae (2)–(5), has an evident physical meaning. For example, in the case of $(1,0) \oplus (0,1)$–representation space there is an analogy with the photon spin states. Namely, the operators X and Y correspond to the right and left polarization states of the photon. For that reason we will call the canonical basis consisting of the vectors $\mid lm; \dot{l}\dot{m}\rangle$ as *a helicity basis*.

As is known, a double covering of the proper orthochronous Lorentz group $\mathfrak{G}_+$, the group $SL(2,\mathbb{C})$, is isomorphic to the Clifford–Lipschitz group $\mathbf{Spin}_+(1,3)$, which, in its turn, is fully defined within a biquaternion algebra $\mathbb{C}_2$, since

$$
\mathbf{Spin}_+(1,3) \simeq \left\{ \begin{pmatrix} \alpha & \beta \\ \gamma & \delta \end{pmatrix} \in \mathbb{C}_2 : \ \det \begin{pmatrix} \alpha & \beta \\ \gamma & \delta \end{pmatrix} = 1 \right\} = SL(2,\mathbb{C}).
$$

Thus, a fundamental representation of the group $\mathfrak{G}_+$ is realized in a spinspace $\mathbb{S}_2$. The spinspace $\mathbb{S}_2$ is a complexification of the minimal left ideal of the algebra $\mathbb{C}_2$: $\mathbb{S}_2 = \mathbb{C} \otimes I_{2,0} = \mathbb{C} \otimes C\!\ell_{2,0} e_{20}$ or $\mathbb{S}_2 = \mathbb{C} \otimes I_{1,1} = \mathbb{C} \otimes C\!\ell_{1,1} e_{11}$ ($\mathbb{C} \otimes I_{0,2} = \mathbb{C} \otimes C\!\ell_{0,2} e_{02}$), where $C\!\ell_{p,q}$ $(p+q=2)$ is a real subalgebra of $\mathbb{C}_2$, $I_{p,q}$ is the minimal left ideal of the algebra $C\!\ell_{p,q}$, e_{pq} is a primitive idempotent.

Further, let $\overset{*}{\mathbb{C}}_2$ be the biquaternion algebra, in which all the coefficients are complex conjugate to the coefficients of the algebra $\mathbb{C}_2$. The algebra $\overset{*}{\mathbb{C}}_2$ is obtained from $\mathbb{C}_2$ under action of the automorphism $\mathcal{A} \rightarrow \mathcal{A}^\star$ (involution) or the antiautomorphism

$\mathcal{A} \to \widetilde{\mathcal{A}}$ (reversal), where $\mathcal{A} \in \mathbb{C}_2$ (see [45, 46]). Let us compose a tensor product of k algebras $\mathbb{C}_2$ and r algebras $\overset{*}{\mathbb{C}}_2$:

$$\mathbb{C}_2 \otimes \mathbb{C}_2 \otimes \cdots \otimes \mathbb{C}_2 \otimes \overset{*}{\mathbb{C}}_2 \otimes \overset{*}{\mathbb{C}}_2 \otimes \cdots \otimes \overset{*}{\mathbb{C}}_2 \simeq \mathbb{C}_{2k} \otimes \overset{*}{\mathbb{C}}_{2r}. \tag{6}$$

The tensor product (6) induces a spinspace

$$\mathbb{S}_2 \otimes \mathbb{S}_2 \otimes \cdots \otimes \mathbb{S}_2 \otimes \dot{\mathbb{S}}_2 \otimes \dot{\mathbb{S}}_2 \otimes \cdots \otimes \dot{\mathbb{S}}_2 = \mathbb{S}_{2^{k+r}} \tag{7}$$

with 'vectors' (spintensors) of the form

$$\xi^{\alpha_1\alpha_2\cdots\alpha_k\dot{\alpha}_1\dot{\alpha}_2\cdots\dot{\alpha}_r} = \sum \xi^{\alpha_1} \otimes \xi^{\alpha_2} \otimes \cdots \otimes \xi^{\alpha_k} \otimes \xi^{\dot{\alpha}_1} \otimes \xi^{\dot{\alpha}_2} \otimes \cdots \otimes \xi^{\dot{\alpha}_r}. \tag{8}$$

The algebras $\mathbb{C}_2$ ($\overset{*}{\mathbb{C}}_2$) and the spinspaces $\mathbb{S}_2$ ($\dot{\mathbb{S}}_2$) correspond to fundamental represesentations $\boldsymbol{\tau}_{\frac{1}{2},0}$ ($\boldsymbol{\tau}_{0,\frac{1}{2}}$) of the Lorentz group $\mathfrak{G}_+$. In general case, the spinspace (7) is reducible, that is, there exists a decomposition of the original spinspace $\mathbb{S}_{2^{k+r}}$ into a direct sum of irreducible subspaces with respect to a representation

$$\underbrace{\boldsymbol{\tau}_{\frac{1}{2},0} \otimes \boldsymbol{\tau}_{\frac{1}{2},0} \otimes \cdots \otimes \boldsymbol{\tau}_{\frac{1}{2},0}}_{k \text{ times}} \otimes \underbrace{\boldsymbol{\tau}_{0,\frac{1}{2}} \otimes \boldsymbol{\tau}_{0,\frac{1}{2}} \otimes \cdots \otimes \boldsymbol{\tau}_{0,\frac{1}{2}}}_{r \text{ times}}. \tag{9}$$

The full representation space $\mathbb{S}_{2^{k+r}}$ contains both symmetric and antisymmetric spintensors (8). A decomposition of the spinspace (7) into irreducible subspaces with respect to the action of $SL(2,\mathbb{C})$ is a particular case of the Weyl theory [54]. In this case an explicit form of the decomposition is defined by the Clebsh-Gordan formula, and each irreducible representation of the group $SL(2,\mathbb{C})$ is given by the Young diagram consisting of only one row. For that reason irreducible subspaces are spaces of symmetric spintensors $\mathrm{Sym}(k,r) \subset \mathbb{S}_{2^{k+r}}$. The dimension of $\mathrm{Sym}(k,r)$ is equal to $(k+1)(r+1)$ or $(2l+1)(2l'+1)$ at $l=\frac{k}{2}$, $l'=\frac{r}{2}$. It is easy to see that the space $\mathrm{Sym}(k,r)$ is a space of Van der Waerden representation (5). The space $\mathrm{Sym}(k,r)$ can be considered as a space of polynomials

$$p(z_0, z_1, \bar{z}_0, \bar{z}_1) = \sum_{\substack{(\alpha_1,\ldots,\alpha_k)\\ (\dot{\alpha}_1,\ldots,\dot{\alpha}_r)}} \frac{1}{k!\,r!} a^{\alpha_1\cdots\alpha_k\dot{\alpha}_1\cdots\dot{\alpha}_r} z_{\alpha_1}\cdots z_{\alpha_k}\bar{z}_{\dot{\alpha}_1}\cdots\bar{z}_{\dot{\alpha}_r} \tag{10}$$

$$(\alpha_i, \dot{\alpha}_i = 0, 1),$$

where the numbers $a^{\alpha_1\cdots\alpha_k\dot{\alpha}_1\cdots\dot{\alpha}_r}$ are unaffected at the permutations of indices. Infinitesimal operators of $\mathfrak{G}_+$ in the helicity basis have a very simple form

$$\begin{aligned}
\mathsf{A}_1 \mid l, m; \dot{l}, \dot{m}\rangle &= -\frac{i}{2}\boldsymbol{\alpha}^l_m \mid l, m-1; \dot{l}, \dot{m}\rangle - \frac{i}{2}\boldsymbol{\alpha}^l_{m+1} \mid l, m+1; \dot{l}\dot{m}\rangle,\\
\mathsf{A}_2 \mid l, m; \dot{l}, \dot{m}\rangle &= \frac{1}{2}\boldsymbol{\alpha}^l_m \mid l, m-1; \dot{l}, \dot{m}\rangle - \frac{1}{2}\boldsymbol{\alpha}^l_{m+1} \mid l, m+1; \dot{l}, \dot{m}\rangle,\\
\mathsf{A}_3 \mid l, m; \dot{l}, \dot{m}\rangle &= -im \mid l, m; \dot{l}, \dot{m}\rangle,
\end{aligned} \tag{11}$$

$$\begin{aligned}
\mathsf{B}_1 \mid l, m; \dot{l}, \dot{m}\rangle &= -\frac{1}{2}\boldsymbol{\alpha}^l_m \mid l, m-1; \dot{l}, \dot{m}\rangle - \frac{1}{2}\boldsymbol{\alpha}^l_{m+1} \mid l, m+1; \dot{l}, \dot{m}\rangle, \\
\mathsf{B}_2 \mid l, m; \dot{l}, \dot{m}\rangle &= -\frac{i}{2}\boldsymbol{\alpha}^l_m \mid l, m-1; \dot{l}, \dot{m}\rangle + \frac{i}{2}\boldsymbol{\alpha}^l_{m+1} \mid l, m+1; \dot{l}, \dot{m}\rangle, \\
\mathsf{B}_3 \mid l, m; \dot{l}, \dot{m}\rangle &= -m \mid l, m; \dot{l}, \dot{m}\rangle,
\end{aligned} \tag{12}$$

$$\begin{aligned}
\widetilde{\mathsf{A}}_1 \mid l, m; \dot{l}, \dot{m}\rangle &= -\frac{i}{2}\boldsymbol{\alpha}^{\dot{l}}_{\dot{m}} \mid l, m; \dot{l}, \dot{m}-1\rangle - \frac{i}{2}\boldsymbol{\alpha}^{\dot{l}}_{\dot{m}+1} \mid l, m; \dot{l}, \dot{m}+1\rangle, \\
\widetilde{\mathsf{A}}_2 \mid l, m; \dot{l}, \dot{m}\rangle &= \frac{1}{2}\boldsymbol{\alpha}^{\dot{l}}_{\dot{m}} \mid l, m; \dot{l}, \dot{m}-1\rangle - \frac{1}{2}\boldsymbol{\alpha}^{\dot{l}}_{\dot{m}+1} \mid l, m; \dot{l}, \dot{m}+1\rangle, \\
\widetilde{\mathsf{A}}_3 \mid l, m; \dot{l}, \dot{m}\rangle &= -i\dot{m} \mid l, m; \dot{l}, \dot{m}\rangle,
\end{aligned} \tag{13}$$

$$\begin{aligned}
\widetilde{\mathsf{B}}_1 \mid l, m; \dot{l}, \dot{m}\rangle &= \frac{1}{2}\boldsymbol{\alpha}^{\dot{l}}_{\dot{m}} \mid l, m; \dot{l}, \dot{m}-1\rangle + \frac{1}{2}\boldsymbol{\alpha}^{\dot{l}}_{\dot{m}+1} \mid l, m; \dot{l}, \dot{m}+1\rangle, \\
\widetilde{\mathsf{B}}_2 \mid l, m; \dot{l}, \dot{m}\rangle &= \frac{i}{2}\boldsymbol{\alpha}^{\dot{l}}_{\dot{m}} \mid l, m; \dot{l}, \dot{m}-1\rangle - \frac{i}{2}\boldsymbol{\alpha}^{\dot{l}}_{\dot{m}+1} \mid l, m; \dot{l}, \dot{m}+1\rangle, \\
\widetilde{\mathsf{B}}_3 \mid l, m; \dot{l}, \dot{m}\rangle &= -\dot{m} \mid l, m; \dot{l}, \dot{m}\rangle,
\end{aligned} \tag{14}$$

where

$$\boldsymbol{\alpha}^l_m = \sqrt{(l+m)(l-m+1)}.$$

3 Hyperspherical functions

As noted previously, the explicit form of the isomorphism $SL(2,\mathbb{C}) \simeq SU(2)\otimes SU(2)$ can be obtained via the complexification of $SU(2)$. The group $SL(2,\mathbb{C})$ is a group of all complex matrices

$$\begin{pmatrix} \alpha & \beta \\ \gamma & \delta \end{pmatrix}$$

of 2-nd order with the determinant $\alpha\delta - \gamma\beta = 1$. The group $SU(2)$ is one of the real forms of $SL(2,\mathbb{C})$. The transition from $SU(2)$ to $SL(2,\mathbb{C})$ is realized via the complexification of three real parameters φ, θ, ψ (Euler angles). Let $\theta^c = \theta - i\tau$, $\varphi^c = \varphi - i\epsilon$, $\psi^c = \psi - i\varepsilon$ be complex Euler angles, where

$$\begin{array}{lllllllll}
0 & \leq & \operatorname{Re}\theta^c = \theta & \leq & \pi, & -\infty & < & \operatorname{Im}\theta^c = \tau & < & +\infty, \\
0 & \leq & \operatorname{Re}\varphi^c = \varphi & < & 2\pi, & -\infty & < & \operatorname{Im}\varphi^c = \epsilon & < & +\infty, \\
-2\pi & \leq & \operatorname{Re}\psi^c = \psi & < & 2\pi, & -\infty & < & \operatorname{Im}\psi^c = \varepsilon & < & +\infty.
\end{array}$$

Replacing in $SU(2)$ the angles φ, θ, ψ by the complex angles $\varphi^c, \theta^c, \psi^c$ we come to the following matrix

$$\mathfrak{g} = \begin{pmatrix} \cos\frac{\theta^c}{2} e^{\frac{i(\varphi^c+\psi^c)}{2}} & i\sin\frac{\theta^c}{2} e^{\frac{i(\varphi^c-\psi^c)}{2}} \\ i\sin\frac{\theta^c}{2} e^{\frac{i(\psi^c-\varphi^c)}{2}} & \cos\frac{\theta^c}{2} e^{-\frac{i(\varphi^c+\psi^c)}{2}} \end{pmatrix} =$$

$$\begin{pmatrix} \left[\cos\frac{\theta}{2}\cosh\frac{\tau}{2}+i\sin\frac{\theta}{2}\sinh\frac{\tau}{2}\right] e^{\frac{\epsilon+\varepsilon+i(\varphi+\psi)}{2}} & \left[\cos\frac{\theta}{2}\sinh\frac{\tau}{2}+i\sin\frac{\theta}{2}\cosh\frac{\tau}{2}\right] e^{\frac{\epsilon-\varepsilon+i(\varphi-\psi)}{2}} \\ \left[\cos\frac{\theta}{2}\sinh\frac{\tau}{2}+i\sin\frac{\theta}{2}\cosh\frac{\tau}{2}\right] e^{\frac{\varepsilon-\epsilon+i(\psi-\varphi)}{2}} & \left[\cos\frac{\theta}{2}\cosh\frac{\tau}{2}+i\sin\frac{\theta}{2}\sinh\frac{\tau}{2}\right] e^{\frac{-\epsilon-\varepsilon-i(\varphi+\psi)}{2}} \end{pmatrix}, \quad (15)$$

since $\cos\frac{1}{2}(\theta - i\tau) = \cos\frac{\theta}{2}\cosh\frac{\tau}{2} + i\sin\frac{\theta}{2}\sinh\frac{\tau}{2}$, and $\sin\frac{1}{2}(\theta - i\tau) = \sin\frac{\theta}{2}\cosh\frac{\tau}{2} - i\cos\frac{\theta}{2}\sinh\frac{\tau}{2}$. It is easy to verify that the matrix (15) coincides with a matrix of the fundamental representation of the group $SL(2,\mathbb{C})$ (in Euler parametrization):

$$\mathfrak{g}(\varphi,\, \epsilon,\, \theta,\, \tau,\, \psi,\, \varepsilon) =$$

$$\begin{pmatrix} e^{i\frac{\varphi}{2}} & 0 \\ 0 & e^{-i\frac{\varphi}{2}} \end{pmatrix}\begin{pmatrix} e^{\frac{\epsilon}{2}} & 0 \\ 0 & e^{-\frac{\epsilon}{2}} \end{pmatrix}\begin{pmatrix} \cos\frac{\theta}{2} & i\sin\frac{\theta}{2} \\ i\sin\frac{\theta}{2} & \cos\frac{\theta}{2} \end{pmatrix}\begin{pmatrix} \cosh\frac{\tau}{2} & \sinh\frac{\tau}{2} \\ \sinh\frac{\tau}{2} & \cosh\frac{\tau}{2} \end{pmatrix}\begin{pmatrix} e^{i\frac{\psi}{2}} & 0 \\ 0 & e^{-i\frac{\psi}{2}} \end{pmatrix}\begin{pmatrix} e^{\frac{\varepsilon}{2}} & 0 \\ 0 & e^{-\frac{\varepsilon}{2}} \end{pmatrix}.$$

Moreover, the complexification of $SU(2)$ gives us the most simple and direct way for calculation of matrix elements of the Lorentz group. It is known that these elements have a great importance in quantum field theory and widely used at the study of relativistic amplitudes. The most degenerate representation of such the elements was first obtained by Dolginov, Toptygin and Moskalev [13, 14, 15] via an analytic continuation of representations of the group O_4. Later on, matrix elements were studied on the hyperboloid [50, 49], on the direct product of the hyperboloid and sphere [29, 28]. The matrix elements of the principal series were studied by Ström [42, 43] in the Gel'fand-Naimark basis. However, all the matrix elements, calculated in the GN-basis, have a very complicated form.

Matrix elements of $SL(2,\mathbb{C})$ in the helicity basis (see [48]) are

$$t^l_{mn}(\mathfrak{g}) = e^{-m(\epsilon+i\varphi)-n(\varepsilon+i\psi)}Z^l_{mn} = e^{-m(\epsilon+i\varphi)-n(\varepsilon+i\psi)}\times$$

$$\sum_{k=-l}^{l} i^{m-k}\sqrt{\Gamma(l-m+1)\Gamma(l+m+1)\Gamma(l-k+1)\Gamma(l+k+1)}\times$$

$$\cos^{2l}\frac{\theta}{2}\tan^{m-k}\frac{\theta}{2}\times$$

$$\sum_{j=\max(0,k-m)}^{\min(l-m,l+k)} \frac{i^{2j}\tan^{2j}\dfrac{\theta}{2}}{\Gamma(j+1)\Gamma(l-m-j+1)\Gamma(l+k-j+1)\Gamma(m-k+j+1)}\times$$

$$\sqrt{\Gamma(l-n+1)\Gamma(l+n+1)\Gamma(l-k+1)\Gamma(l+k+1)}\cosh^{2l}\frac{\tau}{2}\tanh^{n-k}\frac{\tau}{2}\times$$

$$\sum_{s=\max(0,k-n)}^{\min(l-n,l+k)} \frac{\tanh^{2s}\dfrac{\tau}{2}}{\Gamma(s+1)\Gamma(l-n-s+1)\Gamma(l+k-s+1)\Gamma(n-k+s+1)}. \tag{16}$$

We will call the functions Z^l_{mn} in (16) as *hyperspherical functions.* Hyperspherical functions Z^l_{mn} can be written via the hypergeometric series [48]:

$$Z^l_{mn} = \cos^{2l}\frac{\theta}{2}\cosh^{2l}\frac{\tau}{2}\sum_{k=-l}^{l} i^{m-k}\tan^{m-k}\frac{\theta}{2}\tanh^{n-k}\frac{\tau}{2}\times$$

$${}_2F_1\left(\begin{matrix} m-l+1, 1-l-k \\ m-k+1 \end{matrix}\middle| i^2\tan^2\frac{\theta}{2}\right){}_2F_1\left(\begin{matrix} n-l+1, 1-l-k \\ n-k+1 \end{matrix}\middle| \tanh^2\frac{\tau}{2}\right). \tag{17}$$

Therefore, matrix elements can be expressed by means of the function (*a generalized hyperspherical function*)

$$\mathfrak{M}^l_{mn}(\mathfrak{g}) = e^{-m(\epsilon+i\varphi)}Z^l_{mn}e^{-n(\varepsilon+i\psi)}, \tag{18}$$

where

$$Z^l_{mn} = \sum_{k=-l}^{l} P^l_{mk}(\cos\theta)\mathfrak{P}^l_{kn}(\cosh\tau), \tag{19}$$

here $P^l_{mn}(\cos\theta)$ is a generalized spherical function on the group $SU(2)$ (see [20]), and $\mathfrak{P}^l_{mn}$ is an analog of the generalized spherical function for the group $QU(2)$ (so–called Jacobi function [51]). $QU(2)$ is a group of quasiunitary unimodular matrices of second order. As well as the group $SU(2)$, the group $QU(2)$ is one of the real forms of $SL(2,\mathbb{C})$ ($QU(2)$ is noncompact). Further, from (17) we see that the function Z^l_{mn} depends on two variables θ and τ. Therefore, using Bateman factorization we can express the hyperspherical functions Z^l_{mn} via Appell functions F_1–F_4 (hypergeometric series of the two variables [4, 8]).

Infinitesimal operators of the group $\mathfrak{G}_+$ can be expressed via the Euler angles as follows (for detailed deriving of these formulae see [48])

$$\mathsf{A}_1 = \cos\psi^c \frac{\partial}{\partial\theta} + \frac{\sin\psi^c}{\sin\theta^c}\frac{\partial}{\partial\varphi} - \cot\theta^c \sin\psi^c \frac{\partial}{\partial\psi}, \tag{20}$$

$$\mathsf{A}_2 = -\sin\psi^c \frac{\partial}{\partial\theta} + \frac{\cos\psi^c}{\sin\theta^c}\frac{\partial}{\partial\varphi} - \cot\theta^c \cos\psi^c \frac{\partial}{\partial\psi}, \tag{21}$$

$$\mathsf{A}_3 = \frac{\partial}{\partial\psi}, \tag{22}$$

$$\mathsf{B}_1 = \cos\psi^c \frac{\partial}{\partial\tau} + \frac{\sin\psi^c}{\sin\theta^c}\frac{\partial}{\partial\epsilon} - \cot\theta^c \sin\psi^c \frac{\partial}{\partial\varepsilon}, \tag{23}$$

$$\mathsf{B}_2 = -\sin\psi^c \frac{\partial}{\partial\tau} + \frac{\cos\psi^c}{\sin\theta^c}\frac{\partial}{\partial\epsilon} - \cot\theta^c \cos\psi^c \frac{\partial}{\partial\varepsilon}, \tag{24}$$

$$\mathsf{B}_3 = \frac{\partial}{\partial\varepsilon}. \tag{25}$$

It is easy to verify that operators A_i, B_i, defined by the formulae (22), (25), (20), (23) and (21), (24), are satisfy the commutation relations (1).

Between generalized hyperspherical functions $\mathfrak{M}^l_{mn}$ (and also the hyperspherical functions Z^l_{mn}) there exists a wide variety of recurrence relations. Part of them relates the hyperspherical functions of one and the same order (with identical l), other part relates the functions of different orders (for more details see [48]). In virtue of the Van der Waerden representation (5) the recurrence formulae for the hyperspherical functions of one and the same order follow from the equalities

$$\mathsf{X}_-\mathfrak{M}^l_{mn} = \alpha_n \mathfrak{M}^l_{m,n-1}, \quad \mathsf{X}_+\mathfrak{M}^l_{mn} = \alpha_{n+1}\mathfrak{M}^l_{m,n+1}, \tag{26}$$

$$\mathsf{Y}_-\mathfrak{M}^{\dot{l}}_{\dot{m}\dot{n}} = \alpha_{\dot{n}} \mathfrak{M}^{\dot{l}}_{\dot{m},\dot{n}-1}, \quad \mathsf{Y}_+\mathfrak{M}^{\dot{l}}_{\dot{m}\dot{n}} = \alpha_{\dot{n}+1}\mathfrak{M}^{\dot{l}}_{\dot{m},\dot{n}+1}, \tag{27}$$

where

$$\alpha_n = \sqrt{(l+n)(l-n+1)}, \quad \alpha_{\dot{n}} = \sqrt{(\dot{l}+\dot{n})(\dot{l}-\dot{n}+1)}.$$

Using the formulae (20), (23) and (21), (24) we obtain

$$\mathsf{X}_+ = \frac{e^{-i\psi^c}}{2}\left[i\frac{\partial}{\partial\theta} - \frac{1}{\sin\theta^c}\frac{\partial}{\partial\varphi} + \cot\theta^c\frac{\partial}{\partial\psi} - \frac{\partial}{\partial\tau} - \frac{i}{\sin\theta^c}\frac{\partial}{\partial\epsilon} + i\cot\theta^c\frac{\partial}{\partial\varepsilon}\right], \tag{28}$$

$$\mathsf{X}_- = \frac{e^{i\psi^c}}{2}\left[i\frac{\partial}{\partial\theta} + \frac{1}{\sin\theta^c}\frac{\partial}{\partial\varphi} - \cot\theta^c\frac{\partial}{\partial\psi} - \frac{\partial}{\partial\tau} + \frac{i}{\sin\theta^c}\frac{\partial}{\partial\epsilon} - i\cot\theta^c\frac{\partial}{\partial\varepsilon}\right], \tag{29}$$

$$\mathsf{Y}_+ = \frac{e^{-i\dot{\psi}^c}}{2}\left[i\frac{\partial}{\partial\theta} - \frac{1}{\sin\dot{\theta}^c}\frac{\partial}{\partial\varphi} + \cot\dot{\theta}^c\frac{\partial}{\partial\psi} + \frac{\partial}{\partial\tau} + \frac{i}{\sin\dot{\theta}^c}\frac{\partial}{\partial\epsilon} - i\cot\dot{\theta}^c\frac{\partial}{\partial\varepsilon}\right], \tag{30}$$

$$\mathsf{Y}_- = \frac{e^{i\dot{\psi}^c}}{2}\left[i\frac{\partial}{\partial\theta} + \frac{1}{\sin\dot{\theta}^c}\frac{\partial}{\partial\varphi} - \cot\dot{\theta}^c\frac{\partial}{\partial\psi} + \frac{\partial}{\partial\tau} - \frac{i}{\sin\dot{\theta}^c}\frac{\partial}{\partial\epsilon} + i\cot\dot{\theta}^c\frac{\partial}{\partial\varepsilon}\right]. \tag{31}$$

Further, substituting the function $\mathfrak{M}^l_{mn} = e^{-m(\epsilon - i\varphi)} Z^l_{mn}(\theta, \tau) e^{-n(\varepsilon - i\psi)}$ into the relations (26) and taking into account the operators (28)–(29), and also using the symmetry of the functions $\dot{Z}^l_{mn}(\theta, \tau)$, we find that

$$i\frac{\partial Z^l_{mn}}{\partial\theta} - \frac{\partial Z^l_{mn}}{\partial\tau} - \frac{2i(n - m\cos\theta^c)}{\sin\theta^c} Z^l_{mn} = 2\alpha_m Z^l_{m-1,n}, \tag{32}$$

$$i\frac{\partial Z^l_{mn}}{\partial\theta} - \frac{\partial Z^l_{mn}}{\partial\tau} + \frac{2i(n - m\cos\theta^c)}{\sin\theta^c} Z^l_{mn} = 2\alpha_{m+1} Z^l_{m+1,n}. \tag{33}$$

Analogously, substituting the function $\mathfrak{M}^{\dot{l}}_{\dot{m}\dot{n}} = e^{-\dot{m}(\epsilon - i\varphi)} Z^{\dot{l}}_{\dot{m}\dot{n}}(\theta, \tau) e^{-\dot{n}(\varepsilon - i\psi)}$ into the relations (27), we obtain

$$i\frac{\partial Z^{\dot{l}}_{\dot{m}\dot{n}}}{\partial\theta} + \frac{\partial Z^{\dot{l}}_{\dot{m}\dot{n}}}{\partial\tau} + \frac{2i(\dot{n} - \dot{m}\cos\dot{\theta}^c)}{\sin\dot{\theta}^c} Z^{\dot{l}}_{\dot{m}\dot{n}} = 2\alpha_{\dot{m}} Z^{\dot{l}}_{\dot{m}-1,\dot{n}}, \tag{34}$$

$$i\frac{\partial Z^{\dot{l}}_{\dot{m}\dot{n}}}{\partial\theta} + \frac{\partial Z^{\dot{l}}_{\dot{m}\dot{n}}}{\partial\tau} - \frac{2i(\dot{n} - \dot{m}\cos\dot{\theta}^c)}{\sin\dot{\theta}^c} Z^{\dot{l}}_{\dot{m}\dot{n}} = 2\alpha_{\dot{m}+1} Z^{\dot{l}}_{\dot{m}+1,\dot{n}}. \tag{35}$$

4 Gel'fand-Yaglom equations in the helicity basis

As a direct consequence of the isomorphism $SL(2, \mathbb{C}) \simeq SU(2) \otimes SU(2)$ we have equations

$$\begin{aligned} \sum_{i=1}^{3} \Lambda_i \frac{\partial\psi}{\partial x_i} - i\sum_{i=1}^{3} \Lambda_i \frac{\partial\psi}{\partial x^*_i} + \kappa^c\psi &= 0, \\ \sum_{i=1}^{3} \Lambda^*_i \frac{\partial\dot{\psi}}{\partial\widetilde{x}_i} + i\sum_{i=1}^{3} \Lambda^*_i \frac{\partial\dot{\psi}}{\partial\widetilde{x}^*_i} + \dot{\kappa}^c\dot{\psi} &= 0, \end{aligned} \tag{36}$$

The equations (36) are defined in a 3-dimensional complex space $\mathbb{C}^3$. In turn, the space $\mathbb{C}^3$ is isometric to a 6-dimensional bivector space $\mathbb{R}^6$ (a parameter space of the Lorentz group [26, 33]). The bivector space $\mathbb{R}^6$ is a tangent space of a group manifold $\mathfrak{L}$ of the Lorentz group, that is, the group manifold $\mathfrak{L}$ in the each point is equivalent to the space $\mathbb{R}^6$. There exists a close relationship between the metrics of the Minkowski spacetime $\mathbb{R}^{1,3}$ and the metrics of $\mathbb{R}^6$ defined by the formulae (see [33])

$$g_{ab} \longrightarrow g_{\alpha\beta\gamma\delta} \equiv g_{\alpha\gamma}g_{\beta\delta} - g_{\alpha\delta}g_{\beta\gamma}, \tag{37}$$

where $g_{\alpha\beta}$ is a metric tensor of the spacetime $\mathbb{R}^{1,3}$, and collective indices are skewsymmetric pairs $\alpha\beta \to a$, $\gamma\delta \to b$. In more detail, if

$$g_{\alpha\beta} = \begin{pmatrix} -1 & 0 & 0 & 0 \\ 0 & -1 & 0 & 0 \\ 0 & 0 & -1 & 0 \\ 0 & 0 & 0 & 1 \end{pmatrix},$$

then in virtue of (37) for the metric tensor of $\mathbb{R}^6$ we obtain

$$g_{ab} = \begin{pmatrix} -1 & 0 & 0 & 0 & 0 & 0 \\ 0 & -1 & 0 & 0 & 0 & 0 \\ 0 & 0 & -1 & 0 & 0 & 0 \\ 0 & 0 & 0 & 1 & 0 & 0 \\ 0 & 0 & 0 & 0 & 1 & 0 \\ 0 & 0 & 0 & 0 & 0 & 1 \end{pmatrix}, \tag{38}$$

where the order of collective indices in $\mathbb{R}^6$ is $23 \to 0$, $10 \to 1$, $20 \to 2$, $30 \to 3$, $31 \to 4$, $12 \to 5$. As it shown in [26], the Lorentz transformations can be represented by linear transformations of the space $\mathbb{R}^6$. Let us write an invariance condition of the system (36). Let $\mathfrak{g} : x' = \mathfrak{g}^{-1}x$ be a transformation of the bivector space $\mathbb{R}^6$, that is, $x' = \sum_{b=1}^{6} g_{ba}x_b$, where $x = (x_1, x_2, x_3, \widetilde{x}_1, \widetilde{x}_2, \widetilde{x}_3)$ and g_{ba} is the metric tensor (38). We can write the tensor (38) in the form $g_{ab} = \begin{pmatrix} g^{-}_{ik} & \\ & g^{+}_{ik} \end{pmatrix}$, then $x' = \sum_{k=1}^{3} g^{-}_{ki}x_k$, $\widetilde{x}' = \sum_{k=1}^{3} g^{+}_{ki}\widetilde{x}_k$. Replacing ψ via $T_{\mathfrak{g}}^{-1}\psi'$, and differentiation on x_k $(\widetilde{x}_k)$ by differentiation on x'_k $(\widetilde{x}'_k)$ via the formulae

$$\frac{\partial}{\partial x_k} = \sum g^{-}_{ik}\frac{\partial}{\partial x'_i}, \quad \frac{\partial}{\partial \widetilde{x}_k} = \sum g^{+}_{ik}\frac{\partial}{\partial \widetilde{x}'_i},$$

we obtain

$$\sum_i \left[g^{-}_{i1}\Lambda_1 T_{\mathfrak{g}}^{-1}\frac{\partial \psi'}{\partial x'_i} + g^{-}_{i2}\Lambda_2 T_{\mathfrak{g}}^{-1}\frac{\partial \psi'}{\partial x'_i} + g^{-}_{i3}\Lambda_3 T_{\mathfrak{g}}^{-1}\frac{\partial \psi'}{\partial x'_i} - \right.$$
$$\left. -ig^{-}_{i1}\Lambda_1 T_{\mathfrak{g}}^{-1}\frac{\partial \psi'}{\partial x^{*\prime}_i} - ig^{-}_{i2}\Lambda_2 T_{\mathfrak{g}}^{-1}\frac{\partial \psi'}{\partial x^{*\prime}_i} - ig^{-}_{i3}\Lambda_3 T_{\mathfrak{g}}^{-1}\frac{\partial \psi'}{\partial x^{*\prime}_i} \right] + \kappa^c T_{\mathfrak{g}}^{-1}\psi' = 0,$$
$$\sum_i \left[g^{+}_{i1}\Lambda^*_1 \overset{*}{T}{}_{\mathfrak{g}}^{-1}\frac{\partial \dot{\psi}'}{\partial \widetilde{x}'_i} + g^{+}_{i2}\Lambda^*_2 \overset{*}{T}{}_{\mathfrak{g}}^{-1}\frac{\partial \dot{\psi}'}{\partial \widetilde{x}'_i} + g^{+}_{i3}\Lambda^*_3 \overset{*}{T}{}_{\mathfrak{g}}^{-1}\frac{\partial \dot{\psi}'}{\partial \widetilde{x}'_i} + \right.$$
$$\left. +ig^{+}_{i1}\Lambda^*_1 \overset{*}{T}{}_{\mathfrak{g}}^{-1}\frac{\partial \dot{\psi}'}{\partial \widetilde{x^*}_i{}'} + ig^{+}_{i2}\Lambda^*_2 \overset{*}{T}{}_{\mathfrak{g}}^{-1}\frac{\partial \dot{\psi}'}{\partial \widetilde{x^*}_i{}'} + ig^{+}_{i3}\Lambda^*_3 \overset{*}{T}{}_{\mathfrak{g}}^{-1}\frac{\partial \dot{\psi}'}{\partial \widetilde{x^*}_i{}'} \right] + \dot{\kappa}^c \overset{*}{T}{}_{\mathfrak{g}}^{-1}\dot{\psi}' = 0.$$

For coincidence of the latter system with (36) we must multiply this system by $T_{\mathfrak{g}}$ $(\overset{*}{T}_{\mathfrak{g}})$ from the left:

$$\begin{aligned} \sum_i \sum_k g^{-}_{ik} T_{\mathfrak{g}}\Lambda_k T_{\mathfrak{g}}^{-1}\frac{\partial \psi'}{\partial x'_i} - i\sum_i \sum_k g^{-}_{ik} T_{\mathfrak{g}}\Lambda_k T_{\mathfrak{g}}^{-1}\frac{\partial \psi'}{\partial x^{*\prime}_i} + \kappa^c\psi' &= 0, \\ \sum_i \sum_k g^{+}_{ik} \overset{*}{T}_{\mathfrak{g}}\Lambda^*_k \overset{*}{T}{}_{\mathfrak{g}}^{-1}\frac{\partial \dot{\psi}'}{\partial \widetilde{x}'_i} + i\sum_i \sum_k g^{+}_{ik} \overset{*}{T}_{\mathfrak{g}}\Lambda^*_k \overset{*}{T}{}_{\mathfrak{g}}^{-1}\frac{\partial \dot{\psi}'}{\partial \widetilde{x^*}_i{}'} + \dot{\kappa}^c\dot{\psi}' &= 0. \end{aligned}$$

The requirement of invariance means that for any transformation $\mathfrak{g}$ between the matrices Λ_k (Λ_k^*) we must have the relations

$$\sum_k g_{ik}^{-} T_{\mathfrak{g}} \Lambda_k T_{\mathfrak{g}}^{-1} = \Lambda_i, \qquad \sum_k g_{ik}^{+} \overset{*}{T}_{\mathfrak{g}} \Lambda_k^* \overset{*}{T}_{\mathfrak{g}}^{-1} = \Lambda_i^*, \tag{39}$$

where Λ_i^* are the matrices of the equations in the dual representation space, κ^c is a complex number, $\partial/\partial\widetilde{x}_i$ mean covariant derivatives in the dual space.

Let us find commutation relations between the matrices Λ_i, Λ_i^* and infinitesimal operators (11), (12), (13), (14) defined in the helicity basis. First of all, let us present transformations $T_{\mathfrak{g}}$ ($\overset{*}{T}_{\mathfrak{g}}$) in the infinitesimal form, $\mathsf{I}+\mathsf{A}_i\xi+\ldots$, $\mathsf{I}+\mathsf{B}_i\xi+\ldots$, $\mathsf{I}+\widetilde{\mathsf{A}}_i\xi+\ldots$, $\mathsf{I}+\widetilde{\mathsf{B}}_i\xi+\ldots$. Substituting these transformations into invariance conditions (39) we obtain with an accuracy of the terms of second order the following commutation relations

$$\begin{array}{lll} [\mathsf{A}_1,\Lambda_1] = 0, & [\mathsf{A}_1,\Lambda_2] = \Lambda_3, & [\mathsf{A}_1,\Lambda_3] = -\Lambda_2,\\ [\mathsf{A}_2,\Lambda_1] = -\Lambda_3, & [\mathsf{A}_2,\Lambda_2] = 0, & [\mathsf{A}_2,\Lambda_3] = \Lambda_1,\\ [\mathsf{A}_3,\Lambda_1] = \Lambda_2, & [\mathsf{A}_3,\Lambda_2] = -\Lambda_1, & [\mathsf{A}_3,\Lambda_3] = 0. \end{array} \tag{40}$$

$$\begin{array}{lll} [\mathsf{B}_1,\Lambda_1] = 0, & [\mathsf{B}_1,\Lambda_2] = -i\Lambda_3, & [\mathsf{B}_1,\Lambda_3] = i\Lambda_2,\\ [\mathsf{B}_2,\Lambda_1] = i\Lambda_3, & [\mathsf{B}_2,\Lambda_2] = 0, & [\mathsf{B}_2,\Lambda_3] = -i\Lambda_1,\\ [\mathsf{B}_3,\Lambda_1] = -i\Lambda_2, & [\mathsf{B}_3,\Lambda_2] = i\Lambda_1, & [\mathsf{B}_3,\Lambda_3] = 0. \end{array} \tag{41}$$

$$\begin{array}{lll} \left[\widetilde{\mathsf{A}}_1,\Lambda_1^*\right] = 0, & \left[\widetilde{\mathsf{A}}_1,\Lambda_2^*\right] = \Lambda_3^*, & \left[\widetilde{\mathsf{A}}_1,\Lambda_3^*\right] = -\Lambda_2^*,\\ \left[\widetilde{\mathsf{A}}_2,\Lambda_1^*\right] = -\Lambda_3^*, & \left[\widetilde{\mathsf{A}}_2,\Lambda_2^*\right] = 0, & \left[\widetilde{\mathsf{A}}_2,\Lambda_3^*\right] = \Lambda_1^*,\\ \left[\widetilde{\mathsf{A}}_3,\Lambda_1^*\right] = \Lambda_2^*, & \left[\widetilde{\mathsf{A}}_3,\Lambda_2^*\right] = -\Lambda_1^*, & \left[\widetilde{\mathsf{A}}_3,\Lambda_3^*\right] = 0. \end{array} \tag{42}$$

$$\begin{array}{lll} \left[\widetilde{\mathsf{B}}_1,\Lambda_1^*\right] = 0, & \left[\widetilde{\mathsf{B}}_1,\Lambda_2^*\right] = i\Lambda_3^*, & \left[\widetilde{\mathsf{B}}_1,\Lambda_3^*\right] = -i\Lambda_2^*,\\ \left[\widetilde{\mathsf{B}}_2,\Lambda_1^*\right] = -i\Lambda_3^*, & \left[\widetilde{\mathsf{B}}_2,\Lambda_2^*\right] = 0, & \left[\widetilde{\mathsf{B}}_2,\Lambda_3^*\right] = i\Lambda_1^*,\\ \left[\widetilde{\mathsf{B}}_3,\Lambda_1^*\right] = i\Lambda_2^*, & \left[\widetilde{\mathsf{B}}_3,\Lambda_2^*\right] = -i\Lambda_1^*, & \left[\widetilde{\mathsf{B}}_3,\Lambda_3^*\right] = 0. \end{array} \tag{43}$$

Further, using the latter relations and taking into account (2), it is easy to establish commutation relations between Λ_3, Λ_3^* and generators $\mathsf{Y}_\pm$, Y_3, $\mathsf{X}_\pm$, X_3:

$$\begin{array}{ll} [[\Lambda_3,\mathsf{X}_-],\mathsf{X}_+] = 2\Lambda_3, & [[\Lambda_3^*,\mathsf{Y}_-],\mathsf{Y}_+] = 2\Lambda_3^*,\\ [\Lambda_3,\mathsf{X}_3] = 0, & [\Lambda_3^*,\mathsf{Y}_3] = 0, \end{array} \tag{44}$$

Using the relations (44) we will find an explicit form of the matrices Λ_3 and Λ_3^*, and after this we will find Λ_1, Λ_2 and Λ_1^*, Λ_2^*. The wave function ψ is transformed within some representation $T_{\mathfrak{g}}$ of the group $\mathfrak{G}_+$. We assume that $T_{\mathfrak{g}}$ is decomposed into irreducible representations. The components of the function ψ we will numerate by the indices l and m, where l is a weight of irreducible representation, m is a number of the component in the representation of the weight l. In the case when a representation with one and the same weight l at the decomposition of ψ occurs more than one time, then with the aim to distinguish these representations we will add the index k, which indicates a number of the representation of the weight l. Denoting $\zeta_{lm;\dot{l}\dot{m}} = | lm; \dot{l}\dot{m}\rangle$ and coming to the helicity basis, we obtain a following decomposition for the wave function:

$$\psi(x_1, x_2, x_3) = \sum_{l,m,k} \psi^k_{lm;\dot{l}\dot{m}}(x_1, x_2, x_3)\zeta^k_{lm;\dot{l};\dot{m}},$$

where x_1, x_2, x_3 are the coordinates of the complex space $\mathbb{C}^{32}$. Analogously, for the dual representation we have

$$\dot{\psi}(\widetilde{x}_1, \widetilde{x}_2, \widetilde{x}_3) = \sum_{\dot{l},\dot{m},\dot{k}} \psi^{\dot{k}}_{\dot{l}\dot{m};lm}(\widetilde{x}_1, \widetilde{x}_2, \widetilde{x}_3)\zeta^{\dot{k}}_{\dot{l}\dot{m};lm}.$$

The transformation Λ_3 in the helicity basis has a form

$$\Lambda_3 \zeta^k_{lm;\dot{l}\dot{m}} = \sum_{l',m',k'} c^{k'k}_{l'l,m'm} \zeta^{k'}_{l'm';\dot{l}\dot{m}}.$$

Calculating the commutators $[\Lambda_3, \mathsf{X}_3]$, $[[\Lambda_3, \mathsf{X}_-], \mathsf{X}_+]$, we find the numbers $c^{k'k}_{l'l,m'm}$:

$$\Lambda_3 : \left\{ \begin{array}{rcl} c^{k'k}_{l-1,l,m} & = & c^{k'k}_{l-1,l}\sqrt{l^2 - m^2}, \\ c^{k'k}_{l,l,m} & = & c^{k'k}_{ll} m, \\ c^{k'k}_{l+1,l,m} & = & c^{k'k}_{l+1,l}\sqrt{(l+1)^2 - m^2}. \end{array} \right. \tag{45}$$

All other elements of the matrix Λ_3 are equal to zero. Let define now elements of the matrices Λ_1 and Λ_2. For the transformations Λ_1 and Λ_2 in the helicity basis we have

$$\begin{aligned} \Lambda_1 \zeta^k_{lm;\dot{l}\dot{m}} &= \sum_{l',m',k'} a^{k'k}_{l'l,m'm} \zeta^{k'}_{l'm';\dot{l}\dot{m}}, \\ \Lambda_2 \zeta^k_{lm;\dot{l}\dot{m}} &= \sum_{l',m',k'} b^{k'k}_{l'l,m'm} \zeta^{k'}_{l'm';\dot{l}\dot{m}}. \end{aligned}$$

Using the relations $\Lambda_1 = [\mathsf{A}_2, \Lambda_3]$ (or $\Lambda_1 = i\,[\mathsf{B}_2, \Lambda_3]$) and (11) (or (12)), and also (45), we find the elements $a^{k'k}_{l'l,m'm}$ of the matrix Λ_1. Analogously, from the relations

[2] Recall that the wave function $\psi(x_1, x_2, x_3)$ is defined on the group manifold $\mathfrak{L}$, that is, ψ is a function on the Lorentz group.

$\Lambda_2 = -[\mathsf{A}_1, \Lambda_3]$ (or $\Lambda_2 = -i[\mathsf{B}_1, \Lambda_3]$) and (11) (or (12)), (45) we obtain the elements $b^{k'k}_{l'l,m'm}$ of Λ_2. Thus,

$$\Lambda_1: \begin{cases} a^{k'k}_{l-1,l,m-1,m} &= -\dfrac{c_{l-1,l}}{2}\sqrt{(l+m)(l+m-1)}, \\ a^{k'k}_{l,l,m-1,m} &= \dfrac{c_{ll}}{2}\sqrt{(l+m)(l-m+1)}, \\ a^{k'k}_{l+1,l,m-1,m} &= \dfrac{c_{l+1,l}}{2}\sqrt{(l-m+1)(l-m+2)}, \\ a^{k'k}_{l-1,l,m+1,m} &= \dfrac{c_{l-1,l}}{2}\sqrt{(l-m)(l-m-1)}, \\ a^{k'k}_{l,l,m+1,m} &= \dfrac{c_{ll}}{2}\sqrt{(l+m+1)(l-m)}, \\ a^{k'k}_{l+1,l,m+1,m} &= -\dfrac{c_{l+1,l}}{2}\sqrt{(l+m+1)(l+m+2)}. \end{cases} \tag{46}$$

$$\Lambda_2: \begin{cases} b^{k'k}_{l-1,l,m-1,m} &= -\dfrac{ic_{l-1,l}}{2}\sqrt{(l+m)(l+m-1)}, \\ b^{k'k}_{l,l,m-1,m} &= \dfrac{ic_{ll}}{2}\sqrt{(l+m)(l-m+1)}, \\ b^{k'k}_{l+1,l,m-1,m} &= \dfrac{ic_{l+1,l}}{2}\sqrt{(l-m+1)(l-m+2)}, \\ b^{k'k}_{l-1,l,m+1,m} &= -\dfrac{ic_{l-1,l}}{2}\sqrt{(l-m)(l-m-1)}, \\ b^{k'k}_{l,l,m+1,m} &= -\dfrac{ic_{ll}}{2}\sqrt{(l+m+1)(l-m)}, \\ b^{k'k}_{l+1,l,m+1,m} &= \dfrac{ic_{l+1,l}}{2}\sqrt{(l+m+1)(l+m+2)}. \end{cases} \tag{47}$$

Coming to the dual representations, we find elements of the matrices Λ_1^*, Λ_2^* and Λ_3^*. The transformations Λ_i^* in the dual helicity basis are

$$\begin{aligned} \Lambda_1^*\zeta^{\dot{k}}_{\dot{l}\dot{m};lm} &= \sum_{\dot{l}',\dot{m}',\dot{k}'} d^{\dot{k}'\dot{k}}_{\dot{l}'\dot{l},\dot{m}'\dot{m}}\zeta^{\dot{k}'}_{\dot{l}'\dot{m}';lm}, \\ \Lambda_2^*\zeta^{\dot{k}}_{\dot{l}\dot{m};lm} &= \sum_{\dot{l}',\dot{m}',\dot{k}'} e^{\dot{k}'\dot{k}}_{\dot{l}'\dot{l},\dot{m}'\dot{m}}\zeta^{\dot{k}'}_{\dot{l}'\dot{m}';lm}, \\ \Lambda_3^*\zeta^{\dot{k}}_{\dot{l}\dot{m};lm} &= \sum_{\dot{l}',\dot{m}',\dot{k}'} f^{\dot{k}'\dot{k}}_{\dot{l}'\dot{l},\dot{m}'\dot{m}}\zeta^{\dot{k}'}_{\dot{l}'\dot{m}';lm}. \end{aligned}$$

Calculating the commutators $[\Lambda_3^*, \mathsf{Y}_3]$, $[[\Lambda_3^*, \mathsf{Y}_-], \mathsf{Y}_+]$ with respect to the vectors $\zeta^{\dot{k}}_{\dot{l}\dot{m};lm}$ of the dual basis, we find elements of the matrix Λ_3^*. Using the relations $\Lambda_1^* = \left[\widetilde{\mathsf{A}}_2, \Lambda_3^*\right]$ (or $\Lambda_1^* = -i\left[\widetilde{\mathsf{B}}_2, \Lambda_3^*\right]$) and (13) (or (14)), we find elements $d^{\dot{k}'\dot{k}}_{\dot{l}'\dot{l},\dot{m}'\dot{m}}$ of the matrix Λ_1^*. And also from the relations $\Lambda_2^* = -\left[\widetilde{\mathsf{A}}_1, \Lambda_3^*\right]$ (or $\Lambda_2^* = i\left[\widetilde{\mathsf{B}}_1, \Lambda_3^*\right]$) we obtain elements $e^{\dot{k}'\dot{k}}_{\dot{l}'\dot{l},\dot{m}'\dot{m}}$ of Λ_2^*. All calculations are analogous to the calculations presented for the case of Λ_i. In the result we have

$$\Lambda_1^* : \begin{cases} d^{\dot{k}'\dot{k}}_{\dot{l}-1,\dot{l},\dot{m}-1,\dot{m}} &= -\frac{c_{\dot{l}-1,\dot{l}}}{2}\sqrt{(\dot{l}+\dot{m})(\dot{l}-\dot{m}-1)}, \\ d^{\dot{k}'\dot{k}}_{\dot{l},\dot{l},\dot{m}-1,\dot{m}} &= \frac{c_{\dot{l}\dot{l}}}{2}\sqrt{(\dot{l}+\dot{m})(\dot{l}-\dot{m}+1)}, \\ d^{\dot{k}'\dot{k}}_{\dot{l}+1,\dot{l},\dot{m}-1,\dot{m}} &= \frac{c_{\dot{l}+1,\dot{l}}}{2}\sqrt{(\dot{l}-\dot{m}+1)(\dot{l}-\dot{m}+2)}, \\ d^{\dot{k}'\dot{k}}_{\dot{l}-1,\dot{l},\dot{m}+1,\dot{m}} &= \frac{c_{\dot{l}-1,\dot{l}}}{2}\sqrt{(\dot{l}-\dot{m})(\dot{l}-\dot{m}-1)}, \\ d^{\dot{k}'\dot{k}}_{\dot{l},\dot{l},\dot{m}+1,\dot{m}} &= \frac{c_{\dot{l}\dot{l}}}{2}\sqrt{(\dot{l}+\dot{m}+1)(\dot{l}-\dot{m})}, \\ d^{\dot{k}'\dot{k}}_{\dot{l}+1,\dot{l},\dot{m}+1,\dot{m}} &= -\frac{c_{\dot{l}+1,\dot{l}}}{2}\sqrt{(\dot{l}+\dot{m}+1)(\dot{l}+\dot{m}+2)}. \end{cases} \tag{48}$$

$$\Lambda_2^* : \begin{cases} e^{\dot{k}'\dot{k}}_{\dot{l}-1,\dot{l},\dot{m}-1,\dot{m}} &= -\frac{ic_{\dot{l}-1,\dot{l}}}{2}\sqrt{(\dot{l}+\dot{m})(\dot{l}-\dot{m}-1)}, \\ e^{\dot{k}'\dot{k}}_{\dot{l},\dot{l},\dot{m}-1,\dot{m}} &= \frac{ic_{\dot{l}\dot{l}}}{2}\sqrt{(\dot{l}+\dot{m})(\dot{l}-\dot{m}+1)}, \\ e^{\dot{k}'\dot{k}}_{\dot{l}+1,\dot{l},\dot{m}-1,\dot{m}} &= \frac{ic_{\dot{l}+1,\dot{l}}}{2}\sqrt{(\dot{l}-\dot{m}+1)(\dot{l}-\dot{m}+2)}, \\ e^{\dot{k}'\dot{k}}_{\dot{l}-1,\dot{l},\dot{m}+1,\dot{m}} &= \frac{-ic_{\dot{l}-1,\dot{l}}}{2}\sqrt{(\dot{l}-\dot{m})(\dot{l}-\dot{m}-1)}, \\ e^{\dot{k}'\dot{k}}_{\dot{l},\dot{l},\dot{m}+1,\dot{m}} &= \frac{-ic_{\dot{l}\dot{l}}}{2}\sqrt{(\dot{l}+\dot{m}+1)(\dot{l}-\dot{m})}, \\ e^{\dot{k}'\dot{k}}_{\dot{l}+1,\dot{l},\dot{m}+1,\dot{m}} &= -\frac{ic_{\dot{l}+1,\dot{l}}}{2}\sqrt{(\dot{l}+\dot{m}+1)(\dot{l}+\dot{m}+2)}. \end{cases} \tag{49}$$

$$\Lambda_3^* : \begin{cases} f^{\dot{k}'\dot{k}}_{\dot{l}-1,\dot{l},\dot{m}} &= c^{\dot{k}'\dot{k}}_{\dot{l}-1,\dot{l}}\sqrt{\dot{l}^2-\dot{m}^2}, \\ f^{\dot{k}'\dot{k}}_{\dot{l}\dot{l},\dot{m}} &= c^{\dot{k}'\dot{k}}_{\dot{l}\dot{l}}\dot{m}, \\ f^{\dot{k}'\dot{k}}_{\dot{l}+1,\dot{l},\dot{m}} &= c^{\dot{k}'\dot{k}}_{\dot{l}+1,\dot{l}}\sqrt{(\dot{l}+1)^2-\dot{m}^2}. \end{cases} \tag{50}$$

Let us construct in $\mathbb{C}^3$ a two–dimensional complex sphere from the quantities $z_k = x_k + iy_k$, $\overset{*}{z}_k = x_k - iy_k$ as follows (see **Fig. 1**)

$$\mathbf{z}^2 = z_1^2 + z_2^2 + z_3^2 = \mathbf{x}^2 - \mathbf{y}^2 + 2i\mathbf{x}\mathbf{y} = r^2 \tag{51}$$

and its complex conjugate (dual) sphere

$$\overset{*}{\mathbf{z}}{}^2 = \overset{*}{z}_1{}^2 + \overset{*}{z}_2{}^2 + \overset{*}{z}_3{}^2 = \mathbf{x}^2 - \mathbf{y}^2 - 2i\mathbf{x}\mathbf{y} = \overset{*}{r}{}^2. \tag{52}$$

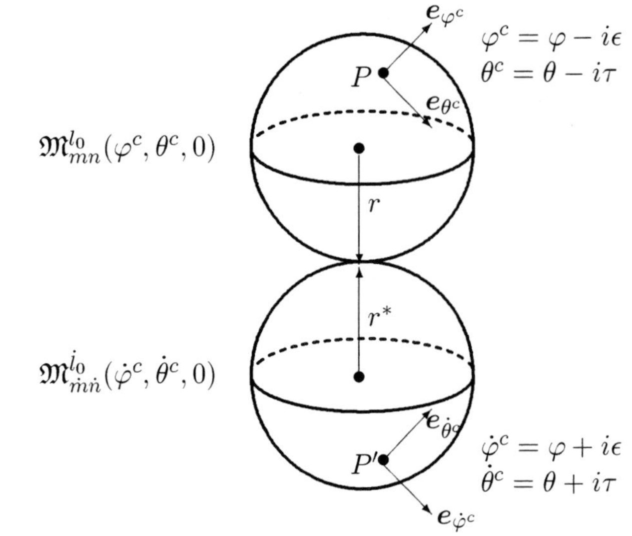

Figure 1. Two–dimensional complex sphere $z_1^2 + z_2^2 + z_3^2 = r^2$ in three–dimensional complex space $\mathbb{C}^3$. The space $\mathbb{C}^3$ is isometric to the bivector space $\mathbb{R}^6$. The dual (complex conjugate) sphere $\overset{*}{z}_1{}^2 + \overset{*}{z}_2{}^2 + \overset{*}{z}_3{}^2 = \overset{*}{r}{}^2$ is a mirror image of the complex sphere with respect to the hyperplane. The hyperspherical functions $\mathfrak{M}^l_{mn}(\varphi^c, \theta^c, 0)$ $(\mathfrak{M}^{\dot{l}}_{\dot{m}\dot{n}}(\dot{\varphi}^c, \dot{\theta}^c, 0))$ are defined on the surface of the complex (dual) sphere.

It is well-known that both quantities $\mathbf{x}^2 - \mathbf{y}^2$, $\mathbf{xy}$ are invariant with respect to the Lorentz transformations, since a surface of the complex sphere is invariant (Casimir operators of the Lorentz group are constructed from such quantities). It is easy to see that three–dimensional complex space $\mathbb{C}^3$ is isometric to a real space $\mathbb{R}^{3,3}$ with a basis $\{i\mathbf{e}_1, i\mathbf{e}_2, i\mathbf{e}_3, \mathbf{e}_4, \mathbf{e}_5, \mathbf{e}_6\}$. At this point a metric tensor of $\mathbb{R}^{3,3}$ has the form (38). Hence it immediately follows that $\mathbb{C}^3$ is isometric to the bivector space $\mathbb{R}^6$. Therefore, with the each point of the Minkowski spacetime $\mathbb{R}^{1,3}$ we can associate the two–dimensional complex sphere and its conjugate.

Let us introduce now hyperspherical coordinates on the surfaces of the complex and dual spheres

$$\begin{array}{rclrcl} z_1 & = & r\sin\theta^c\cos\varphi^c, & z_1^* & = & r^*\sin\dot{\theta}^c\cos\dot{\varphi}^c, \\ z_2 & = & r\sin\theta^c\sin\varphi^c, & z_2^* & = & r^*\sin\dot{\theta}^c\sin\dot{\varphi}^c, \\ z_3 & = & r\cos\theta^c, & z_3^* & = & r^*\cos\dot{\theta}^c, \end{array} \tag{53}$$

where θ^c, φ^c are the complex Euler angles.

Let us show that solutions of the equations (36) can be found in the form of the series in generalized hyperspherical functions considered in the previous section. With this end in view let us transform the system (36) as follows. First of all, let us define the derivatives $\dfrac{\partial}{\partial x_i}, \dfrac{\partial}{\partial x_i^*}$ on the surface of the complex sphere (51) and write them in the hyperspherical coordinates (53) as

$$\frac{\partial}{\partial x_1} = -\frac{\sin\varphi^c}{r\sin\theta^c}\frac{\partial}{\partial\varphi} + \frac{\cos\varphi^c\cos\theta^c}{r}\frac{\partial}{\partial\theta} + \cos\varphi^c\sin\theta^c\frac{\partial}{\partial r}, \tag{54}$$

$$\frac{\partial}{\partial x_2} = \frac{\cos\varphi^c}{r\sin\theta^c}\frac{\partial}{\partial\varphi} + \frac{\sin\varphi^c\cos\theta^c}{r}\frac{\partial}{\partial\theta} + \sin\varphi^c\sin\theta^c\frac{\partial}{\partial r}, \tag{55}$$

$$\frac{\partial}{\partial x_3} = -\frac{\sin\theta^c}{r}\frac{\partial}{\partial\theta} + \cos\theta^c\frac{\partial}{\partial r}. \tag{56}$$

$$\frac{\partial}{\partial x_1^*} = i\frac{\partial}{\partial x_1} = -\frac{\sin\varphi^c}{r\sin\theta^c}\frac{\partial}{\partial\epsilon} + \frac{\cos\varphi^c\sin\theta^c}{r}\frac{\partial}{\partial\tau} + i\cos\varphi^c\sin\theta^c\frac{\partial}{\partial r}, \tag{57}$$

$$\frac{\partial}{\partial x_2^*} = i\frac{\partial}{\partial x_2} = \frac{\cos\varphi^c}{r\sin\theta^c}\frac{\partial}{\partial\epsilon} + \frac{\sin\varphi^c\cos\theta^c}{r}\frac{\partial}{\partial\tau} + i\sin\varphi^c\sin\theta^c\frac{\partial}{\partial r}, \tag{58}$$

$$\frac{\partial}{\partial x_3^*} = i\frac{\partial}{\partial x_3} = -\frac{\sin\theta^c}{r}\frac{\partial}{\partial\tau} + i\cos\theta^c\frac{\partial}{\partial r}. \tag{59}$$

Analogously, on the surface of the dual sphere we have

$$\frac{\partial}{\partial\widetilde{x}_1} = -\frac{\sin\dot{\varphi}^c}{r^*\sin\dot{\theta}^c}\frac{\partial}{\partial\varphi} + \frac{\cos\dot{\varphi}^c\cos\dot{\theta}^c}{r^*}\frac{\partial}{\partial\theta} + \cos\dot{\varphi}^c\sin\dot{\theta}^c\frac{\partial}{\partial r^*}, \tag{60}$$

$$\frac{\partial}{\partial\widetilde{x}_2} = \frac{\cos\dot{\varphi}^c}{r^*\sin\dot{\theta}^c}\frac{\partial}{\partial\varphi} + \frac{\sin\dot{\varphi}^c\cos\dot{\theta}^c}{r^*}\frac{\partial}{\partial\theta} + \sin\dot{\varphi}^c\sin\dot{\theta}^c\frac{\partial}{\partial r^*}, \tag{61}$$

$$\frac{\partial}{\partial\widetilde{x}_3} = -\frac{\sin\dot{\theta}^c}{r^*}\frac{\partial}{\partial\theta} + \cos\dot{\theta}^c\frac{\partial}{\partial r^*}. \tag{62}$$

$$\frac{\partial}{\partial\widetilde{x}_1^*} = -i\frac{\partial}{\partial\widetilde{x}_1} = \frac{\sin\dot{\varphi}^c}{r^*\sin\dot{\theta}^c}\frac{\partial}{\partial\epsilon} - \frac{\cos\dot{\varphi}^c\cos\dot{\theta}^c}{r^*}\frac{\partial}{\partial\tau} - i\cos\dot{\varphi}^c\sin\dot{\theta}^c\frac{\partial}{\partial r^*}, \tag{63}$$

$$\frac{\partial}{\partial\widetilde{x}_2^*} = -i\frac{\partial}{\partial\widetilde{x}_2} = -\frac{\cos\dot{\varphi}^c}{r^*\sin\dot{\theta}^c}\frac{\partial}{\partial\epsilon} - \frac{\sin\dot{\varphi}^c\cos\dot{\theta}^c}{r^*}\frac{\partial}{\partial\tau} - i\sin\dot{\varphi}^c\sin\dot{\theta}^c\frac{\partial}{\partial r^*}, \tag{64}$$

$$\frac{\partial}{\partial\widetilde{x}_3^*} = -i\frac{\partial}{\partial\widetilde{x}_3} = \frac{\sin\dot{\theta}^c}{r^*}\frac{\partial}{\partial\tau} - i\cos\dot{\theta}^c\frac{\partial}{\partial r^*}. \tag{65}$$

Substituting the functions $\psi = T_{\mathfrak{g}}^{-1}\psi'$ ($\dot{\psi} = \overset{*}{T}{}_{\mathfrak{g}}^{-1}\dot{\psi}'$) and the derivatives (54)–(59), (60)–(65) into the system (36), and then multiplying by $T_{\mathfrak{g}} = T(\varphi^c, \theta^c, 0)$ ($\overset{*}{T}_{\mathfrak{g}} = \overset{*}{T}(\dot{\varphi}^c, \dot{\theta}^c, 0)$) from the left and taking into account the invariance conditions

(39), we obtain

$$\frac{1}{r\sin\theta^c}\Lambda_1 T_{\mathfrak{g}}\frac{\partial(T_{\mathfrak{g}}^{-1}\psi')}{\partial\varphi}+\frac{i}{r\sin\theta^c}\Lambda_1 T_{\mathfrak{g}}\frac{\partial(T_{\mathfrak{g}}^{-1}\psi')}{\partial\epsilon}-$$
$$-\frac{1}{r}\Lambda_2 T_{\mathfrak{g}}\frac{\partial(T_{\mathfrak{g}}^{-1}\psi')}{\partial\theta}-\frac{i}{r}\Lambda_2 T_{\mathfrak{g}}\frac{\partial(T_{\mathfrak{g}}^{-1}\psi')}{\partial\tau}+2\Lambda_3 T_{\mathfrak{g}}\frac{\partial(T_{\mathfrak{g}}^{-1}\psi')}{\partial r}+\kappa^c\psi'=0,$$
$$\frac{1}{r^*\sin\dot{\theta}^c}\Lambda_1^*\overset{*}{T}_{\mathfrak{g}}\frac{\partial(\overset{*}{T}{}_{\mathfrak{g}}^{-1}\dot{\psi}')}{\partial\varphi}-\frac{i}{r^*\sin\dot{\theta}^c}\Lambda_1^*\overset{*}{T}_{\mathfrak{g}}\frac{\partial(\overset{*}{T}{}_{\mathfrak{g}}^{-1}\dot{\psi}')}{\partial\epsilon}-$$
$$-\frac{1}{r^*}\Lambda_2^*\overset{*}{T}_{\mathfrak{g}}\frac{\partial(\overset{*}{T}{}_{\mathfrak{g}}^{-1}\dot{\psi}')}{\partial\theta}+\frac{i}{r^*}\Lambda_2^*\overset{*}{T}_{\mathfrak{g}}\frac{\partial(\overset{*}{T}{}_{\mathfrak{g}}^{-1}\dot{\psi}')}{\partial\tau}+2\Lambda_3^*\overset{*}{T}_{\mathfrak{g}}\frac{\partial(\overset{*}{T}{}_{\mathfrak{g}}^{-1}\dot{\psi}')}{\partial r^*}+\dot{\kappa}^c\dot{\psi}'=0. \quad (66)$$

The matrices $T_{\mathfrak{g}}^{-1}$, $\overset{*}{T}{}_{\mathfrak{g}}^{-1}$ depend on φ, ϵ, θ, τ. By this reason we must differentiate in $T_{\mathfrak{g}}^{-1}\psi'$ ($\overset{*}{T}{}_{\mathfrak{g}}^{-1}\dot{\psi}'$) the both factors. After differentiation, the system (66) takes a form

$$\frac{1}{r\sin\theta^c}\Lambda_1\frac{\partial\psi'}{\partial\varphi}+\frac{i}{r\sin\theta^c}\Lambda_1\frac{\partial\psi'}{\partial\epsilon}-\frac{1}{r}\Lambda_2\frac{\partial\psi'}{\partial\theta}-$$
$$-\frac{i}{r}\Lambda_2\frac{\partial\psi'}{\partial\tau}+2\Lambda_3\frac{\partial\psi'}{\partial r}+\left[\frac{1}{r\sin\theta^c}\Lambda_1 T_{\mathfrak{g}}\frac{\partial T_{\mathfrak{g}}^{-1}}{\partial\varphi}+\right.$$
$$\left.+\frac{i}{r\sin\theta^c}\Lambda_1 T_{\mathfrak{g}}\frac{\partial T_{\mathfrak{g}}^{-1}}{\partial\epsilon}-\frac{1}{r}\Lambda_2 T_{\mathfrak{g}}\frac{\partial T_{\mathfrak{g}}^{-1}}{\partial\theta}-\frac{i}{r}\Lambda_2 T_{\mathfrak{g}}\frac{\partial T_{\mathfrak{g}}^{-1}}{\partial\tau}+\kappa^c I\right]\psi'=0,$$
$$\frac{1}{r^*\sin\dot{\theta}^c}\Lambda_1^*\frac{\partial\dot{\psi}'}{\partial\varphi}-\frac{i}{r^*\sin\dot{\theta}^c}\Lambda_1^*\frac{\partial\dot{\psi}'}{\partial\epsilon}-\frac{1}{r^*}\Lambda_2^*\frac{\partial\dot{\psi}'}{\partial\theta}+$$
$$+\frac{i}{r^*}\Lambda_2^*\frac{\partial\dot{\psi}'}{\partial\tau}+2\Lambda_3^*\frac{\partial\dot{\psi}'}{\partial r^*}+\left[\frac{1}{r^*\sin\dot{\theta}^c}\Lambda_1^*\overset{*}{T}_{\mathfrak{g}}\frac{\partial\overset{*}{T}{}_{\mathfrak{g}}^{-1}}{\partial\varphi}-\right.$$
$$\left.-\frac{i}{r^*\sin\dot{\theta}^c}\Lambda_1^*\overset{*}{T}_{\mathfrak{g}}\frac{\partial\overset{*}{T}{}_{\mathfrak{g}}^{-1}}{\partial\epsilon}-\frac{1}{r^*}\Lambda_2^*\overset{*}{T}_{\mathfrak{g}}\frac{\partial\overset{*}{T}{}_{\mathfrak{g}}^{-1}}{\partial\theta}+\frac{i}{r^*}\Lambda_2^*\overset{*}{T}_{\mathfrak{g}}\frac{\partial\overset{*}{T}{}_{\mathfrak{g}}^{-1}}{\partial\tau}+\dot{\kappa}^c I\right]\dot{\psi}'=0. \quad (67)$$

Let us show that the products $T_{\mathfrak{g}}\frac{\partial T_{\mathfrak{g}}^{-1}}{\partial\varphi}$, ..., $\overset{*}{T}_{\mathfrak{g}}\frac{\partial\overset{*}{T}{}_{\mathfrak{g}}^{-1}}{\partial\tau}$ are expressed via linear combinations of the operators A_i, B_i, $\widetilde{\mathsf{A}}_i$, $\widetilde{\mathsf{B}}_i$. Indeed, for the fundamental representation $\mathfrak{g}(\varphi,\epsilon,\theta,\tau,0,0)=T_{\mathfrak{g}}(\varphi^c,\theta^c,0)$ and dual fundamental representation $\dot{\mathfrak{g}}(\varphi,\epsilon,\theta,\tau,0,0)=$

$\overset{*}{T}_{\mathfrak{g}}(\dot{\varphi}^c, \dot{\theta}^c, 0)$ of the group $\mathfrak{G}_+$ we obtain

$$\mathfrak{g}^T \frac{\partial\left(\mathfrak{g}^T\right)^{-1}}{\partial\varphi} = \frac{i}{2}\begin{pmatrix} -\cos\theta^c & i\sin\theta^c \\ -i\sin\theta^c & \cos\theta^c \end{pmatrix} = -\mathsf{A}_3\cos\theta^c - \mathsf{A}_2\sin\theta^c, \tag{68}$$

$$\mathfrak{g}^T \frac{\partial\left(\mathfrak{g}^T\right)^{-1}}{\partial\epsilon} = \frac{1}{2}\begin{pmatrix} -\cos\theta^c & i\sin\theta^c \\ -i\sin\theta^c & \cos\theta^c \end{pmatrix} = -\mathsf{B}_3\cos\theta^c - \mathsf{B}_2\sin\theta^c, \tag{69}$$

$$\mathfrak{g}^T \frac{\partial\left(\mathfrak{g}^T\right)^{-1}}{\partial\theta} = \frac{1}{2}\begin{pmatrix} 0 & -i \\ -i & 0 \end{pmatrix} = \mathsf{A}_1, \tag{70}$$

$$\mathfrak{g}^T \frac{\partial\left(\mathfrak{g}^T\right)^{-1}}{\partial\tau} = \frac{1}{2}\begin{pmatrix} 0 & 1 \\ 1 & 0 \end{pmatrix} = -\mathsf{B}_1. \tag{71}$$

$$\dot{\mathfrak{g}}^T \frac{\partial\left(\dot{\mathfrak{g}}^T\right)^{-1}}{\partial\varphi} = \frac{i}{2}\begin{pmatrix} \cos\dot{\theta}^c & i\sin\dot{\theta}^c \\ -i\sin\dot{\theta}^c & -\cos\dot{\theta}^c \end{pmatrix} = \widetilde{\mathsf{A}}_3\cos\dot{\theta}^c - \widetilde{\mathsf{A}}_2\sin\dot{\theta}^c, \tag{72}$$

$$\dot{\mathfrak{g}}^T \frac{\partial\left(\dot{\mathfrak{g}}^T\right)^{-1}}{\partial\epsilon} = \frac{1}{2}\begin{pmatrix} -\cos\dot{\theta}^c & -i\sin\dot{\theta}^c \\ i\sin\dot{\theta}^c & \cos\dot{\theta}^c \end{pmatrix} = \widetilde{\mathsf{B}}_3\cos\dot{\theta}^c - \widetilde{\mathsf{B}}_2\sin\dot{\theta}^c, \tag{73}$$

$$\dot{\mathfrak{g}}^T \frac{\partial\left(\dot{\mathfrak{g}}^T\right)^{-1}}{\partial\theta} = \frac{1}{2}\begin{pmatrix} 0 & i \\ i & 0 \end{pmatrix} = -\widetilde{\mathsf{A}}_1, \tag{74}$$

$$\dot{\mathfrak{g}}^T \frac{\partial\left(\dot{\mathfrak{g}}^T\right)^{-1}}{\partial\tau} = \frac{1}{2}\begin{pmatrix} 0 & 1 \\ 1 & 0 \end{pmatrix} = \widetilde{\mathsf{B}}_1. \tag{75}$$

It is easy to verify that relations (68)–(75) take place for any representation $\mathfrak{g} \to T_{\mathfrak{g}}$ of the group $\mathfrak{G}_+$. Substituting these relations into the system (67), we obtain

$$\begin{aligned}
&\frac{1}{r\sin\theta^c}\Lambda_1\frac{\partial\psi'}{\partial\varphi} + \frac{i}{r\sin\theta^c}\Lambda_1\frac{\partial\psi'}{\partial\epsilon} - \frac{1}{r}\Lambda_2\frac{\partial\psi'}{\partial\theta} - \frac{i}{r}\Lambda_2\frac{\partial\psi'}{\partial\tau} + \\
&\qquad + 2\Lambda_3\frac{\partial\psi'}{\partial r} + \frac{1}{r}\left[-(\Lambda_1\mathsf{A}_2 + \Lambda_2\mathsf{A}_1) + i(\Lambda_2\mathsf{B}_1 - \Lambda_1\mathsf{B}_2) + \right. \\
&\qquad\qquad \left. + \cot\theta^c(\Lambda_1\mathsf{A}_3 + i\Lambda_1\mathsf{B}_3)\right]\psi' + \kappa^c\psi = 0, \\
&\frac{1}{r^*\sin\dot{\theta}^c}\Lambda_1^*\frac{\partial\dot{\psi}'}{\partial\varphi} - \frac{i}{r^*\sin\dot{\theta}^c}\Lambda_1^*\frac{\partial\dot{\psi}'}{\partial\epsilon} - \frac{1}{r^*}\Lambda_2^*\frac{\partial\dot{\psi}'}{\partial\theta} + \frac{i}{r^*}\Lambda_2^*\frac{\partial\dot{\psi}'}{\partial\tau} + \\
&\qquad 2\Lambda_3^*\frac{\partial\dot{\psi}'}{\partial r^*} + \frac{1}{r^*}\left[\Lambda_2^*\widetilde{\mathsf{A}}_1 - \Lambda_1^*\widetilde{\mathsf{A}}_2 + i(\Lambda_1^*\widetilde{\mathsf{B}}_2 + \Lambda_2^*\widetilde{\mathsf{B}}_1) + \right. \\
&\qquad\qquad \left. + \cot\dot{\theta}^c(\Lambda_1^*\widetilde{\mathsf{A}}_3 - i\Lambda_1^*\widetilde{\mathsf{B}}_3)\right]\dot{\psi}' + \dot{\kappa}^c\dot{\psi} = 0.
\end{aligned} \tag{76}$$

Now we are in a position that allows us to separate variables in relativistically invariant system. We decompose the each component ψ^k_{lm} of the wave function ψ into the series on the generalized hyperspherical functions. This procedure gives rise to separation of variables, that is, it reduces the relativistically invariant system to a system of ordinary differential equations. Preliminarily, let us calculate elements of the matrices $V = \Lambda_2\mathsf{B}_1 - \Lambda_1\mathsf{B}_2$, $U = \Lambda_2^*\widetilde{\mathsf{A}}_1 - \Lambda_1^*\widetilde{\mathsf{A}}_2$, $G = \Lambda_1\mathsf{A}_2 + \Lambda_2\mathsf{A}_1$, $W = \Lambda_1^*\widetilde{\mathsf{B}}_2 + \Lambda_2^*\widetilde{\mathsf{B}}_1$. First of all, let us find elements of the matrix $V = \Lambda_2\mathsf{B}_1 - \Lambda_1\mathsf{B}_2$. Using the relations (41), we can write $V = 2i\Lambda_3 + \mathsf{B}_2\Lambda_1 - \mathsf{B}_1\Lambda_2$. The action of the transformation V in the helicity basis has a form

$$V\zeta^k_{l,m;\dot{l}\dot{m}} = \sum_{l',m',k'} v^{k'k}_{l'l,m'm}\zeta^{k'}_{l',m';\dot{l}\dot{m}}$$

Taking into account (46)–(45) and (12), we obtain

$$V: \quad \begin{cases} v^{k'k}_{l-1,l,m\,m} &= ic^{k'k}_{l-1,l}(l+1)\sqrt{l^2-m^2}, \\ v^{k'k}_{l\,l,m\,m} &= ic^{k'k}_{l\,l}m, \\ v^{k'k}_{l+1,l,m\,m} &= -ic^{k'k}_{l+1,l}l\sqrt{(l+1)^2-m^2}. \end{cases} \tag{77}$$

All other elements $v^{k'k}_{l'l,m'm}$ are equal to zero. Analogously, using the relations (42) and the operators (13), we find that elements of the matrix $U = \Lambda_2^*\widetilde{\mathsf{A}}_1 - \Lambda_1^*\widetilde{\mathsf{A}}_2$ are

$$U: \quad \begin{cases} u^{\dot{k}'\dot{k}}_{\dot{l}-1,\dot{l},\dot{m}\dot{m}} &= -c^{\dot{k}'\dot{k}}_{\dot{l}-1,\dot{l}}(\dot{l}+1)\sqrt{\dot{l}^2-\dot{m}^2}, \\ u^{\dot{k}'\dot{k}}_{\dot{l}\dot{l},\dot{m}\dot{m}} &= -c^{\dot{k}'\dot{k}}_{\dot{l}\dot{l}}\dot{m}, \\ u^{\dot{k}'\dot{k}}_{\dot{l}+1,\dot{l},\dot{m}\dot{m}} &= c^{\dot{k}'\dot{k}}_{\dot{l}+1,\dot{l}}\dot{l}\sqrt{(\dot{l}+1)^2-\dot{m}^2}. \end{cases} \tag{78}$$

Further, using (40), (11) and (43), (14), it is easy to verify that all the elements of the matrices $G = \Lambda_1\mathsf{A}_2 + \Lambda_2\mathsf{A}_1$ and $W = \Lambda_1^*\widetilde{\mathsf{B}}_2 + \Lambda_2^*\widetilde{\mathsf{B}}_1$ are equal to zero.

The system (76) in the components $\psi^k_{lm;\dot{l}\dot{m}}$ is written as follows

$$
\begin{aligned}
&\frac{1}{r\sin\theta^c}\sum_{l',m',k'} a^{kk'}_{ll',mm'}\frac{\partial\psi^{k'}_{l'm';\dot{l}\dot{m}}}{\partial\varphi}+\frac{i}{r\sin\theta^c}\sum_{l',m',k'} a^{kk'}_{ll',mm'}\frac{\partial\psi^{k'}_{l'm';\dot{l}\dot{m}}}{\partial\epsilon}-\\
&-\frac{1}{r}\sum_{l',m',k'} b^{kk'}_{ll',mm'}\frac{\partial\psi^{k'}_{l'm';\dot{l}\dot{m}}}{\partial\theta}-\frac{i}{r}\sum_{l',m',k'} b^{kk'}_{ll',mm'}\frac{\partial\psi^{k'}_{l'm';\dot{l}\dot{m}}}{\partial\tau}+2\sum_{l',m',k'} c^{kk'}_{ll',mm'}\frac{\partial\psi^{k'}_{l'm';\dot{l}\dot{m}}}{\partial r}+\\
&+\frac{i}{r}\sum_{l',m',k'} v^{kk'}_{ll',mm'}\psi^{k'}_{l'm';\dot{l}\dot{m}}+\frac{2i}{r}\cot\theta^c\sum_{l',m',k'} m'a^{kk'}_{ll',mm'}\psi^{k'}_{l'm';\dot{l}\dot{m}}+\kappa^c\psi^k_{lm;\dot{l}\dot{m}}=0,\\
&\frac{1}{r^*\sin\dot{\theta}^c}\sum_{\dot{l}',\dot{m}',\dot{k}'} d^{\dot{k}\dot{k}'}_{\dot{l}\dot{l}',\dot{m}\dot{m}'}\frac{\partial\psi^{\dot{k}'}_{\dot{l}'\dot{m}';lm}}{\partial\varphi}-\frac{i}{r^*\sin\dot{\theta}^c}\sum_{\dot{l}',\dot{m}',\dot{k}'} d^{\dot{k}\dot{k}'}_{\dot{l}\dot{l}',\dot{m}\dot{m}'}\frac{\partial\psi^{\dot{k}'}_{\dot{l}'\dot{m}';lm}}{\partial\epsilon}-\\
&-\frac{1}{r^*}\sum_{\dot{l}',\dot{m}',\dot{k}'} e^{\dot{k}\dot{k}'}_{\dot{l}\dot{l}',\dot{m}\dot{m}'}\frac{\partial\psi^{\dot{k}'}_{\dot{l}'\dot{m}';lm}}{\partial\theta}+\frac{i}{r^*}\sum_{\dot{l}',\dot{m}',\dot{k}'} e^{\dot{k}\dot{k}'}_{\dot{l}\dot{l}',\dot{m}\dot{m}'}\frac{\partial\psi^{\dot{k}'}_{\dot{l}'\dot{m}';lm}}{\partial\tau}+2\sum_{\dot{l}',\dot{m}',\dot{k}'} f^{\dot{k}\dot{k}'}_{\dot{l}\dot{l}',\dot{m}\dot{m}'}\frac{\partial\psi^{\dot{k}'}_{\dot{l}'\dot{m}';lm}}{\partial r^*}+\\
&+\frac{1}{r^*}\sum_{\dot{l}',\dot{m}',\dot{k}'} u^{\dot{k}\dot{k}'}_{\dot{l}\dot{l}',\dot{m}\dot{m}'}\psi^{\dot{k}'}_{\dot{l}'\dot{m}';lm}-\frac{2i}{r^*}\cot\dot{\theta}^c\sum_{\dot{l}',\dot{m}',\dot{k}'} \dot{m}'d^{\dot{k}\dot{k}'}_{\dot{l}\dot{l}',\dot{m}\dot{m}'}\psi^{\dot{k}'}_{\dot{l}'\dot{m}';lm}+\dot{\kappa}^c\psi^{\dot{k}}_{\dot{l}\dot{m};lm}=0,
\end{aligned}
\tag{79}
$$

where the coefficients $a^{kk'}_{ll',mm'}$, $b^{kk'}_{ll',mm'}$, $c^{kk'}_{ll',mm'}$, $v^{kk'}_{ll',mm'}$, $d^{\dot{k}\dot{k}'}_{\dot{l}\dot{l}',\dot{m}\dot{m}'}$, $e^{\dot{k}\dot{k}'}_{\dot{l}\dot{l}',\dot{m}\dot{m}'}$, $f^{\dot{k}\dot{k}'}_{\dot{l}\dot{l}',\dot{m}\dot{m}'}$, $u^{\dot{k}\dot{k}'}_{\dot{l}\dot{l}',\dot{m}\dot{m}'}$ are defined by the formulae (46), (47), (45), (77), (48), (49), (50), (78).

With a view to separate the variables in (79) let us assume that

$$
\begin{aligned}
\psi^k_{lm;\dot{l}\dot{m}} &= \boldsymbol{f}^{l_0}_{lmk}(r)\mathfrak{M}^{l_0}_{mn}(\varphi,\epsilon,\theta,\tau,0,0),\\
\psi^{\dot{k}}_{\dot{l}\dot{m};lm} &= \boldsymbol{f}^{\dot{l}_0}_{\dot{l}\dot{m}\dot{k}}(r^*)\mathfrak{M}^{\dot{l}_0}_{\dot{m}\dot{n}}(\varphi,\epsilon,\theta,\tau,0,0),
\end{aligned}
\tag{80}
$$

where $l_0\geq l$, $-l_0\leq m$, $n\leq l_0$ $\dot{l}_0\geq\dot{l}$, $-\dot{l}_0\leq\dot{m}$, $\dot{n}\leq\dot{l}_0$. Substituting the functions (80) into the system (79) and taking into account values of the coefficients $a^{kk'}_{ll',mm'}$, $b^{kk'}_{ll',mm'}$, $c^{kk'}_{ll',mm'}$, $v^{kk'}_{ll',mm'}$, $d^{\dot{k}\dot{k}'}_{\dot{l}\dot{l}',\dot{m}\dot{m}'}$, $e^{\dot{k}\dot{k}'}_{\dot{l}\dot{l}',\dot{m}\dot{m}'}$, $f^{\dot{k}\dot{k}'}_{\dot{l}\dot{l}',\dot{m}\dot{m}'}$, $u^{\dot{k}\dot{k}'}_{\dot{l}\dot{l}',\dot{m}\dot{m}'}$ let us collect together the terms with identical radial functions. In the result we obtain

$$\sum_{k'} c^{kk'}_{l,l-1}\left\{\left[2\sqrt{l^2-m^2}\frac{\partial \boldsymbol{f}^{l_0}_{l-1,m,k'}}{\partial r}-\right.\right.$$

$$\left.-\frac{1}{r}(l+1)\sqrt{l^2-m^2}\boldsymbol{f}^{l_0}_{l-1,m,k'}\right]\mathfrak{M}^{l_0}_{mn}(\varphi,\epsilon,\theta,\tau,0,0)+$$

$$+\frac{1}{2r}\sqrt{(l+m)(l+m-1)}\boldsymbol{f}^{l_0}_{l-1,m-1,k'}\left[-\frac{1}{\sin\theta^c}\frac{\partial\mathfrak{M}^{l_0}_{m-1,n}}{\partial\varphi}-\frac{i}{\sin\theta^c}\frac{\partial\mathfrak{M}^{l_0}_{m-1,n}}{\partial\epsilon}+\right.$$

$$\left.+i\frac{\partial\mathfrak{M}^{l_0}_{m-1,n}}{\partial\theta}-\frac{\partial\mathfrak{M}^{l_0}_{m-1,n}}{\partial\tau}-\frac{2i(m-1)\cos\theta^c}{\sin\theta^c}\mathfrak{M}^{l_0}_{m-1,n}\right]+$$

$$+\frac{1}{2r}\sqrt{(l-m)(l-m-1)}\boldsymbol{f}^{l_0}_{l-1,m+1,k'}\left[\frac{1}{\sin\theta^c}\frac{\partial\mathfrak{M}^{l_0}_{m+1,n}}{\partial\varphi}+\frac{i}{\sin\theta^c}\frac{\partial\mathfrak{M}^{l_0}_{m+1,n}}{\partial\epsilon}+\right.$$

$$\left.\left.+i\frac{\partial\mathfrak{M}^{l_0}_{m+1,n}}{\partial\theta}-\frac{\partial\mathfrak{M}^{l_0}_{m+1,n}}{\partial\tau}+\frac{2i(m+1)\cos\theta^c}{\sin\theta^c}\mathfrak{M}^{l_0}_{m+1,n}\right]\right\}+$$

$$+\sum_{k'} c^{kk'}_{ll}\left\{\left[2m\frac{\partial \boldsymbol{f}^{l_0}_{l,m,k'}}{\partial r}-\frac{1}{r}m\boldsymbol{f}^{l_0}_{l,m,k'}\right]\mathfrak{M}^{l_0}_{m,n}(\varphi,\epsilon,\theta,\tau,0,0)+\right.$$

$$+\frac{1}{2r}\sqrt{(l+m)(l-m+1)}\boldsymbol{f}^{l_0}_{l,m-1,k'}\left[\frac{1}{\sin\theta^c}\frac{\partial\mathfrak{M}^{l_0}_{m-1,n}}{\partial\varphi}+\frac{i}{\sin\theta^c}\frac{\partial\mathfrak{M}^{l_0}_{m-1,n}}{\partial\epsilon}-\right.$$

$$\left.-i\frac{\partial\mathfrak{M}^{l_0}_{m-1,n}}{\partial\theta}+\frac{\partial\mathfrak{M}^{l_0}_{m-1,n}}{\partial\tau}+\frac{2i(m-1)\cos\theta^c}{\sin\theta^c}\mathfrak{M}^{l_0}_{m-1,n}\right]+$$

$$+\frac{1}{2r}\sqrt{(l+m+1)(l-m)}\boldsymbol{f}^{l_0}_{l,m+1,k'}\left[\frac{1}{\sin\theta^c}\frac{\partial\mathfrak{M}^{l_0}_{m+1,n}}{\partial\varphi}+\frac{i}{\sin\theta^c}\frac{\partial\mathfrak{M}^{l_0}_{m+1,n}}{\partial\epsilon}+\right.$$

$$\left.\left.+i\frac{\partial\mathfrak{M}^{l_0}_{m+1,n}}{\partial\theta}-\frac{\partial\mathfrak{M}^{l_0}_{m+1,n}}{\partial\tau}+\frac{2i(m+1)\cos\theta^c}{\sin\theta^c}\mathfrak{M}^{l_0}_{m+1,n}\right]\right\}+$$

$$+\sum_{k'} c^{kk'}_{l,l+1}\left\{\left[2\sqrt{(l+1)^2-m^2}\frac{\partial \boldsymbol{f}^{l_0}_{l+1,m,k'}}{\partial r}+\right.\right.$$

$$\left.+\frac{1}{r}l\sqrt{(l+1)^2-m^2}\boldsymbol{f}^{l_0}_{l+1,m,k'}\right]\mathfrak{M}^{l_0}_{m,n}(\varphi,\epsilon,\theta,\tau,0,0)+$$

$$+\frac{1}{2r}\sqrt{(l-m+1)(l-m+2)}\boldsymbol{f}^{l_0}_{l+1,m-1,k'}\left[\frac{1}{\sin\theta^c}\frac{\partial\mathfrak{M}^{l_0}_{m-1,n}}{\partial\varphi}+\frac{i}{\sin\theta^c}\frac{\partial\mathfrak{M}^{l_0}_{m-1,n}}{\partial\epsilon}-\right.$$

$$-i\frac{\partial \mathfrak{M}^{l_0}_{m-1,n}}{\partial \theta}+\frac{\partial \mathfrak{M}^{l_0}_{m-1,n}}{\partial \tau}+\frac{2i(m-1)\cos\theta^c}{\sin\theta^c}\mathfrak{M}^{l_0}_{m-1,n}\Bigg]+$$

$$+\frac{1}{2r}\sqrt{(l+m+1)(l+m+2)}\boldsymbol{f}^{l_0}_{l+1,m+1,k'}\Bigg[-\frac{1}{\sin\theta^c}\frac{\partial \mathfrak{M}^{l_0}_{m+1,n}}{\partial \varphi}-\frac{i}{\sin\theta^c}\frac{\partial \mathfrak{M}^{l_0}_{m+1,n}}{\partial \epsilon}-$$

$$-i\frac{\partial \mathfrak{M}^{l_0}_{m+1,n}}{\partial \theta}+\frac{\partial \mathfrak{M}^{l_0}_{m+1,n}}{\partial \tau}-\frac{2i(m+1)\cos\theta^c}{\sin\theta^c}\mathfrak{M}^{l_0}_{m+1,n}\Bigg]\Bigg\}+$$

$$+\kappa^c\boldsymbol{f}^{l_0}_{lm}\mathfrak{M}^{l_0}_{m,n}(\varphi,\epsilon,\theta,\tau,0,0)=0,$$

$$\sum_{\dot{k}'}c^{\dot{k}\dot{k}'}_{\dot{l},\dot{l}-1}\Bigg\{\Bigg[2\sqrt{\dot{l}^2-\dot{m}^2}\frac{\partial \boldsymbol{f}^{\dot{l}_0}_{\dot{l}-1,\dot{m},\dot{k}'}}{\partial r^*}-$$

$$-\frac{1}{r^*}(\dot{l}+1)\sqrt{\dot{l}^2-\dot{m}^2}\boldsymbol{f}^{\dot{l}_0}_{\dot{l}-1,\dot{m},\dot{k}'}\Bigg]\mathfrak{M}^{\dot{l}_0}_{\dot{m},\dot{n}}(\varphi,\epsilon,\theta,\tau,0,0)+$$

$$+\frac{1}{2r^*}\sqrt{(\dot{l}+\dot{m})(\dot{l}+\dot{m}-1)}\boldsymbol{f}^{\dot{l}_0}_{\dot{l}-1,\dot{m}-1,\dot{k}'}\Bigg[-\frac{1}{\sin\dot{\theta}^c}\frac{\partial \mathfrak{M}^{\dot{l}_0}_{\dot{m}-1,\dot{n}}}{\partial \varphi}+\frac{i}{\sin\dot{\theta}^c}\frac{\partial \mathfrak{M}^{\dot{l}_0}_{\dot{m}-1,\dot{n}}}{\partial \epsilon}+$$

$$+i\frac{\partial \mathfrak{M}^{\dot{l}_0}_{\dot{m}-1,\dot{n}}}{\partial \theta}+\frac{\partial \mathfrak{M}^{\dot{l}_0}_{\dot{m}-1,\dot{n}}}{\partial \tau}+\frac{2i(\dot{m}-1)\cos\dot{\theta}^c}{\sin\dot{\theta}^c}\mathfrak{M}^{\dot{l}_0}_{\dot{m}-1,\dot{n}}\Bigg]+$$

$$+\frac{1}{2r^*}\sqrt{(\dot{l}-\dot{m})(\dot{l}-\dot{m}-1)}\boldsymbol{f}^{\dot{l}_0}_{\dot{l}-1,\dot{m}+1,\dot{k}'}\Bigg[\frac{1}{\sin\dot{\theta}^c}\frac{\partial \mathfrak{M}^{\dot{l}_0}_{\dot{m}+1,\dot{n}}}{\partial \varphi}-\frac{i}{\sin\dot{\theta}^c}\frac{\partial \mathfrak{M}^{\dot{l}_0}_{\dot{m}+1,\dot{n}}}{\partial \epsilon}+$$

$$+i\frac{\partial \mathfrak{M}^{\dot{l}_0}_{\dot{m}+1,\dot{n}}}{\partial \theta}+\frac{\partial \mathfrak{M}^{\dot{l}_0}_{\dot{m}+1,\dot{n}}}{\partial \tau}-\frac{2i(\dot{m}+1)\cos\dot{\theta}^c}{\sin\dot{\theta}^c}\mathfrak{M}^{\dot{l}_0}_{\dot{m}+1,\dot{n}}\Bigg]\Bigg\}+$$

$$+\sum_{\dot{k}'}c^{\dot{k}\dot{k}'}_{\dot{l}\dot{l}}\Bigg\{\Bigg[2\dot{m}\frac{\partial \boldsymbol{f}^{\dot{l}_0}_{\dot{l},\dot{m},\dot{k}'}}{\partial r^*}-\frac{1}{r^*}\dot{m}\boldsymbol{f}^{\dot{l}_0}_{\dot{l},\dot{m},\dot{k}'}\Bigg]\mathfrak{M}^{\dot{l}_0}_{\dot{m},\dot{n}}(\varphi,\epsilon,\theta,\tau,0,0)+$$

$$+\frac{1}{2r^*}\sqrt{(\dot{l}+\dot{m})(\dot{l}-\dot{m}+1)}\boldsymbol{f}^{\dot{l}_0}_{\dot{l},\dot{m}-1,\dot{k}'}\left[\frac{1}{\sin\dot{\theta}^c}\frac{\partial\mathfrak{M}^{\dot{l}_0}_{\dot{m}-1,\dot{n}}}{\partial\varphi}-\frac{i}{\sin\dot{\theta}^c}\frac{\partial\mathfrak{M}^{\dot{l}_0}_{\dot{m}-1,\dot{n}}}{\partial\epsilon}-\right.$$

$$\left.-i\frac{\partial\mathfrak{M}^{\dot{l}_0}_{\dot{m}-1,\dot{n}}}{\partial\theta}-\frac{\partial\mathfrak{M}^{\dot{l}_0}_{\dot{m}-1,\dot{n}}}{\partial\tau}-\frac{2i(\dot{m}-1)\cos\dot{\theta}^c}{\sin\dot{\theta}^c}\mathfrak{M}^{\dot{l}_0}_{\dot{m}-1,\dot{n}}\right]+$$

$$+\frac{1}{2r^*}\sqrt{(\dot{l}+\dot{m}+1)(\dot{l}-\dot{m})}\boldsymbol{f}^{\dot{l}_0}_{\dot{l},\dot{m}+1,\dot{k}'}\left[\frac{1}{\sin\dot{\theta}^c}\frac{\partial\mathfrak{M}^{\dot{l}_0}_{\dot{m}+1,\dot{n}}}{\partial\varphi}-\frac{i}{\sin\dot{\theta}^c}\frac{\partial\mathfrak{M}^{\dot{l}_0}_{\dot{m}+1,\dot{n}}}{\partial\epsilon}+\right.$$

$$\left.\left.+i\frac{\partial\mathfrak{M}^{\dot{l}_0}_{\dot{m}+1,\dot{n}}}{\partial\theta}+\frac{\partial\mathfrak{M}^{\dot{l}_0}_{\dot{m}+1,\dot{n}}}{\partial\tau}-\frac{2i(\dot{m}+1)\cos\dot{\theta}^c}{\sin\dot{\theta}^c}\mathfrak{M}^{\dot{l}_0}_{\dot{m}+1,\dot{n}}\right]\right\}+$$

$$+\sum_{\dot{k}'}c^{\dot{k}\dot{k}'}_{\dot{l},\dot{l}+1}\left\{\left[2\sqrt{(\dot{l}+1)^2-\dot{m}^2}\frac{\partial\boldsymbol{f}^{\dot{l}_0}_{\dot{l}+1,\dot{m},\dot{k}'}}{\partial r^*}+\right.\right.$$

$$\left.+\frac{1}{r^*}\sqrt{(\dot{l}+1)^2-\dot{m}^2}\boldsymbol{f}^{\dot{l}_0}_{\dot{l}+1,\dot{m},\dot{k}'}\right]\mathfrak{M}^{\dot{l}_0}_{\dot{m},\dot{n}}(\varphi,\epsilon,\theta,\tau,0,0)+$$

$$+\frac{1}{2r^*}\sqrt{(\dot{l}-\dot{m}+1)(\dot{l}-\dot{m}+2)}\boldsymbol{f}^{\dot{l}_0}_{\dot{l}+1,\dot{m}-1,\dot{k}'}\left[\frac{1}{\sin\dot{\theta}^c}\frac{\partial\mathfrak{M}^{\dot{l}_0}_{\dot{m}-1,\dot{n}}}{\partial\varphi}-\frac{i}{\sin\dot{\theta}^c}\frac{\partial\mathfrak{M}^{\dot{l}_0}_{\dot{m}-1,\dot{n}}}{\partial\epsilon}-\right.$$

$$\left.-i\frac{\partial\mathfrak{M}^{\dot{l}_0}_{\dot{m}-1,\dot{n}}}{\partial\theta}-\frac{\partial\mathfrak{M}^{\dot{l}_0}_{\dot{m}-1,\dot{n}}}{\partial\tau}-\frac{2i(\dot{m}-1)\cos\dot{\theta}^c}{\sin\dot{\theta}^c}\mathfrak{M}^{\dot{l}_0}_{\dot{m}-1,\dot{n}}\right]+$$

$$+\frac{1}{2r^*}\sqrt{(\dot{l}+\dot{m}+1)(\dot{l}+\dot{m}+2)}\boldsymbol{f}^{\dot{l}_0}_{\dot{l}+1,\dot{m}+1,\dot{k}'}\left[-\frac{1}{\sin\dot{\theta}^c}\frac{\partial\mathfrak{M}^{\dot{l}_0}_{\dot{m}+1,\dot{n}}}{\partial\varphi}+\frac{i}{\sin\dot{\theta}^c}\frac{\partial\mathfrak{M}^{\dot{l}_0}_{\dot{m}+1,\dot{n}}}{\partial\epsilon}-\right.$$

$$\left.\left.-i\frac{\partial\mathfrak{M}^{\dot{l}_0}_{\dot{m}+1,\dot{n}}}{\partial\theta}-\frac{\partial\mathfrak{M}^{\dot{l}_0}_{\dot{m}+1,\dot{n}}}{\partial\tau}+\frac{2i(\dot{m}+1)\cos\dot{\theta}^c}{\sin\dot{\theta}^c}\mathfrak{M}^{\dot{l}_0}_{\dot{m}+1,\dot{n}}\right]\right\}+$$

$$+\dot{\kappa}^c\boldsymbol{f}^{\dot{l}_0}_{\dot{l}\dot{m}}\mathfrak{M}^{\dot{l}_0}_{\dot{m}\dot{n}}(\varphi,\epsilon,\theta,\tau,0,0)=0. \quad (81)$$

The each equation of the obtained system contains three generalized hyperspherical functions $\mathfrak{M}^{l_0}_{mn}$, $\mathfrak{M}^{l_0}_{m-1,n}$, $\mathfrak{M}^{l_0}_{m+1,n}$ and their conjugate. We apply the recurrence relations (32)–(33), (34)–(35) to square brackets containing the functions $\mathfrak{M}^{l_0}_{m-1,n}$ and $\mathfrak{M}^{l_0}_{m+1,n}$. First of all, let us recall that $\mathfrak{M}^{l_0}_{m\pm1,n}(\varphi,\epsilon,\theta,\tau,0,0)=e^{-n(\epsilon+i\varphi)}Z^{l_0}_{m\pm1,n}(\theta,\tau)$ and $\mathfrak{M}^{\dot{l}_0}_{\dot{m}\pm1,\dot{n}}(\varphi,\epsilon,\theta,\tau,0,0)=e^{-\dot{n}(\epsilon-i\varphi)}Z^{\dot{l}_0}_{\dot{m}\pm1,\dot{n}}(\theta,\tau)$. Therefore, $\dfrac{\partial\mathfrak{M}^{l_0}_{m\pm1,n}}{\partial\varphi}=-in\mathfrak{M}^{l_0}_{m\pm1,n}$, $\dfrac{\partial\mathfrak{M}^{l_0}_{m\pm1,n}}{\partial\epsilon}=-n\mathfrak{M}^{l_0}_{m\pm1,n}$ and $\dfrac{\partial\mathfrak{M}^{\dot{l}_0}_{\dot{m}\pm1,\dot{n}}}{\partial\varphi}=i\dot{n}\mathfrak{M}^{\dot{l}_0}_{\dot{m}\pm1,\dot{n}}$,

$\dfrac{\partial \mathfrak{M}^{\dot{l}_0}_{\dot{m}\pm1,\dot{n}}}{\partial \epsilon} = -\dot{n}\mathfrak{M}^{\dot{l}_0}_{\dot{m}\pm1,\dot{n}}$. Let us replace the square brackets in the system (81) via the recurrence relations and cancel all the equations by $\mathfrak{M}^{l_0}_{mn}$ ($\mathfrak{M}^{\dot{l}_0}_{\dot{m}\dot{n}}$). In such a way, we see that the relativistically invariant system is reduced to a system of ordinary differential equations:

$$\sum_{k'} c^{kk'}_{l,l-1}\left[2\sqrt{l^2-m^2}\frac{d\boldsymbol{f}^{l_0}_{l-1,m,k'}(r)}{dr}-\frac{1}{r}(l+1)\sqrt{l^2-m^2}\boldsymbol{f}^{l_0}_{l-1,m,k'}(r)+\right.$$
$$+\frac{1}{r}\sqrt{(l+m)(l+m-1)}\sqrt{(l_0+m)(l_0-m+1)}\boldsymbol{f}^{l_0}_{l-1,m-1,k'}(r)+$$
$$\left.+\frac{1}{r}\sqrt{(l-m)(l-m-1)}\sqrt{(l_0+m+1)(l_0-m)}\boldsymbol{f}^{l_0}_{l-1,m+1,k'}(r)\right]+$$

$$\sum_{k'} c^{kk'}_{ll}\left[2m\frac{d\boldsymbol{f}^{l_0}_{l,m,k'}(r)}{dr}-\frac{1}{r}m\boldsymbol{f}^{l_0}_{l,m,k'}(r)-\right.$$
$$-\frac{1}{r}\sqrt{(l+m)(l-m+1)}\sqrt{(l_0+m)(l_0-m+1)}\boldsymbol{f}^{l_0}_{l,m-1,k'}(r)+$$
$$\left.+\frac{1}{r}\sqrt{(l+m+1)(l-m)}\sqrt{(l_0+m+1)(l_0-m)}\boldsymbol{f}^{l_0}_{l,m+1,k'}(r)\right]+$$

$$\sum_{k'} c^{kk'}_{l,l+1}\left[2\sqrt{(l+1)^2-m^2}\frac{d\boldsymbol{f}^{l_0}_{l+1,m,k'}(r)}{dr}+\frac{1}{r}l\sqrt{(l+1)^2-m^2}\boldsymbol{f}^{l_0}_{l+1,m,k'}(r)-\right.$$
$$-\frac{1}{r}\sqrt{(l-m+1)(l-m+2)}\sqrt{(l_0+m)(l_0-m+1)}\boldsymbol{f}^{l_0}_{l+1,m-1,k'}(r)-$$
$$\left.-\frac{1}{r}\sqrt{(l+m+1)(l+m+2)}\sqrt{(l_0+m+1)(l_0-m)}\boldsymbol{f}^{l_0}_{l+1,m+1,k'}(r)\right]+$$
$$+\kappa^c\boldsymbol{f}^{l_0}_{lmk}(r)=0,$$

$$\sum_{\dot{k}'} c^{\dot{k}\dot{k}'}_{\dot{l},\dot{l}-1}\left[2\sqrt{\dot{l}^2-\dot{m}^2}\frac{d\boldsymbol{f}^{\dot{l}_0}_{\dot{l}-1,\dot{m},\dot{k}'}(r^*)}{dr^*}-\frac{1}{r^*}(\dot{l}+1)\sqrt{\dot{l}^2-\dot{m}^2}\boldsymbol{f}^{\dot{l}_0}_{\dot{l}-1,\dot{m},\dot{k}'}(r^*)+\right.$$
$$+\frac{1}{r^*}\sqrt{(\dot{l}+\dot{m})(\dot{l}+\dot{m}-1)}\sqrt{(\dot{l}_0+\dot{m})(\dot{l}_0-\dot{m}+1)}\boldsymbol{f}^{\dot{l}_0}_{\dot{l}-1,\dot{m}-1,\dot{k}'}(r^*)+$$
$$\left.+\frac{1}{r^*}\sqrt{(\dot{l}-\dot{m})(\dot{l}-\dot{m}-1)}\sqrt{(\dot{l}_0+\dot{m}+1)(\dot{l}_0-\dot{m})}\boldsymbol{f}^{\dot{l}_0}_{\dot{l}-1,\dot{m}+1,\dot{k}'}(r^*)\right]+$$

$$\sum_{\dot{k}'} c^{\dot{k}\dot{k}'}_{\dot{l},\dot{l}}\left[2\dot{m}\frac{d\boldsymbol{f}^{\dot{l}_0}_{\dot{l},\dot{m},\dot{k}'}(r^*)}{dr^*}-\frac{1}{r^*}\dot{m}\boldsymbol{f}^{\dot{l}_0}_{\dot{l},\dot{m},\dot{k}'}(r^*)-\right.$$
$$-\frac{1}{r^*}\sqrt{(\dot{l}+\dot{m})(\dot{l}-\dot{m}-1)}\sqrt{(\dot{l}_0+\dot{m})(\dot{l}_0-\dot{m}+1)}\boldsymbol{f}^{\dot{l}_0}_{\dot{l},\dot{m}-1,\dot{k}'}(r^*)+$$
$$\left.+\frac{1}{r^*}\sqrt{(\dot{l}+\dot{m}+1)(\dot{l}-\dot{m})}\sqrt{(\dot{l}_0+\dot{m}+1)(\dot{l}_0-\dot{m})}\boldsymbol{f}^{\dot{l}_0}_{\dot{l},\dot{m}+1,\dot{k}'}(r^*)\right]+$$

$$\sum_{\dot{k}'} c^{\dot{k}\dot{k}'}_{\dot{l},\dot{l}+1}\left[2\sqrt{(\dot{l}+1)^2-\dot{m}^2}\frac{d\boldsymbol{f}^{\dot{l}_0}_{\dot{l}+1,\dot{m},\dot{k}'}(r^*)}{dr^*}+\frac{1}{r^*}\dot{l}\sqrt{(\dot{l}+1)^2-\dot{m}^2}\boldsymbol{f}^{\dot{l}_0}_{\dot{l}+1,\dot{m},\dot{k}'}(r^*)-\right.$$
$$-\frac{1}{r^*}\sqrt{(\dot{l}-\dot{m}+1)(\dot{l}-\dot{m}+2)}\sqrt{(\dot{l}_0+\dot{m})(\dot{l}_0-\dot{m}+1)}\boldsymbol{f}^{\dot{l}_0}_{\dot{l}+1,\dot{m}-1,\dot{k}'}(r^*)-$$
$$\left.-\frac{1}{r^*}\sqrt{(\dot{l}+\dot{m}+1)(\dot{l}+\dot{m}+2)}\sqrt{(\dot{l}_0+\dot{m}+1)(\dot{l}_0-\dot{m})}\boldsymbol{f}^{\dot{l}_0}_{\dot{l}+1,\dot{m}+1,\dot{k}'}(r^*)\right]+$$
$$+\dot{\kappa}^c\boldsymbol{f}^{\dot{l}_0}_{\dot{l}\dot{m}\dot{k}'}(r^*)=0. \quad (82)$$

This system is solvable for any spin in a class of the Bessel functions. Therefore, according to (80) relativistic wave functions are expressed via the products of cylindrical and hyperspherical functions. For that reason the wave function directly depends on the parameters of the group $\mathfrak{G}_+$. It follows that all the physical fields are the functions on the Lorentz (Poincaré) group. Such a description corresponds to a quantum field theory on the Poincaré group introduced by Lurçat [30] (see also [6, 27, 10, 5, 44, 16, 21] and references therein). Moreover, it allows us to widely use a harmonic analysis on the Poincaré group [36, 23, 24] (or, on the product $SU(2)\otimes SU(2)$) at the study of relativistic amplitudes.

5 Dirac equation

The first non-trivial case of the systems (36) and (82) corresponds the spin $l = 1/2$. In turn, this value of spin corresponds to the fundamental representation $\boldsymbol{\tau}_{\frac{1}{2},0}$ of the group $\mathfrak{G}_+$. The first physical field, constructed within the fundamental representation, is a Dirac field (electron-positron field). This field is defined by a P-invariant direct sum $(1/2,0)\oplus(0,1/2)$. Using (45)–(50), we find that matrices Λ_i, Λ^*_i of the system (36) at $l=1/2$ have the form

$$\Lambda_1=\frac{1}{2}c_{\frac{1}{2}\frac{1}{2}}\begin{pmatrix}1&0\\0&-1\end{pmatrix},\quad \Lambda_2=\frac{1}{2}c_{\frac{1}{2}\frac{1}{2}}\begin{pmatrix}0&1\\1&0\end{pmatrix},\quad \Lambda_3=\frac{1}{2}c_{\frac{1}{2}\frac{1}{2}}\begin{pmatrix}0&-i\\i&0\end{pmatrix},$$
$$\Lambda^*_1=\frac{1}{2}\dot{c}_{\frac{1}{2}\frac{1}{2}}\begin{pmatrix}1&0\\0&-1\end{pmatrix},\quad \Lambda^*_2=\frac{1}{2}\dot{c}_{\frac{1}{2}\frac{1}{2}}\begin{pmatrix}0&1\\1&0\end{pmatrix},\quad \Lambda^*_3=\frac{1}{2}\dot{c}_{\frac{1}{2}\frac{1}{2}}\begin{pmatrix}0&-i\\i&0\end{pmatrix}.$$

It is easy to see that these matrices coincide with the Pauli matrices σ_i when $c_{\frac{1}{2}\frac{1}{2}} = 2$. As is known, σ_i define spinor representations of the biquaternion algebra $\mathbb{C}_2$. The representation $\boldsymbol{\tau}_{\frac{1}{2},0} \oplus \boldsymbol{\tau}_{0,\frac{1}{2}}$ of the group $\mathfrak{G}_+$ is associated with the Clifford algebra $\mathbb{C}_2 \oplus \overset{*}{\mathbb{C}}_2$, a spinor representation of which is defined by the direct sum of the representations of $\mathbb{C}_2$ and $\overset{*}{\mathbb{C}}_2$:

$$\gamma_0 = \begin{pmatrix} \sigma_0 & 0 \\ 0 & -\sigma_0 \end{pmatrix}, \quad \gamma_1 = \begin{pmatrix} 0 & \sigma_1 \\ -\sigma_1 & 0 \end{pmatrix}, \quad \gamma_2 = \begin{pmatrix} 0 & \sigma_2 \\ -\sigma_2 & 0 \end{pmatrix}, \quad \gamma_3 = \begin{pmatrix} 0 & \sigma_3 \\ -\sigma_3 & 0 \end{pmatrix}. \tag{83}$$

γ-matrices (83) form a so-called canonical basis, which, in turn, is a spinor representation basis of the spacetime algebra $\mathcal{C}\ell_{1,3}$. In this basis the Dirac equation has the form

$$i\gamma_\mu \frac{\partial \boldsymbol{\psi}}{\partial x_\mu} - m\boldsymbol{\psi} = 0, \tag{84}$$

where $\boldsymbol{\psi} = (\psi_1, \psi_2, \dot{\psi}_1, \dot{\psi}_2)^T$ is a Dirac bispinor. In such a way, we see that in the case of $l = 1/2$ the system (36), which acts on the tangent bundle of the group manifold $\mathfrak{L}$, is equivalent to the Dirac equation defined in the spacetime $\mathbb{R}^{1,3}$. Coming to the helicity basis, we will find solutions of the equation (84), that is, we will present components of the Dirac bispinor in terms of the functions on the Lorentz group (the indices k and $\dot{k}$ we can omit, since representations $\boldsymbol{\tau}_{\frac{1}{2},0}$ and $\boldsymbol{\tau}_{0,\frac{1}{2}}$ occur only one time):

$$\begin{aligned}
\psi_1 &= \psi_{\frac{1}{2},\frac{1}{2};\frac{1}{2},\frac{1}{2}} = \boldsymbol{f}^l_{\frac{1}{2},\frac{1}{2}}(r)\mathfrak{M}^l_{\frac{1}{2},n}(\varphi,\epsilon,\theta,\tau,0,0), \\
\psi_2 &= \psi_{\frac{1}{2},-\frac{1}{2};\frac{1}{2},\frac{1}{2}} = \boldsymbol{f}^l_{\frac{1}{2},-\frac{1}{2}}(r)\mathfrak{M}^l_{-\frac{1}{2},n}(\varphi,\epsilon,\theta,\tau,0,0), \\
\dot{\psi}_1 &= \dot{\psi}_{\frac{1}{2},\frac{1}{2};\frac{1}{2},\frac{1}{2}} = \boldsymbol{f}^{\dot{l}}_{\frac{1}{2},\frac{1}{2}}(r^*)\mathfrak{M}^{\dot{l}}_{\frac{1}{2},\dot{n}}(\varphi,\epsilon,\theta,\tau,0,0), \\
\dot{\psi}_2 &= \dot{\psi}_{\frac{1}{2},\frac{1}{2};\frac{1}{2},-\frac{1}{2}} = \boldsymbol{f}^{\dot{l}}_{\frac{1}{2},-\frac{1}{2}}(r^*)\mathfrak{M}^{\dot{l}}_{-\frac{1}{2},\dot{n}}(\varphi,\epsilon,\theta,\tau,0,0),
\end{aligned}$$

Repeating for the case $l = 1/2$ all the transformations presented for the general relativistically invariant system, we come to the following system of ordinary differential equations (the system (82) at $l = 1/2$):

$$\begin{aligned}
2\frac{d\boldsymbol{f}^l_{\frac{1}{2},\frac{1}{2}}(r)}{dr} \quad - \quad \frac{1}{r}\boldsymbol{f}^l_{\frac{1}{2},\frac{1}{2}}(r) \quad - \quad \frac{2\left(l+\frac{1}{2}\right)}{r}\boldsymbol{f}^l_{\frac{1}{2},-\frac{1}{2}}(r) \quad + \quad \kappa^c \boldsymbol{f}^l_{\frac{1}{2},\frac{1}{2}}(r) &= 0,\\
-2\frac{d\boldsymbol{f}^l_{\frac{1}{2},-\frac{1}{2}}(r)}{dr} \quad + \quad \frac{1}{r}\boldsymbol{f}^l_{\frac{1}{2},-\frac{1}{2}}(r) \quad + \quad \frac{2\left(l+\frac{1}{2}\right)}{r}\boldsymbol{f}^l_{\frac{1}{2},\frac{1}{2}}(r) \quad + \quad \kappa^c \boldsymbol{f}^l_{\frac{1}{2},-\frac{1}{2}}(r) &= 0,\\
2\frac{d\boldsymbol{f}^{\dot{l}}_{\frac{1}{2},\frac{1}{2}}(r^*)}{dr^*} \quad - \quad \frac{1}{r^*}\boldsymbol{f}^{\dot{l}}_{\frac{1}{2},\frac{1}{2}}(r^*) \quad - \quad \frac{2\left(\dot{l}+\frac{1}{2}\right)}{r^*}\boldsymbol{f}^{\dot{l}}_{\frac{1}{2},-\frac{1}{2}}(r^*) \quad + \quad \dot{\kappa}^c \boldsymbol{f}^{\dot{l}}_{\frac{1}{2},\frac{1}{2}}(r^*) &= 0,\\
-2\frac{d\boldsymbol{f}^{\dot{l}}_{\frac{1}{2},-\frac{1}{2}}(r^*)}{dr^*} + \frac{1}{2r^*}\boldsymbol{f}^{\dot{l}}_{\frac{1}{2},-\frac{1}{2}}(r^*) + \frac{2\left(\dot{l}+\frac{1}{2}\right)}{r^*}\boldsymbol{f}^{\dot{l}}_{\frac{1}{2},\frac{1}{2}}(r^*) + \dot{\kappa}^c \boldsymbol{f}^{\dot{l}}_{\frac{1}{2},-\frac{1}{2}}(r^*) &= 0,
\end{aligned}$$

where $\kappa^c = -im$. Assuming that $\boldsymbol{f}^l_{\frac{1}{2},-\frac{1}{2}}(r) = -\boldsymbol{f}^l_{\frac{1}{2},\frac{1}{2}}(r)$ and $\boldsymbol{f}^{\dot{l}}_{\frac{1}{2},-\frac{1}{2}}(r^*) = -\boldsymbol{f}^{\dot{l}}_{\frac{1}{2},\frac{1}{2}}(r^*)$, we find solutions of this system:

$$\begin{aligned}
\boldsymbol{f}^l_{\frac{1}{2},\frac{1}{2}}(r) &= \sqrt[3]{r}\sum_{k=0}^{\infty}(-1)^k\left(\frac{2}{\sqrt{3}}\right)^{2k}\Gamma(\nu+k+1)J_\nu\left(2\sqrt{\kappa^c}\sqrt[3]{r}\right),\\
\boldsymbol{f}^{\dot{l}}_{\frac{1}{2},\frac{1}{2}}(r^*) &= \sqrt[3]{r^*}\sum_{k=0}^{\infty}(-1)^k\left(\frac{2}{\sqrt{3}}\right)^{2k}\Gamma(\dot{\nu}+k+1)J_{\dot{\nu}}\left(2\sqrt{\dot{\kappa}^c}\sqrt[3]{r^*}\right),
\end{aligned}$$

where $\nu = -(l-1)$, $\dot{\nu} = -(\dot{l}-1)$, $l = \frac{2s+1}{2}$, $\dot{l} = \frac{2\dot{s}+1}{2}$, $s, \dot{s} = 0, 1, 2, \ldots$ and

$$\begin{aligned}
J_{\frac{2s+1}{2}}(z) = \sqrt{\frac{2}{\pi z}}\Bigg[\sin\left(z-\frac{s\pi}{2}\right)\sum_{k=0}^{\frac{s}{2}}\frac{(-1)^k(s+2k)!}{(2k)!(s-2k)!(2z)^{2k}}+\\
+\cos\left(z-\frac{s\pi}{2}\right)\sum_{k=0}^{\left[\frac{s-1}{2}\right]}\frac{(-1)^k(s+2k+1)!}{(2k+1)!(s-2k-1)!(2z)^{2k+1}}\Bigg]. \qquad (85)
\end{aligned}$$

is the Bessel function of half-integer order. Thus, solutions of the Dirac equation are defined by the following functions:

$$\begin{aligned}
\psi_1(r,\varphi^c,\theta^c) &= \boldsymbol{f}^l_{\frac{1}{2},\frac{1}{2}}(r)\mathfrak{M}^l_{\frac{1}{2},n}(\varphi,\epsilon,\theta,\tau,0,0),\\
\psi_2(r,\varphi^c,\theta^c) &= -\boldsymbol{f}^l_{\frac{1}{2},\frac{1}{2}}(r)\mathfrak{M}^l_{-\frac{1}{2},n}(\varphi,\epsilon,\theta,\tau,0,0),\\
\dot{\psi}_1(r^*,\dot{\varphi}^c,\dot{\theta}^c) &= \boldsymbol{f}^{\dot{l}}_{\frac{1}{2},\frac{1}{2}}(r^*)\mathfrak{M}^{\dot{l}}_{\frac{1}{2},\dot{n}}(\varphi,\epsilon,\theta,\tau,0,0),\\
\dot{\psi}_2(r^*,\dot{\varphi}^c,\dot{\theta}^c) &= -\boldsymbol{f}^{\dot{l}}_{\frac{1}{2},\frac{1}{2}}(r^*)\mathfrak{M}^{\dot{l}}_{-\frac{1}{2},\dot{n}}(\varphi,\epsilon,\theta,\tau,0,0),
\end{aligned}$$

where

$$\begin{aligned}
l &= \frac{1}{2},\ \frac{3}{2},\ \frac{5}{2},\ldots; \quad n = -l,\ -l+1,\ \ldots,\ l;\\
\dot{l} &= \frac{1}{2},\ \frac{3}{2},\ \frac{5}{2},\ldots; \quad \dot{n} = -\dot{l},\ -\dot{l}+1,\ \ldots,\ \dot{l},
\end{aligned}$$

$$\mathfrak{M}^{l}_{\pm\frac{1}{2},n}(\varphi,\epsilon,\theta,\tau,0,0)=e^{\mp\frac{1}{2}(\epsilon+i\varphi)}Z^{l}_{\pm\frac{1}{2},n}(\theta,\tau),$$

$$Z^{l}_{\pm\frac{1}{2},n}(\theta,\tau)=\cos^{2l}\frac{\theta}{2}\cosh^{2l}\frac{\tau}{2}\sum_{k=-l}^{l}i^{\pm\frac{1}{2}-k}\tan^{\pm\frac{1}{2}-k}\frac{\theta}{2}\tanh^{n-k}\frac{\tau}{2}\times$$
$${}_2F_1\left(\begin{matrix}\pm\frac{1}{2}-l+1,1-l-k\\ \pm\frac{1}{2}-k+1\end{matrix}\middle| i^2\tan^2\frac{\theta}{2}\right){}_2F_1\left(\begin{matrix}n-l+1,1-l-k\\ n-k+1\end{matrix}\middle|\tanh^2\frac{\tau}{2}\right),$$

$$\mathfrak{M}^{\dot{l}}_{\pm\frac{1}{2},\dot{n}}(\varphi,\epsilon,\theta,\tau,0,0)=e^{\mp\frac{1}{2}(\epsilon-i\varphi)}Z^{\dot{l}}_{\pm\frac{1}{2},\dot{n}}(\theta,\tau),$$

$$Z^{\dot{l}}_{\pm\frac{1}{2},\dot{n}}(\theta,\tau)=\cos^{2\dot{l}}\frac{\theta}{2}\cosh^{2\dot{l}}\frac{\tau}{2}\sum_{\dot{k}=-\dot{l}}^{\dot{l}}i^{\pm\frac{1}{2}-\dot{k}}\tan^{\pm\frac{1}{2}-\dot{k}}\frac{\theta}{2}\tanh^{\dot{n}-\dot{k}}\frac{\tau}{2}\times$$
$${}_2F_1\left(\begin{matrix}\pm\frac{1}{2}-\dot{l}+1,1-\dot{l}-\dot{k}\\ \pm\frac{1}{2}-\dot{k}+1\end{matrix}\middle| i^2\tan^2\frac{\theta}{2}\right){}_2F_1\left(\begin{matrix}\dot{n}-\dot{l}+1,1-\dot{l}-\dot{k}\\ \dot{n}-\dot{k}+1\end{matrix}\middle|\tanh^2\frac{\tau}{2}\right).$$

In like manner, we can define any physical field (both massive and massless) of arbitrary spin in terms of the functions on the Lorentz group. In general case, the matrix Λ_3 admits the following decompositions: $\Lambda_3=\operatorname{diag}\left(C^0\otimes I_1, C^1\otimes I_3,\ldots,C^s\otimes I_{2s+1},\ldots\right)$ for integer spin and $\Lambda_3=\operatorname{diag}\left(C^{\frac{1}{2}}\otimes I_2, C^{\frac{3}{2}}\otimes I_4,\ldots,C^s\otimes I_{2s+1},\ldots\right)$ for half–integer spin. The matrix Λ_3^* has the same decompositions. C^s is a spin block. The each spin block C^s is realized in the space Sym_s. The matrix elements of the corresponded representations are expressed via the hyperspherical functions. If the spin block C^s has non–null roots, then the particle possesses the spin s [19, 3, 34]. The spin block C^s consists of the elements $c^s_{\boldsymbol{\tau}\boldsymbol{\tau}'}$, where $\boldsymbol{\tau}_{l_1,l_2}$ and $\boldsymbol{\tau}_{l_1',l_2'}$ are interlocking irreducible representations of the Lorentz group, that is, such representations, for which $l_1'=l_1\pm\frac{1}{2}$, $l_2'=l_2\pm\frac{1}{2}$. At this point the block C^s contains only the elements $c^s_{\boldsymbol{\tau}\boldsymbol{\tau}'}$ corresponding to such interlocking representations $\boldsymbol{\tau}_{l_1,l_2}$, $\boldsymbol{\tau}_{l_1',l_2'}$ which satisfy the conditions

$$|l_1-l_2|\le s\le l_1+l_2,\quad |l_1'-l_2'|\le s\le l_1'+l_2'.$$

According to a de Broglie theory of fusion [11, 41] interlocking representations give rise to indecomposable RWE. Otherwise, we have decomposable equations. As known, the indecomposable RWE correspond to composite particles. A wide variety of such representations and RWE it seems to be sufficient for description of all particle world.

References

[1] A.I. Akhiezer, V.B. Berestetskii, *Quantum Electrodynamics* (John Wiley & Sons, New York, 1965).

[2] D.V. Ahluwalia, D.J. Ernst, $(j,0) \oplus (0,j)$ *covariant spinors and causal propagators based on Weinberg formalism*, *Int. J. Mod. Phys.* **E**2, 397–422 (1993).

[3] V. Amar, U. Dozzio, *Gel'fand-Yaglom Equations with Charge or Energy Density of Definite Sign*, *Nuovo Cimento* **A11**, 87–99 (1972).

[4] P. Appell, M.J. Kampé de Fériet, *Fonctions hypégeométriques et hypersphéques. Polynomes d'Hermite* (Gauthier–Villars, 1926).

[5] H. Arodź, *Metric tensors, Lagrangian formalism and Abelian gauge field on the Poincaré group*, *Acta Phys. Pol., Ser.* **B7**, 177–190 (1976).

[6] H. Bacry, A. Kihlberg, *Wavefunctions on homogeneous spaces*, *J. Math. Phys.* **10**, 2132–2141 (1969).

[7] V. Bargmann, E.P. Wigner, *Group theoretical discussion of relativistic wave equations*, *Proc. Nat. Acad. USA* **34**, 211–223 (1948).

[8] H. Bateman, A. Erdélyi, *Higher Transcendental Functions*, vol. I (Mc Grow-Hill Book Company, New York, 1953).

[9] H.J. Bhabha, Relativistic Wave Equations for the Elementary Particles, *Rev. Mod. Phys.* **17**, 200–216 (1945).

[10] C.P. Boyer, G.N. Fleming, Quantum field theory on a seven-dimensional homogeneous space of the Poincaré group, *J. Math. Phys.* **15**, 1007–1024 (1974).

[11] L. de Broglie, *Theorie de particules a spin (methode de fusion)* (Paris, 1943).

[12] P.A.M. Dirac, Relativistic Wave Equations, *Proc. Roy. Soc.* (London) **155A**, 447–459 (1936).

[13] A.Z. Dolginov, Relativistic spherical functions, *Zh. Ehksp. Teor. Fiz.* **30**, 746–755 (1956).

[14] A.Z. Dolginov, I.N. Toptygin, Relativistic spherical functions. II, *Zh. Ehksp. Teor. Fiz.* **37**, 1441–1451 (1959).

[15] A.Z. Dolginov, A.N. Moskalev,Relativistic spherical functions. III, *Zh. Ehksp. Teor. Fiz.* **37**, 1697–1707 (1959).

[16] W. Drechsler, Geometro-stohastically quantized fields with internal spin variables, *J. Math. Phys.* **38**, 5531–5558 (1997).

[17] V.V. Dvoeglazov, Extra Dirac equations, *Nuovo Cimento*, **B**111, 483–496 (1996).

[18] M. Fierz, W. Pauli, On Relativistic Wave Equations of Particles of Arbitrary Spin in an Electromagnetic Field, *Proc. Roy. Soc.* (London) **173A**, 211–232 (1939).

[19] I.M. Gel'fand, A.M. Yaglom, General relativistic–invariant equations and infinite–dimensional representations of the Lorentz group,*Zh. Ehksp. Teor. Fiz.* **18**, 703–733 (1948).

[20] I.M. Gel'fand, R.A. Minlos, Z.Ya. Shapiro, *Representations of the Rotation and Lorentz Groups and their Applications* (Pergamon Press, Oxford, 1963).

[21] D.M. Gitman, A.L. Shelepin, Fields on the Poincaré Group: Arbitrary Spin Description and Relativistic Wave Equations, *Int. J. Theor. Phys.* **40**(3), 603–684 (2001).

[22] M. Goto, F. Grosshans, *Semisimple Lie Algebras* (Marcel Dekker, New York, 1978).

[23] N.X. Hai, Harmonic analysis on the Poincaré group, I. Generilized matrix elements, *Commun. Math. Phys.* **12**, 331–350 (1969).

[24] N.X. Hai, Harmonic analysis on the Poincaré group, II. The Fourier transform, *Commun. Math. Phys.* **22**, 301–320 (1971).

[25] H. Joos, Zur darstellungstheorie der inhomogenen Lorentzgrouppe als grundlade quantenmechanische kinematick, *Fortschr. Phys.* **10**, 65 (1962).

[26] V.F. Kagan, *Ueber einige Zahlensysteme, zu denen die Lorentztransformation fürt*, Publ. House of Institute of Mathematics, Moscow (1926).

[27] A. Kihlberg, Fields on a homogeneous space of the Poincaré group, *Ann. Inst. Henri Poincaré* **13**, 57–76 (1970).

[28] G.I. Kuznetsov, M.A. Liberman, A.A. Makarov, Ya.A. Smorodinsky, Helicity and unitary representations of the Lorentz group, *Yad. Fiz.* **10**, 644–656 (1969).

[29] M.A. Liberman, Ya.A. Smorodinsky, M.B. Sheftel, Unitary representations of the Lorentz group and functions with spin, *Yad. Fiz.* **7**, 202–214 (1968).

[30] F. Lurçat, Quantum field theory and the dynamical role of spin, *Physics* **1**, 95 (1964).

[31] E. Majorana, Teoria relativistica di particelle con momento intrinseco arbitrario, *Nuovo Cimento* **9**, 335–344 (1932).

[32] M.A. Naimark, *Linear Representations of the Lorentz Group* (Pergamon, London, 1964).

[33] A.Z. Petrov, *Einstein Spaces* (Pergamon Press, Oxford, 1969).

[34] V.A. Pletyukhov, V.I. Strazhev, On Dirac-like relativistic wave equations, *Russian J. Phys.* **n.12**, 38-41 (1983).

[35] W. Rarita, J. Schwinger, On a theory of particles with half-integral spin, *Phys. Rev.* **60**, 61 (1941).

[36] G. Rideau, On the reduction of the regular representation of the Poincaré group, *Commun. Math. Phys.* **3**, 218–227 (1966).

[37] Yu.B. Rumer, A.I. Fet, *Group Theory and Quantized Fields* (Nauka, Moscow, 1977) [in Russian].

[38] L. Ryder, *Quantum Field Theory* (Cambridge University Press, Cambridge, 1985).

[39] S.S. Schweber, *An Introduction to Relativistic Quantum Field Theory* (Harper & Row, New York, 1961).

[40] Ya.A. Smorodinsky, M. Huszar, Representations of the Lorentz group and the generalization of helicity states, *Teor. Mat. Fiz.* **4**, 3, 328–340 (1970).

[41] G.A. Sokolik, *Groups Methods in the Theory of Elementary Particles* (Atomizdat, Moscow, 1965) [in Russian].

[42] S. Ström, On the matrix elements of a unitary representation of the homogeneous Lorentz group, *Arkiv för Fysik* **29**, 467–483 (1965).

[43] S. Ström, A note on the matrix elements of a unitary representation of the homogeneous Lorentz group, *Arkiv för Fysik* **33**, 465–469 (1967).

[44] M. Toller, Free quantum fields on the Poincaré group, *J. Math. Phys.* **37**, 2694–2730 (1996).

[45] V.V. Varlamov, Fundamental Automorphisms of Clifford Algebras and an Extension of Dąbrowski Pin Groups, *Hadronic J.* **22**, 497–535 (1999).

[46] V.V. Varlamov, Discrete Symmetries and Clifford Algebras, *Int. J. Theor. Phys.* **40**(4), 769–805 (2001).

[47] V.V. Varlamov, *Group Theoretical Description of Space Inversion, Time Reversal and Charge Conjugation*, preprint math-ph/0203059 (2002).

[48] V.V. Varlamov, Hyperspherical Functions and Linear Representations of the Lorentz Group, *Hadronic J.* **25**, 481 (2002).

[49] I.A. Verdiev, L.A. Dadashev, Matrix elements of the Lorentz group unitary representation, *Yad. Fiz.* **6**, 1094–1099 (1967).

[50] N.Ya. Vilenkin, Ya.A. Smorodinsky, Invariant expansions of relativistic amplitudes, *Zh. Ehksp. Teor. Fiz.* **46**, 1793–1808 (1964).

[51] N.Ya. Vilenkin, *Special Functions and the Theory of Group Representations* (AMS, Providence, 1968).

[52] B.L. van der Waerden, *Die Gruppentheoretische Methode in der Quantenmechanik* (Springer, Berlin, 1932).

[53] S. Weinberg, Feinman rules for any spin I & II & III, *Phys. Rev.* **133B**, 1318–1332 & **134B**, 882–896 (1964) & **181B**, 1893–1899 (1969).

[54] H. Weyl, *Classical Groups* (Princeton, 1939).

In: Frontiers in Quantum Physics Research ISBN 1-59454-002-2
Editor: F. Columbus and V. Krasnoholovets, pp. 83-128

Unified Model of Matter - Fields Duality & Bivacuum Mediated Electromagnetic and Gravitational Interactions

Alex Kaivarainen
University of Turku,
Vesilinnantie 5, FIN-20014, Turku, Finland
H2o@karelia.ru
http://www.karelia.ru/~alexk

Abstract

The Bivacuum model is a consequence of new interpretation of Dirac theory, pointing to equal probability of positive and negative energy. Unified Model (UM) represents our efforts for unification of vacuum, matter and fields from few ground postulates (http://arXiv.org/abs/physics/0207027).

A new concept of Bivacuum is introduced, as a dynamic superfluid matrix of the Universe, composed from non mixing sub-quantum particles of the opposite energies, separated by energy gap. Their collective excitations form mesoscopic vortical structures. These structures, named Bivacuum fermions and antifermions, are presented by infinitive number of double cells-dipoles, each cell containing a pair of correlated rotors and antirotors: V(+) and V(-) of the opposite quantized energy, virtual mass, charge and magnetic moments. The absolute values of rotors and antirotors internal rotational kinetic energy and magnetic moments are postulated to be equal and permanent, in contrast to their virtual mass and internal angle velocities. These dipoles of three generation (e mu, tau), due to very small external zero-point translational momentum, form macroscopic virtual Bose condensate with nonlocal properties. The matter origination in form of sub-elementary fermions or antifermions of three generation is a result of Bivacuum dipoles symmetry shift towards the positive or negative energy, correspondingly. The asymmetry of velocity rotation of rotor and antirotor of sub-elementary fermion, corresponding to Hidden harmony and standing waves conditions, is responsible for the rest mass and resulting charge of sub-elementary fermion origination. The triplets formed by [sub-elementary fermion + sub-elementary antifermion] with opposite spins, charge and energy and one uncompensated sub-elementary fermion represent electrons. The triplets formed by [sub-elementary fermion + sub-elementary antifermion] and one uncompensated sub-elementary antifermion represent the positrons. The quarks are the result of certain superposition/fusion of (mu) and (tau) electrons and positrons. The bosons, like photons, are composed from the equal number of sub-elementary particles and antiparticles as a system of two [electron + positron] triplets.

The [corpuscle (C) - wave (W)] duality is a result of quantum beats between the 'actual' (vortex) and 'complementary' (rotor) states of sub-elementary fermions/antifermions. The [C] phase exists as a mass, electric and magnetic asymmetric dipole with spatial image: [vortex + rotor]. The [W] phase exists in form of Cumulative virtual cloud (CVC) of sub-quantum particles. The angular momentum of CVC excites the nonlocal massless virtual spin waves (VirSW) in Bivacuum with properties of Nambu - Goldstone modes.

It is shown, that the Principle of least action is a consequence of introduced 'Harmonization force (HaF)' of asymmetric Bivacuum. The system [Bivacuum + Matter] has a properties of the active medium, tending to Hidden harmony or Golden mean conditions under HaF influence. The mechanism of quantum entanglement between coherent particles, mediated by HaF and Virtual wave-guides (VirWG) is proposed also.

The pace of time for any closed and coherent system of particles is determined by the pace of translational longitudinal and translational contributions to kinetic energy change of this system, related to in-phase changes of electromagnetic and gravitational potentials, correspondingly. Our Unified model has been used also for possible explanation of Bivacuum mediated nonlocal mental interactions: www.emergentmind.org/PDF_files.htm/Kaiv290703.pdf.

For series of related papers see also: http://arXiv.org/find/physics/1/au: +Kaivarainen_A/0/1/0/all/0/1

1 Basic properties of Bivacuum

Unified Model (UM) represents the next stage of our efforts for unification of vacuum, matter, fields and time from few ground postulates (Kaivarainen, 2001b; 2002; 2003a,b). The new concept of Bivacuum has been introduced, as a dynamic cell-type matrix of the Universe with superfluid and nonlocal properties, composed from *microscopic* sub-quantum particles of the opposite energies, separated by energetic gap. We proceed from the important result of Dirac's theory, pointing to equal probability of positive and negative energy in the Universe. The collective quantum excitations of *sub-quantum particles and antiparticles* form the correlated pairs [*actual* rotor (V^+) + *complementary* antirotor (V^-)], which represent *mesoscopic double cells-dipoles*. The dimensions of sub-quantum particles are supposed to be of Plank length (10^{-33} cm).

The notions of quantum mechanics became applicable after collective excitations of subquantum particles acquire the ability to form a standing waves, like Bivacuum rotors and antirotors.

The *macroscopic* structure of Bivacuum is formed by the infinitive number of cells-dipoles, unified in form of virtual Bose condensate (VirBC) with nonlocal properties, due to their infinitive external translational de Broglie wave length. The rotor (V^+) and antirotor (V^-) of cell-dipoles have the opposite quantized energy, virtual mass, spin, charge and magnetic moments.

In symmetric *primordial* Bivacuum, i.e. in the absence of matter and fields, the absolute values of all these parameters in each dipole are equal. The radiuses of *primordial* rotor and antirotor are equal to Compton radius vortex: $[L^+ = L^- = L_0 = \hbar/m_0c]^i_{1,2,3}$, where m_0^i is the rest mass of the electrons of three leptons generation ($i = e, \mu, \tau$). Such a cells-dipoles are named Bivacuum fermions ($\mathrm{BVF}^\updownarrow = \mathbf{V}^+ \upuparrows \mathbf{V}^-$)

and Bivacuum antifermions ($\mathbf{BVF}^{\downarrow} = \mathbf{V}^{+} \downdownarrows \mathbf{V}^{-}$). Their opposite half integer spins $S = \pm\frac{1}{2}\hbar$, notated as ($\uparrow$ *and* $\downarrow$), depend on direction of clockwise or anticlockwise rotation of pairs of [rotor ($\mathbf{V}^{+}$) + antirotor ($\mathbf{V}^{-}$)], forming them. Bivacuum bosons ($\mathrm{BVB}^{\pm} = \mathbf{V}^{+} \Updownarrow \mathbf{V}^{-}$) represent the intermediate state between $\mathrm{BVF}^{\uparrow}$ and $\mathrm{BVF}^{\downarrow}$.

In *secondary* Bivacuum, in presence of matter and fields, the properties of rotors and antirotors do not compensate each other and $\mathrm{BVF}^{\updownarrow}$ and $\mathrm{BVB}^{\pm}$ turns to asymmetric. In such a conditions they acquire very small, but nonzero mass, momentum and charge.

In primordial Bivacuum, i.e. in the absence of matter and fields, the absolute values of quantized energies of rotors $\left(\left|+E_{\mathbf{V}}^{+}\right|_{n}^{e,\mu,\tau}\right)$ and antirotors $\left(\left|-E_{\mathbf{V}}^{-}\right|_{n}^{e,\mu,\tau}\right)$ are equal to each other and totally compensate each other:

$$\left|E_{\mathbf{V}}^{+}\right|_{n=0}^{e,\mu,\tau} = \left|-E_{\mathbf{V}}^{-}\right|_{n=0}^{e,\mu,\tau} = \frac{1}{2}\left|\pm m_{0}\right|^{e,\mu,\tau} c^{2} = \frac{1}{2}\left|\pm\hbar\omega_{0}^{e,\mu,\tau}\right| \tag{1}$$

$$or: \left|\pm E_{\mathbf{V}}^{\pm}\right|_{n}^{e,\mu,\tau} = \left|\pm m_{0}^{e,\mu,\tau}\right| c^{2}\left(\frac{1}{2}+n\right) = \left|\pm\hbar\omega_{0}^{e,\mu,\tau}\right|\left(\frac{1}{2}+n\right)$$

The energy of double cells - dipoles is a sum of [rotor + antirotor] energies. For symmetrical primordial Bivacuum it is equal to zero:

$$\left(E_{\mathbf{V}^{+}+\mathbf{V}^{-}}\right)_{n}^{e,\mu,\tau} = \left(E_{\mathbf{V}}^{+}\right)_{n}^{e,\mu,\tau} + \left(-E_{\mathbf{V}}^{-}\right)_{n}^{e,\mu,\tau} = 0$$

The total energy of primordial Bivacuum, as a sum of energy of all Bivacuum dipoles also is zero. The energetic gap $(A^{e,\mu,\tau})_{n}$, separating rotor and antirotor, i.e. the difference of their energies, in each double cell-dipole is equal to difference of their energy:

$$(A^{e,\mu,\tau})_{n} = [E_{\mathbf{V}}^{+} - (-E_{\mathbf{V}}^{-})]_{n}^{e,\mu,\tau} = \left|\pm m_{0}^{e,\mu,\tau}\right| c^{2}\,(2n+1) = \tag{1a}$$

$$= \left|\pm\hbar\omega_{0}^{e,\mu,\tau}\right|(2n+1) = \frac{\hbar c}{[L_{V^{+}\|V^{-}}]_{n}^{e,\mu,\tau}} \tag{1b}$$

where the effective quantized radius of vortices of pairs of [rotor (V_{n}^{+}) +antirotor (V_{n}^{-}], equal to that of $\mathrm{BVF}^{\updownarrow}$ and $\mathrm{BVB}^{\pm}$, is defined as:

$$[L_{V^{+}\|V^{-}}]_{n}^{e,\mu,\tau} = [L_{BVF^{\updownarrow};\,BVB^{\pm}}]_{n}^{e,\mu,\tau} = \frac{\hbar c}{m_{0}^{e,\mu,\tau}\, c^{2}(2n+1)} \tag{1c}$$

The double cells-dipoles of different excitation states ($\Delta n \neq 0$) are in the process of permanent dynamic exchange interaction, following by absorption and emission of virtual clouds ($\mathrm{VC}^{\pm}$) of sub-quantum particles and antiparticles in a course of transitions between the rotors and antirotors states of different quantum numbers: $n = 0, 1, 2, 3...$

The resulting energy and momentum of primordial Bivacuum keeps constant in a course of strictly correlated spontaneous transitions in two parts of double cells, corresponding to positive (+) and negative (-) energy, because they compensate each other. The sequential excitation/relaxation of double cells of Bivacuum is followed by virtual pressure waves (VPW^+ and VPW^-) excitation with fundamental frequency:

$$\omega_0^{e,\mu,\tau} = m_0^{e,\mu,\tau} c^2/\hbar \tag{1d}$$

Virtual particles and antiparticles in our model are the result of certain combinations of virtual clouds $(\mathbf{VC}_{j,k}^{+} \sim V_j^+ - V_k^+)^i$ and anti clouds $(\mathbf{VC}_{j,k}^{-} \sim V_j^- - V_k^-)^i$, composed from sub-quantum particles. Virtual clouds and anti clouds emission/absorption represents a correlated transitions between different excitation states (j,k) of rotors $(V_{j,k}^+)^i$ and antirotors $(V_{j,k}^-)^i$ of Bivacuum dipoles $[BVF^{\updownarrow}]^i$ and $[BVB^{\pm}]^i$. Three generation of Bivacuum fermions correspond to three lepton generation $(i = e, \mu, \tau)$.

The process of $[creation \rightleftharpoons annihilation]$ of virtual clouds is accompanied by oscillation of *virtual pressure* ($VP^{\pm}$) *in form of positive and negative virtual pressure waves* (VPW^+ *and* VPW^-), forming in certain conditions the autowaves in Bivacuum with properties of active medium.

In primordial Bivacuum the virtual pressure waves: VPW^+ and VPW^-, emitted/absorbed in a course of exchange interaction between $[BVF^{\uparrow}$ and $BVF^{\downarrow}]^i$ of opposite spins, totally compensate each other. However, in asymmetric secondary Bivacuum, in presence of matter and fields such a compensation of virtual clouds $(\mathbf{VC}_{j,k}^{+})$ and anti clouds $(\mathbf{VC}_{j,k}^{-})$ with positive [⟳] or negative [⟲] angular moments (spins) is perturbed and the resulting pressure of virtual particles or antiparticles becomes nonzero.

In contrast to real particles, the virtual ones may exist only in the wave [W] phase, but not in corpuscular [C] phase (see Section 3). It is a reason, why $[VPW^{\pm}]$ and their superposition in form of the virtual autowaves do not obey the laws of relativist mechanics and causality principle.

The correlated **virtual Cooper pairs** of Bivacuum fermions (BVF) with opposite spins ($S = \pm\frac{1}{2}\hbar$) :

$$[\mathbf{BVF}^{\uparrow} \bowtie \mathbf{BVF}^{\downarrow}]_{S=0} \equiv [(\mathbf{V}^+ \upuparrows \mathbf{V}^-) \bowtie (\mathbf{V}^+ \downdownarrows \mathbf{V}^-)]_{S=0} \tag{2}$$

Such pairs are bosons and have a properties of massless *Goldstone bosons* with zero spin: $S = 0$.

Superposition of their virtual clouds $(\mathbf{VC}_{j,k}^{\pm})$, emitted and absorbed in a course of correlated transitions of $[\mathbf{BVF}^{\uparrow} \bowtie \mathbf{BVF}^{\downarrow}]_{S=0}^{j,k}$ between (j) and (k) sublevels compensate the virtual energy of each other - totally in primordial Bivacuum and partly in secondary Bivacuum. The latter case is a reason for the excessive virtual pressure origination: $\mathbf{\Delta VP}^{\pm} = |\mathbf{VP}^{+} - \mathbf{VP}^{-}| \sim \left|\mathbf{VC}_{j,k}^{+} - \mathbf{VC}_{j,k}^{-}\right|_{S=0} \geq 0$:

$$\left(BVF^{\uparrow}\right)^{j,k} \equiv (\mathbf{V}^{+} \Uparrow \mathbf{V}^{-})^{j,k} \tag{2a}$$

$$\Updownarrow \qquad \rightleftharpoons \left| \left(\mathbf{VC}^{+}_{j,k}\right)^{\circlearrowright} - \left(\mathbf{VC}^{-}_{j,k}\right)^{\circlearrowleft} \right|^{j,k}_{S=0} \geq 0$$

$$\left(BVF^{\downarrow}\right)^{j,k} \equiv (\mathbf{V}^{+} \Downarrow \mathbf{V}^{-})^{j,k} \tag{2b}$$

The Goldstone modes in Bivacuum is a result of collective excitations of system of virtual bosons $[BVF^{\uparrow} \bowtie BVF^{\downarrow}]_{S=0}$. The energy distribution in a system of weakly interacting bosons (ideal gas), described by Bose-Einstein statistics, do not work for *Goldstone bosons* of Bivacuum due to strong coupling of pairs of $[BVF^{\uparrow} \bowtie BVF^{\downarrow}]_{S=0}$, forming virtual Bose condensate (VirBC) with nonlocal properties.

Each of Bivacuum fermions, forming Goldstone bosons, has a properties of *Goldstone fermions (Goldstino)*. In the absence of Bivacuum supersymmetry breach (in primordial Bivacuum), the $BVF^{\uparrow}$ and $BVF^{\downarrow}$ are the massless neutral particles.

In secondary Bivacuum, when supersymmetry is broken under the influence of external fields, the repulsion between $\text{BVF}^{\updownarrow}$ of the same spins is a consequence of Pauli principle action, based, in accordance to our UM, on the effect of excluded volume, induced by simultaneous emission of virtual clouds ($\text{VC}^{\updownarrow}$), as a result of their in-phase $[\text{C} \rightarrow W]$ transitions.

In secondary Bivacuum, i.e. in presence of matter and fields, their external group velocity ($\mathbf{v} \geq \mathbf{0}$), inertial mass and resulting charge becomes nonzero (Kaivarainen, 2002). The corresponding de Broglie wave length, which determines the dimensions of virtual Bose condensate, is equal to:

$$\lambda_{\mathbf{VirBC}} = \frac{\mathbf{h}}{\mathbf{m}_{C}^{+}\mathbf{v}^{2}/\mathbf{c}} \leq \infty \tag{3}$$

In accordance to our models of Bivacuum, this length, which determines the region of nonlocality, is infinitive in primordial Bivacuum, and has the huge cosmic scale in secondary Bivacuum.

This means, that the *Pauli repulsion* between the $\text{BVF}^{\updownarrow}$ of parallel spins may be realized on very big distances, determined by linear dimensions of domains of virtual Bose condensate (VirBC), formed by 2D virtual sheets, composed by Bivacuum fermions with opposite spins, forming *virtual Cooper pairs*: $[BVF^{\uparrow} \bowtie BVF^{\downarrow}]_{S=0}$ in Bivacuum. The interaction between such domains depends on their boarding conditions and can be realized by Josephson junctions.

Changing the dimensions of VirBC domains by increasing or decreasing of Bivacuum symmetry shift, which determines the space curvature, is interrelated with attraction - repulsion forces equilibrium in Bivacuum.

The antiparallel orientation of spins of Bivacuum fermions (BVF) in the case of the exchange spin-spin interaction, provides the *attraction* in pairs $[BVF^{\uparrow} \bowtie BVF^{\downarrow}]_{S=0}$ and 'contraction' of VirBC domains.

The parallel spin orientation of BVF provides the *repulsion* between them in accordance to Pauli principle and expansion (swelling) of VirBC domains of Bivacuum. The Fermi-Dirac statistics of energy distribution valid in a system of weakly interacting fermions (ideal gas of fermions) do not work for Bivacuum fermions ($\mathrm{BVF}^{\updownarrow}$) in Bivacuum, where they are strongly correlated.

The contraction and repulsion of Bivacuum in accordance to mechanisms described above, can be responsible for the attractive dark matter and repulsive dark energy. Jack Sarfatti also interrelates the corresponding negative and positive values of cosmological constant ($\pm\Lambda$) with 'exotic' properties of vacuum (http://qedcorp.com/APS/Ukraine.doc to be published in Progress in Quantum Physics Research (Nova).

Our Bivacuum concept has some similarity with Krasnoholvets (2000) approach. He regarded a vacuum also as a cellular space, each cell representing a 'superparticle' with dimension of 10^{-28} cm. Interaction of moving actual particle with superparticles is accompanied by emission and absorption of elementary virtual excitations - *inertons* by particle. The cloud in inertons resembles cumulative virtual cloud (CVC) of sub-quantum particles, representing [W] phase of sub-elementary particles.

2 Creation of sub-elementary particles & antiparticles

The sub-elementary *fermions and antifermions* ($\mathbf{F}_{\updownarrow}^{+}$ and $\mathbf{F}_{\updownarrow}^{-}$) of the opposite charge (+/-) and energy emerge due to stable symmetry violation between the *actual* (V^{+}) and *complementary* (V^{-}) rotors of $\mathrm{BVF}^{\updownarrow}$ cells-dipoles: [$\mathrm{BVF}^{\updownarrow}$ $\rightarrow \mathbf{F}_{\updownarrow}^{\pm}$]. Such a stability of symmetry shift corresponds to realization of Golden mean conditions $\left[\phi = 0.618 = (\mathbf{v}^{\phi}/\mathbf{c})^2\right]$, when the difference between the *rotating* kinetic energies of the actual and complementary states is equal to the energy of rest mass (m_0c^2) of sub-elementary particles:

$$\left(\left|m_C^{+}c^2\right| - \left|m_C^{-}c^2\right|\right)_{rot}^{\phi} = m_0c^2 \tag{3a}$$

The spatial image of [C] phase of sub-elementary particle (Fig. 5.1) represents the [actual rotor + complementary vortex] dipole, corresponding to the [actual mass (m_C^{+}) + complementary mass (m_C^{-})] dipole. The spatial image of [W] phase in form of cumulative virtual cloud (CVC) of sub-quantum particles is a parted hyperboloid (http://arXiv.org/abs/physics/0207027).

The sub-elementary particles: *fermions and antifermions* ($\mathbf{F}_{\updownarrow}^{+}$ and $\mathbf{F}_{\updownarrow}^{-}$) of the opposite charge (+/-) and energy, composing the matter, emerge due to stable symmetry violation between the actual (V^{+}) and complementary (V^{-}) rotors of $\mathrm{BVF}^{\updownarrow}$ cells-dipoles: [$\mathrm{BVF}^{\updownarrow}$ $\rightarrow \mathbf{F}_{\updownarrow}^{\pm}$] (Fig .1).]

Asymmetric [vortex + rotor] dipoles or sub-elementary particles, forming elementary particles, get the ability to move as respect to symmetric ones with external group velocity $v_{gr}^{ext} > 0$. The pulsation between such asymmetric (excited) and

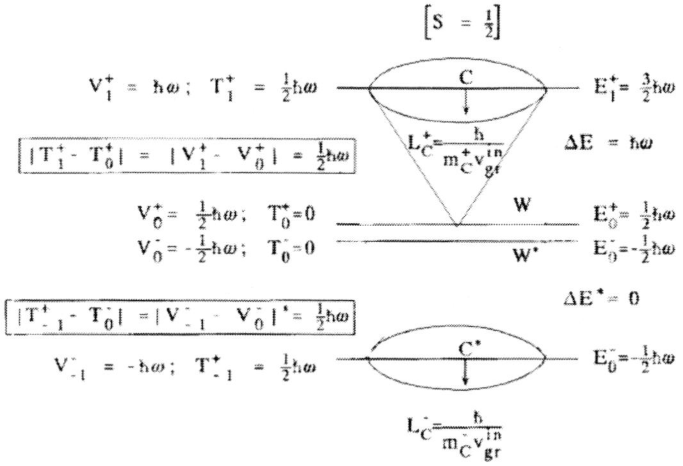

Figure 1: The spatial image of [C] phase of sub-elementary particle in form of [actual rotor + complementary vortex] dipole, corresponding to the [actual mass (m_C^+) + complementary mass (m_C^-)] dipole (http://arXiv.org/abs/physics/0207027).

former symmetric (ground) shape of double cells represents, in accordance to our Unified model, the [corpuscle (C) $\rightleftharpoons$ wave (W)] transitions, described in section 6.

3 Fusion of elementary particles from sub-elementary particles

The triplets of sub-elementary particles/antiparticles: $\langle [F_{\updownarrow}^- \bowtie F_{\updownarrow}^+] + F_{\updownarrow}^{\pm} \rangle^i$, corresponding to three lepton generation $(i = e, \mu, \tau)$ build elementary particles, like electrons, positrons, photons and quarks. The systems of *asymmetric* double cells in form of sub-elementary and elementary particles, atoms and molecules is dissipative and is not more superfluid.

The electron and positron of each generation, in accordance to our model, are the triplets of sub-elementary fermions/antifermions with certain spin orientations:

$$\langle [\mathbf{F}_{\uparrow}^- \bowtie \mathbf{F}_{\downarrow}^+] + \mathbf{F}_{\downarrow}^- \rangle_{e,\mu,\tau} \qquad (electron) \tag{4}$$

$$\langle [\mathbf{F}_{\uparrow}^- \bowtie \mathbf{F}_{\downarrow}^+] + \mathbf{F}_{\uparrow}^+ \rangle_{e,\mu,\tau} \qquad (positron) \tag{4a}$$

formed by pair of [sub-elementary fermion + sub-elementary antifermion] of opposite spins and charges $(\mathbf{F}_{\uparrow}^-$ and $\mathbf{F}_{\downarrow}^+)$ and one sub-elementary fermion $(\mathbf{F}_{\updownarrow}^-)$ or antifermion $(\mathbf{F}_{\updownarrow}^+)$, with two spins $(\pm\frac{1}{2}$, defined as $\uparrow$ and $\downarrow)$, correspondingly. The notation $[\bowtie]$ means that $[C \rightleftharpoons W]$ pulsation of $\mathbf{F}_{\uparrow}^-$ and $\mathbf{F}_{\downarrow}^+$ in pairs $[\mathbf{F}_{\uparrow}^- \bowtie \mathbf{F}_{\downarrow}^+]$ are in-phase with each other and in counterphase with $\mathbf{F}_{\updownarrow}^{\pm}\rangle$. Two sub-elementary fermions in composition of electron $\mathbf{F}_{\uparrow}^- + \mathbf{F}_{\downarrow}^-$ and two sub-elementary antifermions in composition of positron $\mathbf{F}_{\downarrow}^{+} + \mathbf{F}_{\uparrow}^{+}$ have the opposite spins. This means that their $[C \rightleftharpoons W]$ pulsations are counterphase and their are spatially compatible.

The external properties of the electrons and positrons, like mass, spin, charge is determined by uncompensated sub-elementary particle $[\mathbf{F}_{\updownarrow}^{-}>$ *or sub-elementary antiparticle* $[\mathbf{F}_{\updownarrow}^{+}>$.

Photon we introduce, as a superposition of electron and positron in form of three coherent pairs: $\left[3\,\text{sub-elementary fermion}(\mathbf{F}_{\updownarrow}^{-}) + 3\,\text{sub-elementary antifermion}\,(\mathbf{F}_{\updownarrow}^{+})\right]$ with boson properties and resulting spin $J = 1$. *The main difference between bosons and fermions is that the former particles are composed from equal number of standing sub-elementary fermions/antifermions:* $F_{\uparrow}^{-}$ *and* $F_{\downarrow}^{+}$ *and the latter ones - from their non equal number.* In accordance to model, the symmetry of photons - is a factor, which determines their propagation in Bivacuum with light velocity, in contrast to asymmetric fermions. Two possible structure of photon ($S = \pm 1\hbar$), corresponding to its two polarization and spin states can be presented as:

$$\langle 2[\mathbf{F}_{\uparrow}^{-} \bowtie \mathbf{F}_{\downarrow}^{+}] + [\mathbf{F}_{\downarrow}^{+} + \mathbf{F}_{\downarrow}^{-}]\rangle \qquad S = -1 \tag{4b}$$

$$\langle 2[\mathbf{F}_{\uparrow}^{-} \bowtie \mathbf{F}_{\downarrow}^{+}] + [\mathbf{F}_{\uparrow}^{+} + \mathbf{F}_{\uparrow}^{-}]\rangle \qquad S = +1 \tag{4c}$$

The u- quark with charge $\mathbf{Z} = +\frac{2}{3}$ is considered in our model, as the asymmetric superposition of two positron - like structures of heavy μ and/or τ generation:

$$\mathbf{u} \sim [\mathbf{e}^{+} + \mathbf{e}^{+}]^{\mu,\tau} = 2\langle[\mathbf{F}_{\uparrow}^{-} \bowtie \mathbf{F}_{\downarrow}^{+}] + \mathbf{F}_{\downarrow}^{+}\rangle^{\mu,\tau} \tag{4d}$$

The d**- quark** with charge $\mathbf{Z} = -\frac{1}{3}$ can be presented as asymmetric superposition of two electrons and one positron - like structures of μ and/or τ generation:

$$\mathbf{d} \sim [\mathbf{2e}^{-} + \mathbf{1e}^{+}]^{\mu,\tau} = \left\{2\langle[\mathbf{F}_{\uparrow}^{-} \bowtie \mathbf{F}_{\downarrow}^{+}] + \mathbf{F}_{\downarrow}^{-}\rangle + \langle[\mathbf{F}_{\uparrow}^{-} \bowtie \mathbf{F}_{\downarrow}^{+}] + \mathbf{F}_{\uparrow}^{+}\rangle\right\}^{\mu,\tau} \tag{4e}$$

Each of excessive standing sub-elementary particles: $\mathbf{F}^{+}$ and $\mathbf{F}^{-}$ in quark - has an electric charge (Z), equal to +1/3 and -1/3 correspondingly. The electron-positron structure of quarks is formed by sub-elementary particles/antiparticles of $[\mu$ *and/or* $\tau]$ generation, much heavier, than $[e]-$ generation.

In our model, the proton with charge $\mathbf{Z} = +1$:

$$\mathbf{p} = [\mathbf{2u} + \mathbf{d}]^{\mu,\tau} \tag{4f}$$

contains more standing sub-elementary fermions ($\mathbf{12F}^{+}$), than that sub-elementary antifermions ($9\mathbf{F}^{-}$). Each proton contains three excessive standing sub-elementary fermions ($\mathbf{F}^{+}$). The excessive number of ($\mathbf{F}^{+}$) is compensated in the Universe by corresponding number of ($\mathbf{F}^{-}$) in form of excessive number of free electrons. The resulting spin and charge of proton is equal and opposite to that of the electron.

The neutron with charge $\mathbf{Z} = 0$: $\mathbf{n} = [\mathbf{d} + \mathbf{2u}]^{\mu,\tau}$ is composed from the equal number of standing sub-elementary fermions and antifermions: ($12\mathbf{F}^{+}$) and ($12\mathbf{F}^{-}$).

The intermediate transition stage between opposite spin states sub-elementary fermion or antifermion ($S = +\frac{1}{2} \to S = -\frac{1}{2}$) is a sub-elementary boson of two possible polarization ($\mathbf{B}^-$ *and* $\mathbf{B}^+$):

$$\left[\mathbf{F}_{\uparrow}^{-} \rightleftharpoons \mathbf{B}^{-} \rightleftharpoons \mathbf{F}_{\downarrow}^{-}\right]^{e,\mu,\tau} \qquad \left[\mathbf{F}_{\uparrow}^{+} \rightleftharpoons \mathbf{B}^{+} \rightleftharpoons \mathbf{F}_{\downarrow}^{+}\right]^{e,\mu,\tau}$$

Possible mechanism of elementary particles fusion from two kinds of sub-elementary vortex-dipoles ($F_{\updownarrow}^{+}$ and $F_{\updownarrow}^{-}$) and their pairs $[F_{\updownarrow}^{+} \bowtie F_{\updownarrow}^{-}]$ in superfluid Bivacuum with gradient of symmetry shift may have same analogy with suggested by Schester and Dubin (1999), Jin and Dubin (2000) the "vortex crystal" formation.

The symmetry of our Bivacuum as respect to probability of sub-elementary particles and antiparticles creation, makes it principally different from *asymmetric Dirac's vacuum* (1958), with its realm of negative energy saturated with electrons. Positrons in his model represent the 'holes', originated as a result of the electrons jumps to realm of positive energy. Currently it is clear, that the Dirac's model of vacuum is not general enough to explain all know experimental data, for example, the bosons emergency.

The triplets of sub-elementray particles/antiparticles, like electron $<[F_{\updownarrow}^{-} \bowtie F_{\updownarrow}^{+}] + F_{\updownarrow}^{-}>^{i}$ and the counterphase $[C \rightleftharpoons W]$ pulsation of uncompensated $F_{\updownarrow}^{-}>^{i}$ and pair $[F_{\updownarrow}^{-} \bowtie F_{\updownarrow}^{+}]^{i}$, responsible for interaction of particles with Bivacuum, can be presented like:

Consequently, in our UM the matter and antimatter are unified on sub-elementary level, providing stability of elementary particles and antiparticles.

The in-phase $[C \rightleftharpoons W]$ pulsation of pairs $[\mathbf{F}_{\uparrow}^{-} \bowtie \mathbf{F}_{\downarrow}^{+}]$ of triplets provides the dynamic exchange interaction of elementary particles with Bivacuum and modulation of Bivacuum *virtual pressure waves* (VPW$^-$ *and* VPW$^+$).

The structure of triplets is stabilized by exchange of Cumulative Virtual Clouds (CVC) of sub-quantum particles between two sub-elementary fermions or antifermions of the opposite spins: $[\mathbf{F}_{\downarrow}^{+}]$ and $\mathbf{F}_{\uparrow}^{+}\rangle$ or $[\mathbf{F}_{\uparrow}^{-}]$ and $\mathbf{F}_{\downarrow}^{-}\rangle$ in a course of their *counterphase* $[C \rightleftharpoons W]$ pulsation. Stabilization of pair of sub-elementary fermion and antifermion $[\mathbf{F}_{\uparrow}^{-} \bowtie \mathbf{F}_{\downarrow}^{+}]$ or $[\mathbf{F}_{\downarrow}^{+} \bowtie \mathbf{F}_{\uparrow}^{-}]$, pulsing in-phase, occur due to minimization of local Bivacuum energy/symmetry shift, reflecting spatially localized energy conservation.

We assume, that the orientation of cell-dipoles in triplets is normal to each other: the uncompensated $[\mathbf{F}_{\updownarrow}^{\pm}\rangle_z$ is oriented along the $[z]$ axe, coinciding with direction of triplets external momentum, the sub-elementary particle $\mathbf{F}_{\updownarrow}^{+}$ is oriented along axe [y] and sub-elementary antiparticle $\mathbf{F}_{\updownarrow}^{-}$ is oriented along axe [x].

Our 3-dimensional space can be created by triplets of sub-elementary particles and the interaction of their $[C \rightleftharpoons W]$ *pulsation of* $[\mathbf{F}_{\downarrow}^{+} \bowtie \mathbf{F}_{\uparrow}^{-}]$ *with Bivacuum, which determines the number of dimensions of the observed matter and Bivacuum.*

The existence of different 3D structures of virtual autowaves, formed by VPW$^{\pm}$, modulated by external EM, gravitational fields and matter dynamics, are also the

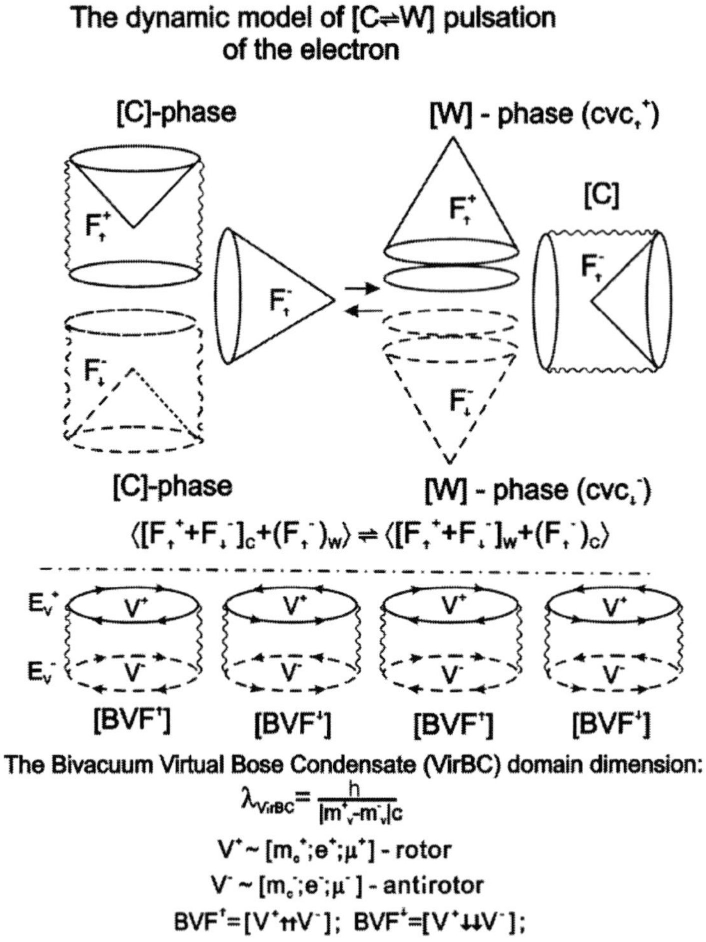

Figure 2: The counterphase $[C \rightleftharpoons W]$ pulsation of uncompensated $F_{\updownarrow}^{-} >$ and pair $[F_{\updownarrow}^{-} \bowtie F_{\updownarrow}^{+}]$ of the electron $<[F_{\updownarrow}^{-} \bowtie F_{\updownarrow}^{+}] + F_{\updownarrow}^{-} >$. The properties of $F_{\updownarrow}^{-} >$ determines the real (measurable) properties of particles, its interaction with other particles and fields. The parameters of sub-elementary particle and sub-elementary antiparticle of pairs compensate each other. The properties of $F_{\updownarrow}^{-} >$ and $[F_{\updownarrow}^{-} \bowtie F_{\updownarrow}^{+}]$ are strongly interrelated and affect the symmetry and dynamics of pairs of Bivacuum fermions ($\mathrm{BVF}^{\uparrow} \Leftrightarrow \mathrm{BVF}^{\downarrow}$), forming virtual Bose condensate with nonlocal properties. The in-phase $[C \rightleftharpoons W]$ pulsation of $[F_{\updownarrow}^{-} \bowtie F_{\updownarrow}^{+}]$ is responsible for interaction of particles with Bivacuum and the Virtual replicas of matter creation (Kaivarainen, 2003: http://www.karelia.ru/~alexk/new_articles/index.html see paper #4.2).

important feature of secondary Bivacuum. The notion of *Virtual Replica (VR)* of condensed matter is introduced, as a multidimensional standing VPW$^{\pm}$, forming the autowaves in Bivacuum under the influence of hierarchy of quantum and molecular dynamics of matter (Kaivarainen, 2001; 2002; 2003a,b).

4 Two Conservation Rules for Bivacuum Fermions (BVF$^{\updownarrow}$) and sub-Elementary Particles (F$^{\pm}_{\updownarrow}$), as a Mass, Magnetic and Electric Dipoles

Two internal conservation rules, responsible for stability of BVF$^{\updownarrow}$ and **sub-elementary** particles and antiparticles ($\mathbf{F}^{+}_{\updownarrow}$ and $\mathbf{F}^{-}_{\updownarrow}$), forming elementary particles of all three generations ($i = e, \mu, \tau$), are postulated in our Unified Model (UM).

I. *Conservation rule of the actual and complementary internal kinetic energies of vortex and antivortex: V^+ and V^- of $BVF^{\updownarrow} = [V^+ \Updownarrow V^-]^i$ and their asymmetric vortex and rotor states of $F^{\pm}_{\updownarrow} = [V^+ \Updownarrow V^-]^{i*}$*, correspondingly, in form of equality of modules of the internal actual $\left|2T^{+}_{kin}\right|^{in}$ and complementary $\left|-2T^{-}_{kin}\right|^{in}$ kinetic energies to the rest mass energy (m_0c^2):

$$\left[\left|2T^{+}_{kin}\right|^{in} = \left|m^{+}_{C}\right| \left(\mathbf{v}^{in}_{gr}\right)^2 = \left|-2T^{-}_{kin}\right|^{in} = \left|-m^{-}_{C}\right| \left(\mathbf{v}^{in}_{ph}\right)^2 = m_0c^2 = const\right]^i \tag{5}$$

where the product if *internal* group ($\mathbf{v}^{in}_{gr}$) and phase (v^{in}_{ph}) velocities is equal to product of *external* group ($\mathbf{v}_{gr} \equiv \mathbf{v}^{ext}_{gr}$) and phase ($\mathbf{v}_{ph} \equiv \mathbf{v}^{ext}_{ph}$) velocities of sub-elementary particle in composition of elementary particle:

$$\mathbf{v}^{in}_{gr}\mathbf{v}^{in}_{ph} = \mathbf{v}_{gr}\mathbf{v}_{ph} = c^2 \tag{5a}$$

From (5), taking into account (5a), we get for the ratio of complementary (m^{-}_{C}) and actual (m^{+}_{C}) mass of sub-elementary particle:

$$\frac{\left|m^{-}_{C}\right|}{\left|m^{+}_{C}\right|} = \left[\frac{\mathbf{v}^{in}_{gr}}{\mathbf{v}^{in}_{ph}}\right]^2 = \left[\frac{\left(\mathbf{v}^{in}_{gr}\right)^2}{c^2}\right]^2 \tag{6}$$

The resulting internal momentum of sub-elementary fermion/antifermion squared ($P_0^2 = m_0^2 c^2$) is permanent and equal to Compton's one:

$$P_0^2 = P^+P^- = (m^{+}_{C}\mathbf{v}^{in}_{gr})(\left|-m^{-}_{C}\right| \mathbf{v}^{in}_{ph}) = (m^{+}_{C}\mathbf{v}_{gr})(\left|-m^{-}_{C}\right| \mathbf{v}_{ph}) = \tag{7}$$

$$= m_0^2 c^2 = \frac{\hbar^2}{L_0^2} = const; \qquad P_0 = m_0c = m_0\omega_0L_0 \tag{7a}$$

where the permanent resulting radius of sub-elementary particle, as a [vortex + rotor] dipole is equal to Compton vorticity radius, determined by particle's rest mass (m_0):

$$L_0 = \frac{\hbar^2}{m_0 c} = (L^+ L^-)^{1/2} \tag{8}$$

where for each sub-elementary particle, the radius of actual vortex is $L^+ = \hbar/(m_C^+ \mathbf{v}_{gr}^{in}) = \hbar/P^+$ and the radius of complementary rotor: $L^- = \hbar/(\left|-m_C^-\right| v_{ph}^{in}) = \hbar/P^-$.

As far from (5) we have:

$$(2T_k^+)^{in} = \frac{(P^+)^2}{m_C^+} = (2T_k^-)^{in} = \frac{(P^-)^2}{\left|-m_C^-\right|} = m_0 c^2 \tag{9}$$

we get for the ratio of cross section of the actual vortex $[S^+ = \pi (L^+)^2]$ and complementary rotor $[S^- = \pi (L^-)^2]$:

$$\frac{S^+}{S^-} = \frac{\pi (L^+)^2}{\pi (L^-)^2} = \frac{(P^-)^2}{(P^+)^2} = \frac{\left|-m_C^-\right|}{m_C^+} = 1 - (\mathbf{v}/c)^2 \tag{10}$$

where, in accordance to our model: $\left|-m_C^-\right| = m_0[1 - (\mathbf{v}/c)^2]^{1/2}$ *and* $m_C^+ = m_0/[1 - (\mathbf{v}/c)^2]^{1/2}$

In *primordial Bivacuum, when sub-elementary particles $F_\updownarrow^\pm$ are absent,* the properties of rotors and antirotors of $\mathrm{BVF}^\updownarrow$ are characterized by equalities:

$$m_C^+ = \left|-m_C^-\right| = m_0 \tag{10a}$$

$$\mathbf{v}_{gr}^{in} = \mathbf{v}_{ph}^{in} = c$$

$$(2T_k^+)^{in} = (2T_k^-)^{in} = (2T_k^0) = P_0^2/m_0 \tag{10b}$$

$$where: \; P_0 = m_0 \omega_0 L_0 = m_0 c \tag{10c}$$

In slightly asymmetric *secondary Bivacuum* in presence of matter and fields, the equalities (5.10a) for $\mathrm{BVF}^\updownarrow$ are perturbed or broken.

II. *Conservation of the absolute values of the internal actual (μ_+) and complementary (μ_-) magnetic moments of vortex and antivortex: V^+ and V^- of Bivacuum fermions: $BVF^\updownarrow = [V^+ \Updownarrow V^-]^i$ and their asymmetric states: vortex and rotor of sub-elementary particles: $F_\updownarrow^\pm = [V^+ \Updownarrow V^-]^{i*}$*, correspondingly, in form of the equality of their modules to the Bohr magneton ($\mu_B^\pm$):

$$|\pm\mu_+| \equiv \frac{1}{2}|e_+| \frac{|\pm\hbar|}{\left|m_C^+\right| \mathbf{v}_{gr}^{in}} = |\pm\mu_-| \equiv \frac{1}{2}|-e_-| \frac{|\pm\hbar|}{\left|-m_C^-\right| \mathbf{v}_{ph}^{in}} = \mu_B \equiv \frac{1}{2}|e| \frac{\hbar}{m_0 c} = const \tag{11}$$

where: e_+ and e_- are the *internal* electric charges of actual vortex and complementary rotor, correspondingly; $|e|$ is a module of the resulting charge of the electron or positron.

The parameters: $|e_\pm|$, $|m_C^\pm|$ and v_{gr}^{in} are not permanent, in contrast to magnetic moments: $|\pm\mu_+| = |\pm\mu_-| = \mu_B$ and ratios:

$$\frac{|e_+|}{|m_C^+|\,\mathbf{v}_{gr}^{in}} = \frac{|-e_-|}{|-m_C^-|\,\mathbf{v}_{ph}^{in}} = const \tag{11a}$$

Such a difference between variable electric and permanent magnetic charges of Bivacuum explains the absence of MONOPOLE, because: $\Delta\mu_\pm = |\pm\mu_+| - |\pm\mu_-| = 0$ always.

For the case of *primordial Bivacuum* (in the absence of matter and fields), when $v = v^{ext} = 0$ and $v_{gr}^{in} = v_{ph}^{in} = c$, we have from (10) and (11) for BVF$^\updownarrow$:

$$|m_C^+| = |-m_C^-| = m_0 \tag{12}$$

$$|e_+| = |e_-| = e \tag{12a}$$

$$\mathbf{v}_{gr}^{in} = \mathbf{v}_{ph}^{in} = c \tag{12b}$$

$$|\pm\mu_+| = |\pm\mu_-| = \mu_B = const \tag{12c}$$

In slightly asymmetric *secondary Bivacuum* in presence of matter and fields, the equalities (12-12b) for BVF$^\updownarrow$ are broken, however 12c remains unchanged, as well, as for sub-elementary particles.

The resulting magnetic moments of sub-elementary fermion/antifermion ($\mu_F^\pm$), equal to the Bohr's magneton (μ_B), we get, as the actual $|\mu_+|$ and complementary $|\mu_-|$ components product average:

$$\mu_F^\pm = (|\mu_+|\,|\mu_-|)^{1/2} = \left[\left(\frac{|e|}{m_0 c}\right)^2 \frac{\hbar^2}{4}\right]^{1/2} = \frac{|e|}{m_0 c}\frac{\hbar}{2} = \mu_B = const \tag{13}$$

where: $|e|^2 = |e_+ e_-|$

For the other hand, the well known formula for the *normal* spin magnetic moment of the electron is:

$$\mu_S = \frac{e}{m_0 c}\mathbf{S} \tag{14}$$

where: $[e/m_0 c]$ is gyromagnetic ratio of the electron.

It follows from our model, that: $\mu_F^\pm = \mu_B = \mu_S^\pm$. Consequently, from eqs. (13 and 14) we get the value of the electron's spin and definition of the Plank constant, leading from our model of sub-elementary particles:

$$\mathbf{S} = \pm\frac{1}{2}\hbar \tag{15}$$

$$where:\ \pm\hbar = \pm\sqrt{|m_C^+|\,|i^2 m_C^-|\,(\mathbf{v}_{gr}^{in}\mathbf{v}_{ph}^{in})(L^+ L^-)} = \pm\sqrt{m_0^2 c^2 L_0^2} \tag{16}$$

From (11) we get, that the *internal resulting electric dipole* ($\mathbf{d}_{el}^{in}$) of sub-elementary particles/antiparticles are related to that of magnetic dipole and the Bohr magneton, as:

$$\left|\mathbf{d}_{el}^{in}\right| = \left[(|e_+|\left|\mathbf{L}^+\right|)(|-e_-|\left|\mathbf{L}^-\right|)\right]^{1/2} = eL_0 = 2\left|\mu_F^{\pm}\right| = 2\mu_B \tag{17}$$

On the distance $r >> L_0 = \frac{\hbar}{m_0 c}$, the electric and magnetic dipole radiations, emitted in a course of in-phase [C $\rightleftharpoons$ W] pulsation of sub-elementary particles or antiparticles should be equal, in accordance with existing theory of dipole radiation.

4.1 The Actual & Complementary Mass and Charge Compensation Principles: *the extension of the Einstein's and Dirac's formalism for free relativistic particles*

The *actual (inertial)* (m_C^+) *& complementary (inertialess)* ($-m_C^- = i^2 m_C^-$) mass compensation principle:

$$\left|m_C^+\right|\left|i^2 m_C^-\right| = m_0^2 \tag{18}$$

$$or: \ \left|m_C^+ m_C^-\right| = m_0^2 \tag{18a}$$

follows from the reverse relativist dependence of the actual and complementary masses on the external rotational-translational group velocity of sub-elementary particles:

$$\left|m_C^+\right| = m_0/[1-(\mathbf{v}/c)^2]^{1/2} \tag{19}$$

$$\left|m_C^-\right| = m_0[1-(\mathbf{v}/c)^2]^{1/2} \tag{19a}$$

From the ratio of (19a) to (19), we get the formula, similar to (10):

$$\frac{\left|m_C^-\right|}{\left|m_C^+\right|} = 1-(\mathbf{v}/c)^2 = \frac{S^+}{S^-} \tag{20}$$

$$or: (\mathbf{v}/c)^2 = 1 - \frac{\left|m_C^-\right|}{\left|m_C^+\right|} \tag{20a}$$

The difference between the absolute values of the actual and complementary masses is:

$$\left|m_C^+\right| - \left|m_C^-\right| = \frac{m_0(\mathbf{v}/\mathbf{c})^2}{[1-(\mathbf{v}/c)^2]^{1/2}} = m_C^+(\mathbf{v}/\mathbf{c})^2 \tag{20b}$$

The eqs. 19 and 19a a can be transformed to following shape:

$$\left(E_C^+\right)^2 = (m_C^+)^2 c^4 = m_0^2 c^4 + (m_C^+ v)^2 c^2 \tag{21}$$

$$\left(E_C^-\right)^2 = (m_C^-)^2 c^4 = m_0^2 c^4 - (m_0 v)^2 c^2 \tag{21a}$$

where: E_C^+ and E_C^- are the actual and complementary energy of wave B, correspondingly.

The first eq. (21) coincides with those, obtained by Dirac. The second (21a) for complementary energy is a new one and reflects the generalization of special theory of relativity and Dirac's theory for relativist particles.

From (11; 5a and 18) follows the internal *actual & complementary charge compensation principle*, symmetric to *mass compensation principle*:

$$|e_+|\,|i^2 e_-| = [i^2 e]^2 \tag{22}$$

$$or: \quad |e_+ e_-| = (e)^2 \tag{22a}$$

The positive *actual* and negative *complementary* internal negative charges: $[e_+]$ and $[i^2 e_-]$, correspond to *vortex* and *rotor* of sub-elementary fermions.

One can see, that the rest mass squared (7) and resulting charge squared (11a) are not dependent on the external group velocity (v), i.e. they are relativist invariants.

5 The Relation Between the External and Internal Parameters of Elementary Particles

Combining eqs.(20 and 6), we get the formula for unification of the internal ($\mathbf{v}_{gr}^{in}$) and external group ($\mathbf{v}_{gr}^{ext} \equiv \mathbf{v}$) velocities of sub-elementary particles, as the asymmetric Bivacuum dipoles:

$$\frac{c}{\mathbf{v}_{gr}^{in}} = \left(\frac{\mathbf{v}_{ph}^{in}}{\mathbf{v}_{gr}^{in}}\right)^{1/2} = \frac{1}{[1-(\mathbf{v}/c)^2]^{1/4}} \tag{22b}$$

Taking into account (5.6; 5.11a and 5.22b) we get the important interrelations between the actual and complementary mass and charge of the asymmetric Bivacuum dipoles and dependence of these parameters ratio on their external group velocity ($\mathbf{v}$):

$$\left(\frac{m_C^+}{m_C^-}\right)^{1/2} = \frac{m_C^+}{m_0} = \frac{\mathbf{v}_{ph}^{in}}{\mathbf{v}_{gr}^{in}} = \left(\frac{c}{\mathbf{v}_{gr}^{in}}\right)^2 = \frac{|e_+|}{|e_-|} = \left(\frac{e_+}{e}\right)^2 = \frac{1}{[1-\mathbf{v}/\mathbf{v}_{ph}]^{1/2}} \tag{22c}$$

one of the consequence of (22c):

$$\frac{e_+}{e} = \frac{1}{[1-\mathbf{v}/\mathbf{v}_{ph}]^{1/4}} \tag{22d}$$

We can see from (22c), that at the external group velocity $\mathbf{v} = 0$, we have the conditions of symmetric Bivacuum double cells-dipoles (BVF), pertinent for primordial Bivacuum in the absence of matter:

$$m_C^+ = m_C^- = m_0; \quad \mathbf{v}_{ph}^{in} = \mathbf{v}_{gr}^{in} = \mathbf{c}; \quad |e_+| = |e_-| = e \quad at \quad \mathbf{v} = \mathbf{0} \tag{22e}$$

6 Corpuscle [C]-Wave [W] duality, as a Background of Electromagnetism and Gravitation

Duality of elementary particles and antiparticles in accordance to Unified model, is a consequence of coherent quantum beats of their sub-elementary particles/anti–particles, as asymmetric Bivacuum dipoles, between two states: the asymmetrically excited state $(BVF^{\updownarrow})^* \equiv \mathbf{F}_{\updownarrow}^{\pm}$ and its symmetric state $(BVF^{\updownarrow})$ - see Fig. 2:

$$\left[\mathbf{F}_{\updownarrow}^{\pm} \overset{\pm CVC}{\rightleftharpoons} BVF^{\updownarrow}\right]^i \tag{23}$$

where: i means *three electron's or positron's* generation: $i = e, \mu, \tau$.

These beats are accompanied by emission *and* absorption of cumulative virtual cloud (CVC) of sub-quantum particles by $BVF^{\updownarrow}$, oscillation of the mass and charge symmetry shift. The $[Corpuscle(C) \rightleftharpoons Wave(W)]$ pulsation of each sub-elementary particle is accompanied by the electric, magnetic and gravitational dipole radiation (Kaivarainen, 2002).

As far the energy of symmetric $BVF^{\updownarrow}$ in (23) is equal to zero, it means that the energy of corpuscular [C] phase, in form of sub-elementary particle $[\mathbf{F}_{\updownarrow}^{\pm}]$ is equal to energy of the wave [W] phase, in form of (CVC): $E_C = E_W$.

The energy of quantum beats in a course of $[C \rightleftharpoons W]$ pulsation of **sub-elementary particle** is equal to difference of energy between the absolute values of actual (vortex) and complementary (rotor) states (see Fig.1).

The energy of sub-elementary de Broglie wave in [C] and [W] phase, its relation to de Broglie wave frequency $(\omega_0 = \omega_{C \rightleftharpoons W})^i$ and the wave length $(\lambda_{C,W})$ can be presented as a sum of one rotational and two translational contributions in few different modes, including spin $(\mathbf{E_S})$, electromagnetic $(\mathbf{E_E})$ and gravitational $(\mathbf{E_G})$ contributions:

$$[E_{C\rightleftharpoons W} = \hbar\omega_{C\rightleftharpoons W} = E_C = E_W]^i = [|m_C^+|\, c^2 - |m_C^-|\, c^2]^i_{tot} = [m_C^+\mathbf{v}^2_{res}]^i \equiv \quad (24)$$

$$[m_C^+\mathbf{v}^2]^i = \left[(m_C^+\mathbf{v}^2_{rot})^\phi + m_C^+\mathbf{v}^2_{\|tr} + m_C^+\mathbf{v}^2_{\perp tr}\right]^i = \left\{m_C^+[(\mathbf{v}^2_{rot})^\phi + \mathbf{v}^2_{\|tr} + \mathbf{v}^{2i}_{\perp tr}]\right\}^i =$$

$$= [(E^S_{C,W})_{rot} + (E_{C,W})_{\|tr} + (E_{C,W})_{\perp tr}]^i = \quad (24a)$$

$$= \left[|m_C^+ - m_C^-|^\phi_{rot}\, c^2 + |m_C^+ - m_C^-|_{\|tr}\, c^2 + |m_C^+ - m_C^-|_{\perp tr}\, c^2\right]^i =$$

$$= \left[m_0\omega_0^2 L_0^2 + \alpha\frac{p^2_{tr}}{m_C^+} + \beta\frac{p^2_{tr}}{m_C^+}\right]^i = \left[2\,(T_k)_{rot} + 2\,(T_k)_{\|tr} + 2(T_k)_{\perp tr}\right]^i = \quad (24b)$$

$$= \left[m_0\omega_0^2 L_0^2 + m_0 c^2\frac{(\mathbf{v}_{\|tr}/c)^2}{[1-(\mathbf{v}/\mathbf{c})^2]^{1/2}} + m_0 c^2\frac{(\mathbf{v}_{\perp tr}/c)^2}{[1-(\mathbf{v}/\mathbf{c})^2]^{1/2}}\right]^i =$$

$$= [\mathbf{E_S} + \mathbf{E_E} + \mathbf{E_G}]^i \quad (24c)$$

where: $\mathbf{E_S}$ is the reversible rotational (spin) contribution to resulting energy of sub-elementary particle; $\mathbf{E_E}$ and $\mathbf{E_G}$ are irreversible electromagnetic and gravitational contributions to resulting energy (25c and 25d).

The longitudinal ($\|$) and transversal ($\perp$) contributions to translational energy of each sub-elementary particle:

$$\left[(E_C = E_W)_{\|,\perp,tr} = \frac{h^2}{m_C^+\lambda^2_{\|,\perp,tr}}\right]^i \quad where: \left[\lambda_{\|,\perp} = \frac{h}{m_C^+\mathbf{v}_{\|,\perp,tr}} = \frac{h}{|m_C^+ - m_C^-|\, c^2/\mathbf{v}_{\|,\perp,tr}}\right]^i \quad (25)$$

where: $p_{\|tr} = m_C^+\mathbf{v}_{\|tr}$ *and* $p_{\perp,tr} = m_C^+\mathbf{v}_{\perp,tr}$ are the external longitudinal and transversal translational momentums of particle; the resulting external group velocity, leading from (24) and (20), *is* :

$$\left[\mathbf{v}^2 = \mathbf{v}^2_{rot} + \mathbf{v}^2_{\|tr} + \mathbf{v}^2_{\perp tr} = \mathbf{c}^2\phi + \alpha\mathbf{v}^2 + \beta\mathbf{v}^2 = \mathbf{c}^2\left(1 - \frac{|m_C^-|}{|m_C^+|}\right)\right]^i \quad (25a)$$

where: $\mathbf{v}^2_{\|,tr} = \alpha\mathbf{v}^2$ and $\mathbf{v}^2_{\perp,tr} = \beta\mathbf{v}^2$ are the longitudinal and transversal external group velocities of elementary particle of $i - generation$ squared (see 25c and 25d).

The external rotational velocity of each sub-elementary particle, which determines the value of the rest mass ($m_0 = |m_C^+ - m_C^-|^\phi)^i$ always is equal to that of Golden mean (GM):

$$\mathbf{v}^2_{rot} = \mathbf{c}^2\phi = \mathbf{v}^2(1 - \alpha - \beta) = const \quad (25b)$$

The longitudinal ($m_C^+\mathbf{v}^2_{\|,tr} = \alpha m_C^+\mathbf{v}^2$) and transversal ($m_C^+\mathbf{v}^2_{\perp,tr} = \beta m_C^+\mathbf{v}^2$) translational contributions to the total actual energy of elementary particle:

$\hbar\omega_{C\rightleftharpoons W} = (m_C^+)_{rot,\,tr}\mathbf{v}^2$, which determine its maximum electromagnetic ($\mathbf{E}_E$) and gravitational ($\mathbf{E}_G$) potentials (Kaivarainen, 2002), can be presented, correspondingly, as:

$$\mathbf{E}_E = \frac{|e_+e_-|}{L^\pm} = \alpha\, m_C^+\mathbf{v}^2 = m_C^+\mathbf{v}^2_{\|,tr} = \frac{m_0\alpha\mathbf{v}^2}{[1-(\mathbf{v}/\mathbf{c})^2]^{1/2}} \tag{25c}$$

$$\mathbf{E}_G = G\frac{|m_C^+\,m_C^-|}{L^\pm} = \beta\, m_C^+\mathbf{v}^2 = m_C^+\mathbf{v}^2_{\perp,tr} = \frac{m_0\beta\mathbf{v}^2}{[1-(\mathbf{v}/\mathbf{c})^2]^{1/2}} \tag{25d}$$

where: $L^\pm = \hbar/[(m_C^+ - m_C^-)c = \hbar/[m_C^+\mathbf{v}(\mathbf{v}/\mathbf{c})]$ is a characteristic distance between the actual and complementary mass and charges of Bivacuum dipoles (sub-elementary fermions). It is hidden, in contrast to external actual de Broglie wave length of particle: $L^+ = \hbar/[m_C^+\mathbf{v}]$. Their ratio: $L^+/L^\pm = \mathbf{v}/\mathbf{c} \to 1\ at\ \mathbf{v} \to \mathbf{c}$;

$\alpha = e^2/\hbar c = (e/Q)^2 = 7.2973506 \cdot 10^{-3}$ is the electromagnetic fine structure constant;

$\beta^e = (m_0^e/M_{Pl})^2 = 1.7385 \cdot 10^{-45}$; $\beta^p = (m_0^p/M_{Pl})^2 = 5.86 \cdot 10^{-39}$

are the introduced in our theory gravitational fine structure constant, different for the electrons and protons; $M_{Pl} = (\hbar c/G)^{1/2} = 2.17671 \cdot 10^{-8}\,\mathrm{kg}$ is a Plank mass; $m_0^e = 9.109534 \cdot 10^{-31}\,kg$ is a rest mass of the electron; $m_0^p = 1.6726485 \cdot 10^{-27}\,kg = m_0^e \cdot 1.8361515 \cdot 10^3\,kg$ is a rest mass of proton.

We can see, that the both: EM and G potentials of elementary particle (25c and 25d) are tending to zero, at $\mathbf{v}^2_{\|tr} \to 0$ and $\mathbf{v}^2_{\perp tr} \to 0$. For the other hand, the translational acceleration, when $\mathbf{v}_{\|tr} \to c$ and $\mathbf{v}_{\perp,tr} \to c$ are accompanied by corresponding increasing of the resulting velocity (25a) and electromagnetic and gravitational potentials of particles.

The resulting actual energy from (24) can be expressed, as a sum of rotational - spin contribution, responsible for he rest mass origination and two translational contributions (25c and 25d), standing for electromagnetic and gravitational maximum potentials:

$$E_{C,W} = (m_C^+)_{res}\mathbf{v}^2 = [\mathbf{E_S} + \mathbf{E_E} + \mathbf{E_G}]^i = \tag{25e}$$

$$= m_0\omega_0^2 L_0^2 + \frac{m_0\alpha\mathbf{v}^2}{[1-(\mathbf{v}/\mathbf{c})^2]^{1/2}} + \frac{m_0\beta\mathbf{v}^2}{[1-(\mathbf{v}/\mathbf{c})^2]^{1/2}}$$

The rotational (spin) contribution to energy is determined by Golden mean conditions (see next section), as a resonant conditions of the exchange interaction elementary particles with Bivacuum virtual pressure waves ($\mathrm{VPW}^\pm$) with fundamental frequency ($\omega_0 = m_0c^2/\hbar$):

$$\left[\left(E^S_{C,W}\right)_{rot} = \left[m_C^+\,\mathbf{v}^2_{rot}\right]^\phi = m_0c^2 = \hbar\omega_0 = m_0\omega_0^2L_0^2 = \frac{p_0^2}{m_0} = \frac{\hbar^2}{m_0L_0^2}\right]^i \tag{26}$$

where the rest mass ($m_0 = |m_C^+ - m_C^-|^\phi$) is determined by difference of the actual vortex mass $|m_C^+|^\phi = m_0/\phi$ and complementary rotor mass $|m_C^-| = \phi m_0$ at Golden mean (GM) conditions $[(\mathbf{v}_{rot}/\mathbf{c})^2 = \phi = 0.618033$ see section 7]; the momentum of rotation, equal to spin momentum of Cumulative virtual clouds ($CVC^{\uparrow}$ or $CVC^{\downarrow}$), corresponding to [W] phase of particles is $p_0 = m_0\omega_0 L_0 = m_0 c$.

The rotational external group velocity of sub-elementary fermion at GM conditions is $\mathbf{v}_{rot}^{\phi} = c\phi^{1/2}$; the frequency of $[C \rightleftharpoons W]$ pulsation for each of $(i = e, \mu, \tau)$ generation of elementary particles is $\omega^i_{C\rightleftharpoons W} = \left[m_C^+ \mathbf{v}_{res}^2\right]^i /\hbar$; the resulting Compton radius of sub-elementary particle is $L_0 = (L_+ L_-)^{1/2} = \hbar/m_0 c$.

The *resulting external group velocity* of particle ($\mathbf{v}_{res}\equiv \mathbf{v}$), taking into account its rotational and translational dynamics, is determined by the ratio of resulting actual and complementary masses from (20 and 25b):

$$\mathbf{v}_{res}\equiv \mathbf{v} = (\mathbf{v}_{rot}^2 + \mathbf{v}_{\parallel tr}^2 + \mathbf{v}_{\perp tr}^2)^{1/2} = c\left[1 - \left|\frac{m_C^-}{m_C^+}\right|\right]^{1/2} \tag{26a}$$

The translational contribution to the total energy of particle is a sum of longitudinal $(\parallel, tr)$ and transversal $(\perp, tr)$ components, corresponding to EM and gravitational (G) contributions in [C] phase of particles from (25c and 25d):

$$\begin{aligned}(E_{C,W})_{tr} &= (E_{C,W})_{\parallel tr} + (E_{C,W})_{\perp tr} = \mathbf{E_E} + \mathbf{E_G} = \\ &= \alpha m_C^+ \mathbf{v}^2 + \beta m_C^+ \mathbf{v}^2 = m_C^+ \mathbf{v}_{\parallel}^2 + m_C^+ \mathbf{v}_{\perp}^2\end{aligned} \tag{26b}$$

Two external **translational** ($\parallel$ and $\perp$) momentums of triplets $\langle[\mathbf{F}_{\uparrow}^- \bowtie \mathbf{F}_{\downarrow}^+] + \mathbf{F}_{\updownarrow}^{\pm}\rangle$ of particles:

$$p_{\parallel,\perp} = m_C^+ \mathbf{v}_{\parallel,\perp} = m_0 \mathbf{v}_{\parallel,\perp}/[1 - (\mathbf{v}/\mathbf{c})^2]^{1/2} \tag{26c}$$

are subdivided to longitudinal $(\parallel)$ and transversal $(\perp)$ ones, as respect to vector of particle's resulting velocity $(\mathbf{v}_{res})$; $\mathbf{v}_{\parallel tr}$ is a longitudinal group velocity of particle's vibrations, induced by oscillation of momentum of uncompensated sub-elementary particle $\mathbf{F}_{\updownarrow}^{\pm}\rangle$, accompanied its$[C \rightleftharpoons W]$ pulsation; $\mathbf{v}_{\perp tr}$ is a transversal group velocity of particle's vibrations.

The virtual electromagnetic photons are emitted and absorbed in a course of $[C \rightleftharpoons W]$ pulsation of sub-elementary particles (fermions) in accordance with known mechanism of the *electric and magnetic dipole radiation, induced by charges acceleration.* The intensity of time-averaged of *electric dipole radiation of each of sub-elementary fermions* may be expressed like:

$$\varepsilon_{E.dip} = \frac{4e^2}{3c^3}\omega^4_{C\rightleftharpoons W}(L^{\pm})^2 = \frac{4}{3c^3}\omega^4_{C\rightleftharpoons W}\, d^2_{\mathbf{F}_{\updownarrow}^{\pm}} \tag{27}$$

where the frequency of $[C \rightleftharpoons W]$ pulsation, equal to that of dipole radiation is

$$\omega_{C\rightleftharpoons W} = \left|m_C^+ - m_C^-\right| c^2/\hbar = m_C^+ \mathbf{v}^2/\hbar$$

is the angle frequency of $[C \rightleftharpoons W]$ pulsation of $[\mathbf{F}_{\updownarrow}^{\pm} >$, equal to frequency of dipole oscillation; $e^2 = |e^- e^+|$ is a resulting charge squared.

The resulting internal dimension of uncompensated sub-elementary particle $[\mathbf{F}_{\updownarrow}^{\pm}>$ is defined as $L^{\pm} = \hbar/(|m_C^+ - m_C^-|\, c)$, equal at Golden mean (GM) conditions to Compton's length: $(L^{\pm})^{\phi} = \hbar/(m_0 c) = L_0$; the electric dipole moment $d_{\mathbf{F}_{\updownarrow}^{\pm}} = eL^{\pm}_{\cdot}$ is equal at GM conditions to

$$\left[d_{\mathbf{F}_{\updownarrow}^{\pm}} = eL^{\pm}_{\cdot}\right]^{\phi} = eL_0 = 2\mu_B$$

where the Bohr's magneton: $\mu_B = e\hbar/(2m_0 c)$.

The intensity of $\varepsilon_{E.dip}$ is maximum in direction, normal to direction of $[C \rightleftharpoons W]$ pulsation and zero along this direction. The set of expressions (24 - 27), in fact, unify the extended special theory of relativity with quantum mechanics, as a consequence of proposed mechanism of corpuscle - wave duality of elementary particles.

Our dynamic presentation of duality explains also the elementary particles, as a permanent sources of electromagnetic and gravitational energies. They are the result of energy redistribution between the negative and positive realms of secondary Bivacuum in a course of $[C \rightleftharpoons W]$ pulsation of unpaired sub-elementary fermions $\mathbf{F}_{\updownarrow}^{\pm}\rangle$, as a mass, electric and magnetic dipoles (Kaivarainen, 2002). The energy of sub-elementary fermions and sub-elementary antifermions in pairs of triplets $\langle[\mathbf{F}_{\uparrow}^{-} \bowtie \mathbf{F}_{\downarrow}^{+}] + \mathbf{F}_{\updownarrow}^{\pm}\rangle$ compensate each other.

7 Quantum Roots of Golden Mean (GM)

It was shown (Kaivarainen, 1995; 2002), that the quantum roots of the famous Golden mean, so widely used in Nature (see huge site of Dan Winter: http://www.soulinvitation.com/indexdw.html), are related to conditions of *Hidden Harmony:* the equality of the internal (*in*) and external (*ext*) group (v_{gr}) and phase ($\mathbf{v}_{ph}$) velocities of each sub-elementary particles/antiparticles, forming the elementary particles:

$$\left[\mathbf{v}_{gr}^{in} = \mathbf{v}_{gr}^{ext} \equiv \mathbf{v}\right] \quad and \quad \left[\mathbf{v}_{ph}^{in} = \mathbf{v}_{ph}^{ext} \equiv \mathbf{v}_{ph}\right] \qquad (27a)$$
$$at: \ \mathbf{v}_{gr}^{in}\,\mathbf{v}_{ph}^{in} = \mathbf{v}_{gr}^{ext}\,\mathbf{v}_{ph}^{ext} = c^2$$

These *Hidden Harmony conditions* turns eq. 22b: $\left(\frac{c}{\mathbf{v}_{gr}^{in}}\right)^2 = \frac{1}{[1-(\mathbf{v}/c)^2]^{1/2}}$ to simple quadratic equation:

$$\phi^2 + \phi - 1 = 0, \ \text{which has a few forms}: \ \phi = \frac{1}{\phi} - 1 \ \ or: \ \frac{\phi}{(1-\phi)^{1/2}} = 1 \qquad (27b)$$

$$where: \ \phi = \left(\frac{\mathbf{v}^2}{c^2}\right)^{ext,\,in} = \left[\frac{\mathbf{v}}{\mathbf{v}_{ph}}\right]^{ext,in} = 0.6180339887 \ \text{(Golden mean)} \qquad (27c)$$

The positive solution of equation (27b) is equal to *Golden mean* ($Psi \equiv \phi = 0.6180339887$).

Taking into account (27b), formula (24) at Hidden Harmony conditions can be transformed to:

$$[|m_C^+|^\phi - |m_C^-|^\phi]^i = \left[m_C^+ \left(\mathbf{v}_{res}^\phi/c\right)^2\right]^i = \frac{m_0\phi}{[1-\phi]^{1/2}} = m_0 = \frac{\hbar\omega_0}{c^2} \qquad (27\text{'c})$$

It is important result, pointing that the *origination of the rest mass* of elementary particles (m_0) is due to mass symmetry shift between the actual and complementary states of unpaired sub-elementary particle, corresponding to Golden mean (Hidden harmony) conditions and fundamental frequency (ω_0) of Bivacuum zero-point oscillations.

It is well known, that Golden mean value is related strongly to *Fibonacci series*:

$$n = 1, 2, 3, 5, 8, 13, 21, 34, 55...$$

where the value of next term of series is defined as a sum of two antecedent terms.

The bigger is number of series (n_j), the closer is its ratio to the next one ($n_{j+1} = n_j + n_{j-1}$) to Golden mean:

$$\frac{n_j}{n_{j+1}} \to 0.6180339887 \quad at \quad j \to \infty$$

At the Golden mean condition (27'c) the formulas for energy ($E_C^\phi = E_W^\phi$), mass $[m_C^+]^\phi$ and $[m_C^-]^\phi$, velocity ($\mathbf{v}^\phi$), the *resulting* momentum (P^ϕ) and de Broglie wave radius ($L^\phi = \lambda^\phi/2\pi$) of sub-elementary particle (eq. 24 and its parameters) turn to the elegant quantitative relations:

$$E_W^\phi = \hbar\omega_{C\rightleftharpoons W}^\phi = \hbar\omega_0 = \left|m_C^+ - m_C^-\right|^\phi c^2 = m_0c^2 = \qquad (27\text{d})$$

$$m_0\omega_0^2L_0^2 = \frac{\hbar^2}{m_0L_0^2} = \left[m_C^+\mathbf{v}^2\right]^\phi = \frac{\hbar^2}{\left[m_C^+(L^+)^2\right]^\phi} = E_C^\phi$$

$$where: \ [m_C^+]^\phi = m_0(c/\mathbf{v}^\phi)^2 = \frac{m_0}{\phi} \simeq 1.618\, m_0;$$

$$[m_C^-]^\phi = \phi m_0 \simeq 0.618\, m_0$$

$$\left[\frac{m_0^2}{\left(m_C^+\right)^2}\right]^\phi = \left[\frac{m_C^-}{m_C^+}\right]^\phi = \phi^2 = 1-\phi \simeq 0.382$$

$$\mathbf{v}^\phi = c\phi^{1/2} = 0.786151377\, c; \qquad \mathbf{v}_{ph}^\phi = \mathbf{c}/\phi^{1/2}$$

$$P^\phi = [m_C^+\mathbf{v}^2]^\phi/c = m_0\, c \equiv P_0; \quad P_0/\left(P^+\right)^\phi = \mathbf{v}^\phi/\mathbf{c} = \phi^{1/2}$$

$$L^\phi = L_0 = \hbar/P_0 = \hbar/m_0\, c; \qquad L_0/\left(L_+\right)^\phi = \left(\lambda_0/\lambda_+\right)^\phi = \mathbf{c}/\mathbf{v}^\phi = 1/\phi^{1/2} \qquad (27\text{e})$$

where: $(P^+)^\phi = m_C^+ \mathbf{v}^\phi$ and $(\lambda_+)^\phi = 2\pi (L_+)^\phi = h/(m_C^+\mathbf{v})^\phi$ are the actual momentum and the actual de Broglie wave length of sub-elementary particle at GM conditions.

We have to point out, that the Hidden Harmony conditions (27a) corresponds to conditions of standing waves: actual (λ_+) and complementary (λ_-). It is a result of superposition of the *internal* circular de Broglie waves (waves B), representing the collective circulation of sub-quantum particles, forming actual vortex and complementary rotor with velocities: $\mathbf{v}_{gr}^{in}$ and $\mathbf{v}_{ph}^{in}$, correspondingly for one side, and the *external* circular waves B in form of rotation of the actual vortex and complementary rotor, as whole, with velocities ($\mathbf{v}_{gr}^{ext}$ and $\mathbf{v}_{ph}^{ext}$) in the opposite direction. The conditions (27a - 27c), taking into account (27d-27e), can be presented as follows:

$$\left[\lambda_+^{in} = \frac{h}{\mathbf{m}_C^+\mathbf{v}_{gr}^{in}} = \frac{h}{\mathbf{m}_C^+\mathbf{v}_{gr}^{ext}} = \lambda_+^{ext}\right]^\phi = \frac{h\,\phi^{1/2}}{\mathbf{m}_0\mathbf{c}}$$

$$\left[\lambda_-^{in} = \frac{h}{\mathbf{m}_C^-\mathbf{v}_{ph}^{in}} = \frac{h}{\mathbf{m}_C^-\mathbf{v}_{ph}^{ext}} = \lambda_-^{ext}\right]^\phi = \frac{h}{\mathbf{m}_0\mathbf{c}\,\phi^{1/2}}$$

Consequently, the ratio of standing de Broglie waves length of the actual vortex and complementary rotor of sub-elementary particles, responsible for the rest mast origination, is also equal to Golden mean:

$$\frac{\lambda_+^{in,ext}}{\lambda_-^{in,ext}} = \frac{L_+^{in,ext}}{L_-^{in,ext}} = \phi \tag{27f}$$

In general case the value of external group velocity $(\mathbf{v})$ is related to value of the quantized mass/energy symmetry shift: $|\Delta m_C|\, c^2 = n m_0 c^2$. In turn, the mass symmetry shift in sub-elementary particles is directly related to quantization of charge symmetry shift in sub-elementary particles or between rotor and antirotor of Bivacuum dipoles ($\mathrm{BVF}^\updownarrow$): $|\Delta e| = n e_0 = n\phi e$. It follows from our theory, that the ratio of charge and mass symmetry shifts is a permanent value:

$$\frac{|\Delta e|}{|\Delta m_C|} = \frac{\phi e}{m_0} \quad or: \quad \frac{|\Delta e|}{\phi e} = \frac{|\Delta m_C|}{m_0} \tag{27g}$$

8 Compensation principle of the local and nonlocal Bivacuum symmetry shifts, as a new explanation of fields, generated by elementary particles

The law of energy conservation keeps the total energy of system [the unpaired sub-elementary particles + secondary Bivacuum] unchanged and equal to zero. It is a consequence of compensation of ***local*** *symmetry shifts of unpaired sub-elementary fermion* $[\mathbf{F}_{\updownarrow}^{\pm}>$ *by* ***nonlocal*** *symmetry shifts of Bivacuum fermions* $(BVF^{\updownarrow})$, *representing the fields, radiated by particles. The local and nonlocal symmetry shifts of paired sub-elementary particles* $[\mathbf{F}_{\uparrow}^{-} \bowtie \mathbf{F}_{\downarrow}^{+}]$ *of triplets* $\langle[\mathbf{F}_{\uparrow}^{-} \bowtie \mathbf{F}_{\downarrow}^{+}]+\mathbf{F}_{\updownarrow}^{\pm}\rangle$ *compensate each other almost totally.*

The local Bivacuum symmetry shift, pertinent for corpuscular [C] phase of each sub-elementary particle, we introduce as a difference between the actual $|m_C^+|$ and complementary $|m_C^-|$ masses, proportional to the energy of this phase (see eq. 24):

$$\Delta m_C(\mathbf{v}) = \Delta m_{C\rightleftharpoons W}(\mathbf{v}) = |m_C^+|-|m_C^-| = m_C^+\mathbf{v}^2/c^2 = m_0+m_C^+\mathbf{v}_{tr\|}^2/c^2+m_C^+\mathbf{v}_{tr\perp}^2/c^2$$

where $(\mathbf{v})$ is the resulting external group velocity of sub-elementary particle (26a), including the rotational contribution and two external translational velocities: longitudinal $(\mathbf{v}_{tr\|})$ and transversal $(\mathbf{v}_{tr\perp}^2)$.

The Bivacuum symmetry shift, related with the rest mass of [C] phase is totally compensated by the local cumulative virtual cloud (CVC) in [W] phase of the same sub-elementary particle.

However, in contrast to local CVC, the local symmetry shifts, related with *translational contributions* to energy of elementary particles, like electrons or positrons, are compensated by nonlocal perturbations of symmetry of infinitive number of Bivacuum fermions ($\mathrm{BVF}^{\uparrow}$ and $\mathrm{BVF}^{\downarrow}$).

It leads from our compensation principle, that *electromagnetic and gravitational fields* in secondary Bivacuum display themselves in slight symmetry shifts between vortex $(\mathbf{V}^+)$ and antivortex $(\mathbf{V}^-)$ of Bivacuum fermions $(\mathbf{BVF}^{\updownarrow}= (\mathbf{V}^+ \Updownarrow \mathbf{V}^-)$, accompanied by the emergency of their longitudinal $(\|)$ and transversal $(\perp)$ vibrations. In accordance to eqs. 22b-22d, the difference between the internal group and phase velocities of $\mathbf{V}^+$ and $\mathbf{V}^-$ means the nonzero external group velocity ($\mathbf{v}$ >0), nonzero mass and charge and, consequently, the external magnetic moment of $\mathbf{BVF}^{\updownarrow}$ of secondary Bivacuum. The $\mathrm{BVF}^{\updownarrow}$ asymmetry and corresponding EM and G - potentials decreases with distance (r) from the local source of asymmetry (unpaired sub-elementary particle) as (1/r).

Three kinds of Bivacuum symmetry oscillations, accompanied $[C \rightleftharpoons W]$ pulsation of unpaired sub-elementary particle $F_{\updownarrow}^{-}>^i$, accompanied by $[emission \rightleftharpoons absorption]$ of cumulative virtual cloud (CVC), are related to oscillation of:

a) energy of rotation (spinning) of sub-elementary particle, which determines its rest mass energy:

$E_S = (m_C^+\mathbf{v}^2)^{\phi} = m_0c^2 = m_0\omega_0^2L_0^2$, reversibly emitted and absorbed in form of CVC in a course of $[C \rightleftharpoons W]$ pulsation;

b) energy of longitudinal translational vibrations, responsible for electromagnetic potential: $(E_{C,W})_{\|tr} = E_E = \alpha m_C^+\mathbf{v}^2 = m_C^+\mathbf{v}_{\|tr}^2$, irreversibly emitted in a course of $[C \rightleftharpoons W]$ pulsation;

c) energy of transversal translational vibrations, responsible for gravitational potential:

$(E_{C,W})_{\perp tr} = E_G = \beta m_C^+ \mathbf{v}^2 = m_C^+ \mathbf{v}^2_{\perp tr}$, irreversibly emitted in a course of $[C \rightleftharpoons W]$ pulsation.

The 1st kind of energy/symmetry oscillation, related with *emission* and *absorp–tion* of the biggest part of CVC energy, is totally local and reversible. However, it is responsible for nonlocal massless virtual spin waves (VirSW) excitation, excited by angular momentum (spin) of cumulative virtual cloud (CVC), representing the [W] phase of sub-elementray particles. Like the collective Nambu-Goldstone modes the spin field is a carrier of the phase/spin, but not the energy.

For the case of longitudinal (b) and transversal (c) translational vibrations of $F_\updownarrow^- >^i$ corresponding energy contributions to the total CVC energy are radiated in space irreversibly.

The energy conservation law and the equality of the resulting energy of system [secondary Bivacuum + fields + matter] to zero demands, that in such a case the *external* to particle Bivacuum symmetry shifts of Bivacuum Fermions (BVF) should compensate the corresponding local symmetry/energy oscillations.

These spatially delocalized perturbations of Bivacuum represent electromagnetic and gravitational fields, generated by corresponding *translational* vibrations of unpaired sub-elementary particles.

Taking this into account, the compensation principle *between the local and non-local symmetry/energy shifts of Bivacuum,* the electromagnetic and gravitational fields, can be presented as:

$$\sum^{n=N} [\mathbf{E}^n_{\mathbf{E}}(F_\updownarrow^\pm)^i + \mathbf{E}^n_{\mathbf{G}}(F_\updownarrow^\pm)^i]_{local} = \tag{28}$$

$$= -\frac{\vec{r}}{r} \sum^{k=\infty} [\, \mathbf{E}^k_{\mathbf{E}(V^+ \rightleftharpoons V^-)^i_\parallel} + \mathbf{E}^k_{\mathbf{G}(V^+ \rightleftharpoons V^-)^i_\perp}]_{\text{nonlocal}} = \tag{28a}$$

$$= -m_0^i c^2 \frac{\vec{r}}{r} \sum^{k=\infty} \ln \left[K_{\mathbf{E}(V^+ \rightleftharpoons V^-)^i} K_{\mathbf{G}(V^+ \rightleftharpoons V^-)^i} \right]^k_{\text{nonlocal}} \tag{28b}$$

where: N is a finite number of charged elementary particles in the closed system under consideration; $k = \infty$ is the infinitive number of asymmetric Bivacuum fermions or antifermions, perturbed by elementary particles; $\mathbf{E}^n_{\mathbf{E}}(F_\updownarrow^\pm)^i$; $\mathbf{E}^n_{\mathbf{G}}(F_\updownarrow^\pm)^i$ are the local electromagnetic and gravitational potentials of elementary particle.

We have to accept, that our perception of macroscopic WORLD is limited by properties of uncompensated sub-elementary particles in corpuscular [C] phase.

Parts of eqs. 28a and 28b represent the sum of two contributions to symmetry shift of sub-elementary particles, induced by translational vibrations (longitudinal and transversal), responsible for electromagnetic and gravitational potentials. Corresponding two kinds of Bivacuum dipoles equilibrium constants can be presented like:

$$K_{\mathbf{E}(V^+\rightleftharpoons V^-)^i} = \exp\left[-\frac{\mathbf{E}^k_{\mathbf{E}(V^+\rightleftharpoons V^-)^i}}{m_0^i c^2}\right] = \exp\left[-\frac{\alpha m^i_{BVF}\,\mathbf{v}^2}{m_0^i c^2}\right] \tag{29}$$

$$K_{\mathbf{G}(V^+\rightleftharpoons V^-)^i} = \exp\left[-\frac{\mathbf{E}^k_{\mathbf{G}(V^+\rightleftharpoons V^-)^i}}{m_0^i c^2}\right] = \exp\left[-\frac{\beta m^i_{BVF}\,\mathbf{v}^2}{m_0^i c^2}\right] \tag{29a}$$

The masses of Bivacuum fermions ($\mathrm{BVF}^\uparrow$ and $\mathrm{BVF}^\downarrow)^i$ of three generation ($i = e,\ \mu, \tau$) are determined by the difference of masses of virtual rotor $(\mathbf{V}^+)^i$ and antirotor $(\mathbf{V}^-)^i$ (see eq.20b):

$$\left[m^i_{BVF} = \left|m^i_{\mathbf{V}^+}\right| - \left|m^i_{\mathbf{V}^-}\right| = \left|m^i_{\mathbf{V}^+}\right| \alpha(\mathbf{v}/\mathbf{c})^2\right]_E \tag{29b}$$

$$\left[m^i_{BVF} = \left|m^i_{\mathbf{V}^+}\right| - \left|m^i_{\mathbf{V}^-}\right| = \left|m^i_{\mathbf{V}^+}\right| \beta(\mathbf{v}/\mathbf{c})^2\right]_G \tag{29c}$$

where: $\alpha\mathbf{v}^2 = \mathbf{v}^2_{\|tr}$ and $\beta\mathbf{v}^2 = \mathbf{v}^2_{\perp tr}$ are the longitudinal and transversal group velocities of sub-elementary fermions in triplets $\langle[\mathbf{F}^-_\uparrow \bowtie \mathbf{F}^+_\downarrow] + \mathbf{F}^\pm_\updownarrow\rangle$. Using these velocities and eq. 19, the values of $\left|m^i_{\mathbf{V}^+}\right|_E$ and $\left|m^i_{\mathbf{V}^+}\right|_G$ can be calculated.

The massless virtual spin waves (VirSW), excited by the angular momentum (spin) of CVC of sub-elementary particles in [W] phase, is compensated by shift of dynamic equilibrium of Bivacuum fermions with opposite spins $(BVF^\uparrow \rightleftharpoons BVF^\downarrow)^i$. However, in this process the equilibrium between parameters of rotors and antirotors $(V^+ \rightleftharpoons V^-)^i$ of $BVF^\updownarrow$ keeps unchanged. *This means that the energy of Bivacuum do not change as a result of VirSW excitation.*

In accordance to known Helmholtz theorem: $\mathbf{F} = \mathbf{rot}\,\mathbf{A} + \mathbf{grad}\,\varphi$, each kind of vector field ($\mathbf{F}$), tending to zero on the infinity, can be presented, as a sum of *rotor* of some vector function, determined in our model by spin equilibrium of Bivacuum fermions and antifermions ($\mathbf{BVF}^\uparrow \bowtie \mathbf{BVF}^\downarrow$), named: $\mathbf{A} \equiv \mathbf{S}$ with its divergence, equal to zero and a gradient of some scalar function, which can be a sum of two or more scalar functions, like ($\varphi = \varphi_E + \varphi_G$):

$$\mathbf{F} = \mathbf{rot}\,\mathbf{S} + \mathbf{grad}\,(\varphi_E + \varphi_G); \qquad \mathbf{div}\,\mathbf{S} = \mathbf{0} \tag{30}$$

where: φ_E *and* φ_G are the scalar potentials of electromagnetic and gravitational fields, as a components of field ($\mathbf{F}$) and $\mathbf{S}$ is a vector potential of this field.

Potentials φ_E *and* φ_G are representing virtual spherical waves. It is known, that general solution of the wave equation for any spherical wave is:

$$\varphi_{E,G} = \frac{1}{r} f_1(ct - r) + \frac{1}{r} f_2(ct + r) \tag{31}$$

where: f_1 and f_2 are arbitrary functions; $\frac{1}{r}f_1(ct-r)$ and $\frac{1}{r}f_2(ct-r)$ are the potentials of diverging wave and converging wave, correspondingly.

9 Different presentations of the local energy contributions, related to spin potential, electric and gravitational potentials at Golden mean conditions

The local rotational - spin contribution in total energy of *sub-elementary particles,* which creates the rest mass and *spin potential,* can be presented for corpuscular [C] and wave [W] phase in form of CVC as (Kaivarainen, 2002):

$$\mathbf{E}^n_{\mathbf{S}}(F^{\pm}_{\updownarrow})^i_{[C]} = \left(\left[m^+_C v^2_{rot}\right]^{\phi} = \left[m^+_C\right]^{\phi} (\phi c)^2_{rot} = \frac{\hbar^2}{\left[m^+_C \left(L^+_C\right)^2\right]^{\phi}} \right)^i_{[C]} \tag{32}$$

$$\mathbf{E}^n_{\mathbf{S}}(F^{\pm}_{\updownarrow})^i_{[W]} = \left(\left|m^+_C - m^-_C\right|^{\phi} c^2 = m_0 c^2 = \frac{\hbar^2}{m_0 L^2_0} = m_0\omega^2_0 L^2_0 = \hbar\omega_0 \right)^i_{[W]} \tag{32a}$$

where the Golden mean angle frequency is $\omega_0 = m_0 c^2/\hbar$; the corresponding Compton radius $L_0 = \hbar/m_0 c$ and the energies of both phase are equal: $\mathbf{E}^n_{\mathbf{S}}(F^{\pm}_{\updownarrow})^i_{[C]} = \mathbf{E}^n_{\mathbf{S}}(F^{\pm}_{\updownarrow})^i_{[W]}$

The energy contribution of longitudinal vibration, induced by $[C \rightleftharpoons W]$ pulsation of uncompensated sub-elementary particles $F^{\pm}_{\updownarrow} >$ of triplets $<[F^-_{\updownarrow} \bowtie F^+_{\updownarrow}] + F^{\pm}_{\updownarrow} >$, is responsible for **electric potential** of elementary particles. At Golden mean conditions it can be presented for [C] and [W] phase as:

$$\mathbf{E}^n_{\mathbf{E}}(F^{\pm}_{\updownarrow})^i_{[C]} = \left[\alpha \left[m^+_C v^2_{res}\right]^{\phi} = \alpha \left[\frac{\mathbf{p}^2}{m^+_C}\right]^{\phi} = \left[m^+_C v^2_{\|tr}\right]^{\phi} = (zc)^2 \left[m^+_C\right]^{\phi} = \alpha\phi \left[m^+_C\right]^{\phi} c^2 \right]^i_{[C]} \tag{33}$$

$$\mathbf{E}^n_{\mathbf{E}}(F^{\pm}_{\updownarrow})^i_{[W]} = \left[\left(\frac{e_+ e_-}{L^{\pm}}\right)^{\phi} = \alpha(m^+_C - m^-_C)^{\phi} c^2 = \alpha m_0 c^2 \right]^i_{[W]} \tag{33a}$$

where: v_{res} is the resulting rotational-translational velocity of sub-elementary fermion (26a); $v^{\phi}_{\|tr} = zc$ is a velocity of longitudinal translational *zero-point vibrations*; $z = (\alpha\phi)^{1/2}$ is a longitudinal zero-point factor; $\alpha = e^2/\hbar c$ is the electromagnetic fine structure constant; $\left(L^{\pm} = \hbar/[(m^+_C - m^-_C)^{\phi} c] = L_0\right)^i$ is a characteristic dimension of asymmetric double cell-dipole separating the actual (e_+) and complementary (e_-) charge of each of three sub-elementary particles of the electron or positron $<[F^-_{\updownarrow} \bowtie F^+_{\updownarrow}] + F^{\pm}_{\updownarrow} >^i$.

The energy contribution of transversal vibration of unpaired sub-elementary particle of triplets $<[F^-_{\updownarrow} \bowtie F^+_{\updownarrow}] + F^{\pm}_{\updownarrow} >^i$, induced by $[C \rightleftharpoons W]$ pulsation, is responsible for **gravitational potential** of elementary particles. At Golden mean conditions the gravitational potential can be presented as:

$$\mathbf{E}^n_{\mathbf{G}}(F^{\pm}_{\updownarrow})^i_{[C]} = \left[\beta\left[m^+_C v^2_{res}\right]^{\phi} = \beta\left[\frac{\mathbf{p}^2}{m^+_C}\right]^{\phi} = \left[m^+_C v^2_{\perp tr}\right]^{\phi} = (xc)^2\left[m^+_C\right]^{\phi}\beta\phi\left[m^+_C\right]^{\phi}c^2\right]^i_{[C]} \quad (34)$$

$$\mathbf{E}^n_{\mathbf{G}}(F^{\pm}_{\updownarrow})^i_{[W]} = \left[G\left(\frac{m^+_C m^-_C}{L^{\pm}}\right)^{\phi} = \beta(m^+_C - m^-_C)^{\phi}c^2 = \beta m_0 c^2 = \Delta m^{\phi}_V c^2\right]^i_{[W]} \quad (34a)$$

where: v_{res} is the resulting rotational-translational velocity of sub-elementary fermion (26a); $v_{\perp tr} = xc$ is a velocity of transversal translational *zero-point vibrations*, responsible for gravitation; $x = (\beta\phi)^{1/2}$ is a transversal zero-point factor.

The electromagnetic and gravitational interaction energy between two particles (1) and (2) can be presented as the square root of product of corresponding potentials of particles:

$$\mathbf{E}^{1,2}_E = [\mathbf{E}^{(1)}_E \mathbf{E}^{(2)}_E]^{1/2}$$

$$\mathbf{E}^{1,2}_G = [\mathbf{E}^{(1)}_G \mathbf{E}^{(2)}_G]^{1/2}$$

The spin-spin interaction (attraction or repulsion, depending on relative spin orientation) energy between two elementary particles in the volume, determined by their de Broglie wave length, can be expressed in similar way:

$$\mathbf{E}^{1,2}_S = [\mathbf{E}^{(1)}_S \mathbf{E}^{(2)}_S]^{1/2}$$

The longitudinal and transversal velocities of elementary particle and corresponding factors at Golden mean conditions are summarized below:

$$\mathbf{v}^{\phi}_{\| tr} = \mathbf{zc} \quad where: \quad z = (\alpha\phi)^{1/2} = 6.71566 \cdot 10^{-2} \quad (34b)$$

$$and \quad z^2 = \left(\frac{\mathbf{v}^{\phi}_{\| tr}}{c}\right)^2 = \alpha\phi = 4.51 \cdot 10^{-3}$$

$$\mathbf{v}^{\phi}_{\perp tr} = xc \quad where: \quad x = (\beta\phi)^{1/2} = 3.27867 \cdot 10^{-23} \quad (34c)$$

$$and \quad x^2 = \left(\frac{\mathbf{v}^{\phi}_{\perp tr}}{c}\right)^2 = \beta^e\phi = 1.07497 \cdot 10^{-45}$$

These minimum values of two kinds of translational velocities correspond to zero-point longitudinal and transversal vibrations of elementary particles, accompanied their $[C \rightleftharpoons W]$ pulsation.

Now, the minimum external resulting group velocity of elementary particle squared, corresponding to GM conditions, using (25a and 34b,c), can be presented as a sum of three contributions:

$$\mathbf{v}^2_{\min} = \mathbf{c}^2\phi + \left(\mathbf{v}^{\phi}_{\| tr}\right)^2 + \left(\mathbf{v}^{\phi}_{\perp tr}\right)^2 = \mathbf{c}^2\phi + \mathbf{c}^2\alpha\phi + \mathbf{c}^2\beta\phi = \mathbf{c}^2(\phi + \alpha\phi + \beta\phi) \quad (34d)$$

10 Interrelation between the electromagnetic and gravitational potentials and their space curvatures at Golden mean conditions

The rotational energy of CVC_{rot} (local spin - potential), of each generation of elementary particles, like electrons of three generation ($i = e, \mu, \tau$) is independent on their translational kinetic energy and always determined by Golden mean conditions:

$$\left[E_S = E_{rot} = \left|m_C^+ - m_C^-\right|^\phi c^2 = m_0 c^2 = m_0 \omega_0^2 L_0^2 = \frac{\hbar^2}{m_0 L_0^2}\right]^i \tag{35}$$

The *curvature*, corresponding to CVC_{rot} and sub-elementary particle in [C] phase, is equal to its Compton radius:

$$\left[L_S = \frac{\hbar}{\left|m_C^+ - m_C^-\right|^\phi c} = \frac{\hbar}{m_0 c} \equiv L_0\right]^i \tag{36}$$

The *energy* of spin field of unpaired sub-elementary particle is local and reversible in the process of $[C \rightleftharpoons W]$ pulsation in contrast to nonlocal action of the *angular momentum* of CVC on$[BVF^\uparrow \rightleftharpoons BVF^\downarrow]$ dynamic equilibrium.

The curvature of Bivacuum, corresponding to symmetry shift, related to *longitudinal* zero-point vibrations of elementary particle, can be find out from the electric potential of the electron:

$$E_E = \alpha \left|m_C^+ - m_C^-\right|^\phi c^2 = \left[m_C^+ v_{\parallel tr}^2\right]^\phi = \alpha m_0 c^2 = \alpha \frac{\hbar c}{L_0} \tag{37}$$

where from 34b: $\left[m_C^+\right]^\phi = m_0/\phi$; $\left[v_{\parallel tr}^2\right]^\phi = \alpha\phi \cdot c^2$

The corresponding to electric potential space curvature at Golden mean conditions $\left|m_C^+ - m_C^-\right|^\phi = m_0$, is equal to radius of the 1st Bohr orbit of the electron (a_B):

$$L_E^\phi = L_\parallel = \frac{1}{\alpha}\frac{\hbar}{m_0 c} = \frac{1}{\alpha} L_0 = \frac{\hbar c}{\left[m_C^+ v_{\parallel tr}^2\right]^\phi} = a_B = 0.5291 \cdot 10^{-10}\, m \tag{38}$$

where $\alpha = e^2/\hbar c = 7.297 \cdot 10^{-3}$ is the electromagnetic fine structure constant.

We get the important result, confirming our Bivacuum and $[C \rightleftharpoons W]$ duality theory, that the circled electromagnetic standing de Broglie wave length ($\lambda_\parallel = 2\pi L_\parallel$), corresponding to *longitudinal* zero-point vibrations of the electron is equal to length of the *1st Bohr orbit in hydrogen atom, containing electron and proton of opposite spins.* This space parameter ($L_\parallel$) characterize also the dimension of the possible coherent cluster of the electrons pairs with opposite spins (Cooper pairs) in state of *microscopic Bose condensation* (μBC), forming for example, in the volume of electric spark. The possibility of mesoscopic molecular Bose condensation (mBC) has been revealed theoretically in water and ice earlier (Kaivarainen, 1995; 2001; 2001a).

From eq.(37) we can see, that the curvature radius, corresponding to circular electromagnetic standing wave of the electron exceeds its Compton radius, equal to spin field curvature (L_S), to about 137 times:

$$\frac{L_E^{\phi}}{L_S} = \frac{a_B}{L_0} = \frac{1}{\alpha} = 137.03604 \tag{39}$$

In similar way, the curvature of Bivacuum, related to *transversal* zero-point vibrations of elementary particles, is defined by the value of gravitational potential at GM conditions:

$$E_G = \beta \left| m_C^+ - m_C^- \right|^{\phi} c^2 = \left[m_C^+ v_{\perp tr}^2 \right]^{\phi} = \beta \frac{\hbar c}{L_0} \tag{39a}$$

where from 34c: $\left[m_C^+ \right]^{\phi} = m_0/\phi;\quad \left[v_{\perp tr}^2 \right]^{\phi} = \beta\phi \cdot c^2$

The corresponding radiuses of the electron's and proton's gravitational curvatures (gravitational standing waves) are:

$$\left(L_G^{\phi} \right)^e = \frac{1}{\beta^e} \frac{\hbar}{m_0^e c} = \frac{\hbar c}{\left[m_C^+ v_{\perp tr}^2 \right]^{\phi}} = \frac{L_0^e}{\beta^e} = \tag{40}$$

$$= a_G^e = \frac{3.86 \cdot 10^{-13}\,\mathrm{m}}{1.7385 \cdot 10^{-45}} = 2.22 \cdot 10^{32}\, m = 2.34 \cdot 10^{16}\, light\, years$$

$$\left(L_G^{\phi} \right)^p = \frac{1}{\beta^p} \frac{\hbar}{m_0^p c} = \frac{\hbar c}{\left[m_C^+ v_{\perp tr}^2 \right]^{\phi}} = \frac{L_0^p}{\beta^p} = \tag{40a}$$

$$= a_G^p = \frac{21 \cdot 10^{-17}\,\mathrm{m}}{5.86 \cdot 10^{-39}} = 4 \cdot 10^{22}\, m = 3.79 \cdot 10^{6}\, light\, years \tag{41}$$

where: $\beta^e = (m_0^e/M_{Pl})^2 = 1.7385 \cdot 10^{-45}$; $\beta^p = (m_0^p/M_{Pl})^2 = 5.86 \cdot 10^{-39}$ are the introduced in our theory gravitational fine structure constant, different for electrons and protons; $M_{Pl} = (\hbar c/G)^{1/2} = 2.17671 \cdot 10^{-8}\, kg$ is a Plank mass; $m_0^e = 9.109534 \cdot 10^{-31}\, kg$ is a rest mass of the electron; $m_0^p = 1.6726485 \cdot 10^{-27}\, kg = m_0^e \cdot 1.8361515 \cdot 10^3\, kg$ is a rest mass of proton; the length of 1 light year is $9.46 \cdot 10^{15}\, m$.

The new parameters (a_G^e and a_G^p), by the analogy with (a_B) can be named as the *1st radius of the circular gravitational standing wave of the electron and proton, correspondingly.*

It is tempting to put forward a conjecture, that the maximum and stable diameter (in future) of our expanding with acceleration Universe is determined by the gravitational curvature radius of the free electrons: $a_G^e = 2.34 \cdot 10^{16}\, light\, years$, coinciding with the curvature radius of the lowest energy of $e-$ neutrino at quantum number at $n = 1$, in accordance to our formula 43b. In such condition of super-low energy/matter density all the matter may change to low energy photons and neutrino, like the relict ones. As a result, the secondary Bivacuum turns again to primordial one with minimum energetic gap between rotors and antirotors of Bivacuum fermions (BVF$^{\updownarrow}$). This means the infinitive virtual Bose condensation of pairs of dipoles $[BVF^{\uparrow} \bowtie BVF^{\downarrow}]$, making Bivacuum highly nonlocal and able for spontaneous but space-time correlated cosmic - large - scale fluctuations of symmetry and energy. It is like the electric breakdown in condensers, when the separation between oppositely charged plates becomes too small.

This breakdown of primordial Bivacuum corresponds to born of the new large-scale Universe after death the old one. In such a model we do not need the concept of Big Bang, starting from singularity.

The gravitational curvature radius of proton, equal to $a_G^p = 3.79 \cdot 10^6\, light\, years$ may have the same importance in cosmology, like the electromagnetic curvature the electron, equal to 1st orbit radius of the hydrogen atom: $L_E^{\phi} = a_B = 0.5291 \cdot 10^{-10}\, m$ in atomic physics. For example very good correlation is existing between a_G^p and the radius of of Local group of galactics, like Milky way, Andromeda galaxy and Magellan clouds, equal approximately to $3 \cdot 10^9$ light years. The radius of Vigro cluster of galactics is also close to a_G^p.

11 Curvature of Bivacuum domains of nonlocality, corresponding to Electromagnetic and Gravitational Fields

It follows from our theory, that nonlocal shift of symmetry between properties of rotors (V^+) and antirotors (V^-), forming $BVF^{\updownarrow} \sim [V^+ \Updownarrow V^-]$ of Bivacuum, as a compensation of local symmetry shifts, induced by *translational vibrations (longitudinal and transversal),* is accompanied by emergency of difference between the actual and complementary mass: $\left|m_C^+ - m_C^-\right| > 0$ and the actual and complementary charge: $|e_+ - e_-| > 0$. This shift in Bivacuum dipoles ($BVF^{\updownarrow}$) in the 'empty' space is much less, than the mass and charge shifts in sub-elementary particles $F_{\updownarrow}^{\pm} >$, forming the matter. However, in accordance to formula (25a), even small decreasing of ratio $\left|m_C^-/m_C^+\right|_{BVF^{\updownarrow}} < 1$ means that the *external rotational - translational* velocity of $BVF^{\updownarrow}$ becomes more, than zero: $\mathbf{v}_{ext} \equiv \mathbf{v} = c\left[1 - \left|\frac{m_C^-}{m_C^+}\right|_{BVF^{\updownarrow}}\right]^{1/2} > 0.$

For the external virtual wave B length of $BVF^{\updownarrow}$ in the vicinity of sub-elementary particle $F_{\updownarrow}^{\pm}$, activated by its translational longitudinal and transversal vibrations at

Golden mean conditions, are correspondingly:

$$\left(L^i_{BVF}\right)_{\parallel} \equiv L_E = \frac{\hbar}{\left(m^i_{BVF}\right)_{\parallel}\left(\mathbf{v}_{ext}\right)^{\phi}_{\parallel}} = \frac{\hbar}{m^i_{BVF}\, c(\alpha\phi)^{1/2}} \tag{42}$$

$$\left(L^i_{BVF}\right)_{\perp} \equiv L_G = \frac{\hbar}{\left(m^i_{BVF}\right)_{\perp}\left(\mathbf{v}_{ext}\right)^{\phi}_{\perp}} = \frac{\hbar}{m^i_{BVF}\, c(\beta\phi)^{1/2}} \tag{42a}$$

where in accordance to relativist dependencies for the actual and complementary mass of rotor and antirotor of $BVF^{\updownarrow} = [\mathbf{V}^{+} \Updownarrow \mathbf{V}^{-}]$ for Golden mean conditions: 34b and 34c, when:

$$(\mathbf{v}^{\phi}_{\parallel}/\mathbf{c})^2 = \alpha\phi = 7.297 \cdot 10^{-3} \cdot 0.618 = 4.51 \cdot 10^{-3}$$

$$(\mathbf{v}^{\phi}_{\perp}/\mathbf{c})^2 = \beta\phi = 5.86 \cdot 10^{-39} \cdot 0.618 = 3.621 \cdot 10^{-39}$$

we get using eq. 20b:

$$\left(m^i_{BVF}\right)_{\parallel} = \left|m^i_{V^+}\right|_{\parallel} - \left|m^i_{V^-}\right|_{\parallel} = \frac{m_0 \alpha\phi}{[1-\alpha\phi]^{1/2}} \tag{42b}$$

$$\left(m^i_{BVF}\right)_{\perp} = \left|m^i_{V^+}\right|_{\perp} - \left|m^i_{V^-}\right|_{\perp} = \frac{m_0 \beta\phi}{[1-\beta\phi]^{1/2}} \tag{42c}$$

Putting 42b and 42c into 42 and 42a, we get the radiuses of domains of virtual Bose condensates (VirBC), formed by Bivacuum fermions (BVF) at Golden mean conditions:

$$\left(L^{\phi}_{E}\right)_{VirBC} = \frac{\hbar}{\left(m^i_{BVF}\right)_{\parallel} c(\alpha\phi)^{1/2}} = \frac{\hbar\,[1-\alpha\phi]^{1/2}}{m_0 c\,(\alpha\phi)^{3/2}} < \infty \tag{42d}$$

$$\left(L^{\phi}_{G}\right)_{VirBC} = \frac{\hbar}{\left(m^i_{BVF}\right)_{\perp} c(\beta\phi)^{1/2}} = \frac{\hbar\,[1-\beta\phi]^{1/2}}{m_0 c\,(\beta\phi)^{3/2}} \cong \frac{\hbar}{m_0 c\,(\beta\phi)^{3/2}} < \infty \tag{42e}$$

These formulas means that the condition of *infinitive* virtual Bose condensation (**VirBC**) of Cooper - like pairs of Bivacuum fermions (BVF), pertinent for primordial Bivacuum, is violated (Kaivarainen, 2002) and nonlocal (instant) electromagnetic and gravitational interactions remains possible only in virtual domains of secondary Bivacuum with dimensions determined by (42d and 42e), correspondingly. The artificial variation of L^i_{BVF} by external fields means Bivacuum engineering due to change the 'interfacial' effects between domains of virtual Bose condensation (**VirBC**), where the nonlocality conditions are violated.

The ratios of radiuses of these nonlocality domains of Bivacuum fermions (42f and 42g) to the radiuses of circular standing waves of the electrons, characterizing their EM potential (eq. 38) and neutrinos, characterizing Gravitational potential of the electrons (eq. 40) are equal to:

$$\frac{\left(L_E^\phi\right)_{VirBC}}{L_E^\phi} = \frac{[1-\alpha\phi]^{1/2}}{\alpha^{1/2}\phi^{3/2}} > 20$$

$$\frac{\left(L_G^\phi\right)_{VirBC}}{L_G^\phi} = \frac{[1-\beta\phi]^{1/2}}{\beta^{1/2}\phi^{3/2}} > 10^{22}$$

VirBC is the *virtual Bose condensation* of virtual Cooper pairs $[\mathbf{BVF}^{\uparrow} \bowtie \mathbf{BVF}^{\downarrow}]$, providing the nonlocal properties of Bivacuum domains. These domains represent multilayer structure. Each of 2D layer is formed by pairs $[\mathbf{BVF}^{\uparrow} \bowtie \mathbf{BVF}^{\downarrow}]$ with defects, represented by $[\mathbf{BVB}^{\pm}]$, as intermediate transition state between $\mathbf{BVF}^{\uparrow}$ **and** $\mathbf{BVF}^{\downarrow}$.

Using eqs. 29 and 29a for $[\mathbf{V}^+ \rightleftharpoons \mathbf{V}^-]$ equilibrium constants, characterizing electromagnetic and gravitational fields at Golden mean conditions in combination with 34b; 34c, 42b and 42c we get:

$$K^\phi_{\mathbf{E}(V^+\rightleftharpoons V^-)^i} = \exp\left[-\frac{\mathbf{E}^k_{\mathbf{E}(V^+\rightleftharpoons V^-)^i}}{m_0^i c^2}\right]^\phi = \exp\left[-\frac{\left(m^i_{BVF}\right)_{\parallel}\mathbf{v}^2_{\parallel tr}}{m_0^i c^2}\right]^\phi_E = \exp\left[-\frac{(\alpha\phi)^2}{[1-\alpha\phi]^{1/2}}\right]$$

$$K^\phi_{\mathbf{G}(V^+\rightleftharpoons V^-)^i} = \exp\left[-\frac{\mathbf{E}^k_{\mathbf{G}(V^+\rightleftharpoons V^-)^i}}{m_0^i c^2}\right]^\phi = \exp\left[-\frac{\left(m^i_{BVF}\right)_{\perp}\mathbf{v}^2_{\perp tr}}{m_0^i c^2}\right]^\phi_G = \exp\left[-\frac{(\beta\phi)^2}{[1-\beta\phi]^{1/2}}\right]$$

Putting these equilibrium constants to eq.(28b), we get for Bivacuum fermions (BVF) energy decay with distance (r), related to electric and gravitational fields, generated by $[\mathrm{C} \rightleftharpoons W]$ pulsation of the electron or positron at Golden mean conditions:

$$\left[\mathbf{E}^i_{\mathbf{E}(V^+\rightleftharpoons V^-)^i}\right]^\phi = \frac{\overrightarrow{r}}{r}\frac{(\alpha\phi)^2}{[1-\alpha\phi]^{1/2}} m_0^i c^2$$

$$\left[\mathbf{E}^i_{\mathbf{G}(V^+\rightleftharpoons V^-)^i}\right]^\phi = \frac{\overrightarrow{r}}{r}\frac{(\beta\phi)^2}{[1-\beta\phi]^{1/2}} m_0^i c^2$$

12 Neutrino and Antineutrino in Unified Model

We put forward a conjecture, that neutrino or antineutrino of three lepton generation, represents a stable non-local Bivacuum symmetry excitations, compensating the local symmetry oscillations, accompanied the creation of the electron's or positron's and their transversal ($\perp$) translational vibrations, responsible for *gravitational potential* of elementary particles. These kind of vibrations, like the longitudinal ones, responsible for electromagnetism, are activated by $[C \rightleftharpoons W]$ pulsation of sub-elementary fermions in triplets $<[F_\uparrow^- \bowtie F_\downarrow^+] + F_\updownarrow^\pm >^i$.

The quantized energy of neutrinos and antineutrinos is related to the rest mass of corresponding generations of the electron and positron $(\pm m_0^{e,\mu,\tau})$ in following manner (see 34):

$$E_{e,\mu,\tau}^{\nu,\widetilde{\nu}} = \left[m_C^+ v_{\perp tr}^2\right]_{e,\mu,\tau}^{\nu,\widetilde{\nu}} = \pm\beta_{e,\mu,\tau}\left(m_0^{e,\mu,\tau}\right)c^2\left(\frac{1}{2}+n\right) = \pm\Delta\left(m_V^{e,\mu,\tau}\right)c^2\left(\frac{1}{2}+n\right) \quad (43)$$

where $(\pm m_0^{e,\mu,\tau})$ are the rest mass of $[i = e, \mu, \tau]$ generations of electrons and positrons; $\beta_{e,\mu,\tau} = \left(m_0^{e,\mu,\tau}/M_{Pl}\right)^2$ is a gravitational fine structure constants, introduced in our theory of gravitation (Kaivarainen, 2002); $\pm\Delta\left(m_V^{e,\mu,\tau}\right)c^2 = \pm\beta_{e,\mu,\tau}$ $\left(m_0^{e,\mu,\tau}\right)c^2$ are the amplitude of Bivacuum symmetry vibrations, corresponding to three neutrino flavors $[e, \mu, \tau]$ at Golden mean conditions.

The charge of neutrino/antineutrino $(\pm e_\nu)$, proportional to symmetry shift: $\pm\Delta\left(m_V^{e,\mu,\tau}\right) = \pm\beta_{e,\mu,\tau}\left(m_0^{e,\mu,\tau}\right)$ (see for comparison 34a) is very close to zero:

$$\pm e_\nu = \pm\beta_{e,\mu,\tau} e_0 \cong 0 \quad (43a)$$

The evidence of neutrino flavor oscillation: $e \rightleftharpoons \mu \rightleftharpoons \tau$ has been recently obtained in Sudbury Neutrino Observatory (SNO, 2002). This means possibility of collective quantum transitions between symmetry shifts of secondary Bivacuum: $\Delta\left(m_V^e\right) \rightleftharpoons \Delta\left(m_V^\mu\right) \rightleftharpoons \Delta\left(m_V^\tau\right)$, as a result of conversions (oscillations) between three basic generation of cell-dipoles $(\mathrm{BVF}^\pm)^i$ with three corresponding resulting mass: $(m_0^e) \rightleftharpoons (m_0^\mu) \rightleftharpoons (m_0^\tau)$, where $m_0^i = \sqrt{(m_C^+ m_C^-)^i}$.

It is known,that neutrinos $(\nu_e; \nu_\mu; \nu_\tau)$ always originates in pairs with positrons $(e^+; \mu^+; \tau^+)$ and antineutrinos $(\widetilde{\nu}_e; \widetilde{\nu}_\mu; \widetilde{\nu}_\tau)$ in pairs with electrons $(e^-; \mu^-; \tau^-)$. In accordance to our approach this confirms our idea of the nonlocal symmetry shift compensation of $\mathbf{BVF}^\updownarrow \equiv [\mathbf{V}^+ \Updownarrow \mathbf{V}^-]$ of the local symmetry shift, accompanied the origination of elementary particles $\langle [F_\uparrow^- \bowtie F_\downarrow^+] + F_\updownarrow^\pm \rangle$, containing the unpaired sub-elementary fermion $F_\updownarrow^\pm \rangle$.

Neutrino and antineutrino may be considered, as a collective excitations of huge domains of Bivacuum in state of virtual Bose condensation (VirBC) with properties of nonlocality. The characteristic radius of such excited states, characterizing the neutrino curvature, is equal to:

$$L_{\nu_{e,\mu,\tau}}^{(n)} = \frac{\hbar}{c[\beta m_0]_{e,\mu,\tau}\left(\frac{1}{2}+n\right)} \quad (43b)$$

Obviously, the neutrino/antineutrino directly participate in gravitational (G) interaction. The energy of G interaction should be dependent on density energy of neutrino and its generation.

13 Harmonization energy (HaE) and force (HaF) of Bivacuum, as a background of Principle of least action & Pace of Time

It is shown, that Principle of least action is a consequence of introduced in our Unified Model (UM) the "Harmonization energy (HaE)" of secondary Bivacuum, driving the matter to Golden Mean conditions and responsible for its evolution on all hierarchic levels (Kaivarainen, 2002; 2003) from elementary particles, atoms, molecules to biopolymers, living organisms and star systems. We introduce the Harmonization energy $(\mathrm{HaE})^i_{x,y,z}$ of Bivacuum (anisotropic in general case, dependent on anisotropy of spatial distribution of kinetic energy of elementary particles), acting on three generations $(i = e, \mu, \tau)$ of elementary particles of matter, as a difference between the total energy of particle $(E^i_{C\rightleftharpoons W} = \hbar\omega^i_0 + E^i_E + E^i_G)_{x,y,z}$ and the basic energy of virtual pressure waves of Bivacuum $(\mathrm{VPW}^{\pm})^i$ with fundamental frequency $\omega^i_0 = m^i_0 c^2/\hbar$ (eq.1d). This frequency corresponds to energy of the rest mass of particle: $E^i_{VPW^\pm} = \hbar\omega^i_0 = m^i_0 c^2$, as it follows from our Unified Model (UM):

$$HaE^i_{x,y,z} = \left|E_{C\rightleftharpoons W} - E^i_{VPW^\pm}\right|^i_{x,y,z} = \left|m^+_C (v^{ext}_{gr})^2 - m_0 c^2\right|^i_{x,y,z} =$$

$$\left(\hbar\left|\omega_{C\rightleftharpoons W} - \omega_0\right|^i = \hbar\left|\omega_{HaE}\right|^i = p^i_{HaE}\, c\right)_{x,y,z} \tag{44}$$

where the frequency of Harmonization energy $|\omega_{HaE}|^i$ is equal to frequency of quantum beats between frequency of $[C \rightleftharpoons W]$ pulsation of elementary particles $(\omega_{C\rightleftharpoons W})$ and fundamental frequency (ω_0) of Bivacuum virtual pressure waves $(\mathrm{VPW}^{\pm})^i$:

$$\left(\left|\omega_{HaE}\right|^i = \left|\omega_{C\rightleftharpoons W} - \omega_0\right|^i = p^i_{HaE}\, c/\hbar\right)_{x,y,z} \tag{44a}$$

The momentum of HaE^i can be expressed, as

$$p^i_{HaE} = \hbar\frac{\left|\omega_{HaE}\right|^i}{c} = \frac{\hbar}{L^i_{HaE}} \tag{44b}$$

Taking into account the expressions for total energy of elementary particle (24-24c), the eq.(44) for Harmonization energy can be expressed via the sum of Electromagnetic and Gravitational potentials $(E_E + E_G)^i$. These potentials are related directly to longitudinal $\left|m^+_C v^2_{\|tr}\right|^i = E^i_E$ and transversal $\left|m^+_C \mathbf{v}^2_{\perp tr}\right|^i = E^i_G$ translational contributions to the total energy of elementary particle $(E^i_{C\rightleftharpoons W} = \hbar\omega^i_0 + E^i_E + E^i_G)_{x,y,z}$:

$$HaE^i_{x,y,z} = [(E_{C,W})_{\|tr} + (E_{C,W})_{\perp tr}]^i_{x,y,z} = (E_E + E_G)^i_{x,y,z} = \hbar \left|\omega_{C\rightleftharpoons W} - \omega_0\right|^i_{x,y,z} = \tag{45a}$$

$$= \left(\left|m_C^+ - m_C^-\right|_{\|tr} c^2 + \left|m_C^+ - m_C^-\right|_{\perp tr} c^2 = \left|m_C^+ \mathbf{v}^2_{\|tr}\right|^i + \left|m_C^+ \mathbf{v}^2_{\perp tr}\right|^i \right)_{x,y,z} = \tag{45b}$$

$$= \left(m_0^i c^2 \frac{\alpha\,(\mathbf{v}/c)^2}{[1-(\mathbf{v}/c)^2]^{1/2}} + m_0^i c^2 \frac{\beta\,(\mathbf{v}/c)^2}{[1-(\mathbf{v}/c)^2]^{1/2}} = HaF^i \lambda^i_{HaE} \right)_{x,y,z} \tag{45c}$$

where: $HaF^i = HaE^i/\lambda^i_{HaE}$ is the Harmonization force of Bivacuum; the virtual Harmonization wavelength, as a carrier of Harmonization energy is:

$$\left[\lambda^i_{HaE} = 2\pi\, L^i_{HaE} = 2\pi c/\left|\omega_{HaE}\right|^i = hc/(E_E + E_G)\right]_{x,y,z} \tag{45c'}$$

The resulting group velocity of particle from (25a), is:

$$\left[\mathbf{v}^2 = \left(\mathbf{v}^2_{rot}\right)^{\phi} + \mathbf{v}^2_{\|tr} + \mathbf{v}^2_{\perp tr} = \mathbf{c}^2\phi + \alpha\mathbf{v}^2 + \beta\mathbf{v}^2\right]_{x,y,z} \tag{45d}$$

The Harmonization energy and force of Bivacuum $(HaE^i \text{ and } HaF^i)_{x,y,z}$ realize the Bivacuum exchange interaction with matter, as a system of triplets of $\langle [F_\uparrow^- \bowtie F_\downarrow^+] + F_\updownarrow^\pm \rangle$ via the pairs $[F_\uparrow^- \bowtie F_\downarrow^+]$. We have to keep in mind, that the frequencies/energies of $[C \rightleftharpoons W]$ pulsation of the paired sub-elementary fermions and that of unpaired $F_\updownarrow^\pm\rangle$, related to their rotational-translational dynamics are strongly correlated and equal to each other:

$$(\omega_{C\rightleftharpoons W})_{F_\updownarrow^\pm\rangle} = \frac{m_C^+ \mathbf{v}^2}{\hbar} = (\omega_{C\rightleftharpoons W})_{[F_\uparrow^- \bowtie F_\downarrow^+]} \tag{45e}$$

The counterphase pulsation of $F_\updownarrow^\pm\rangle$ and $[F_\uparrow^- \bowtie F_\downarrow^+]$ makes them spatially compatible.

The induced resonant influence of Bivacuum fundamental frequency: $\omega_0^i = m_0^i c^2/\hbar$ of virtual pressure waves ($\mathrm{VPW}^\pm$) of Bivacuum, corresponding to Harmonization Energy on frequency of $[C \rightleftharpoons W]$ pulsation and other kind of dynamics of elementary particles, atoms and molecules - could be a physical background of *Principle of Least Action*, as demonstrated below.

The resulting *Action in Lagrangian form* for any elementary particle, as a difference between its external (S^{ext}) and internal (S^{in}) actions, taking into account (44 and 45a), can be presented as a difference between the internal and external actions, equal to to product of Harmonization energy of Bivacuum (HaE) on the time interval:

$$\left[S = S^{ext} - S^{in} = \hbar\left|\omega_{HaE}\right| t = \left|m_C^+(v_{gr}^{ext})^2 - m_0 c^2\right| t = (E_E + E_G)\, t = (HaE)\; t\right]_{x,y,z}^{i} \tag{46}$$

The *Principle of Least Action*, demanding that variation of action should be minimum $(\delta S = 0)$, turns (46) to important form:

$$\delta S = \left[\delta\left(HaE^i\right) t + \left(HaE^i\right)\delta t\right]_{x,y,z}^{i} = 0 \tag{47}$$

or, using (45a,b), we come to few formulations of notion of dimensionless *pace of time*:

$$\left[\frac{\delta t}{t}\right]_{x,y,z}^{i} = -\left[\frac{\delta\left(HaE^i\right)}{HaE^i}\right]_{x,y,z} = \left[-\frac{\delta\omega_{HaE}^i}{\omega_{HaE}^i}\right]_{x,y,z} = \left[-\frac{\delta[(T_k)_{\parallel tr} + (T_k)_{\perp tr}]^i}{\left[(T_k)_{\parallel tr} + (T_k)_{\perp tr}\right]^i}\right]_{x,y,z} = \tag{48}$$

$$= \left[-\frac{\delta\left(E_E + E_G\right)^i}{\left(E_E + E_G\right)^i}\right]_{x,y,z} = \left[-\frac{\delta[m_C^+(v_{gr}^{ext})_{tr}^2]}{m_C^+(v_{gr}^{ext})_{tr}^2}\right]_{x,y,z} = \left[-\frac{\delta T_{kin}}{T_{kin}}\right]_{x,y,z}$$

This formula interrelates the dimensionless pace of time $[\delta t/t]^i = d\ln t^i$ *for free particles* of three generations $(i = e, \mu, \tau)$ with pace of changes of translational kinetic energy contributions, responsible for electromagnetic and gravitational potentials of these particles. It is important to note here, that similar result we get from analysis of uncertainty principle for free particle in coherent form:

$$T_{kin} t = \hbar \quad \rightarrow \quad \left[\frac{\delta t}{t} = -\frac{\delta T_{kin}}{T_{kin}}\right]_{x,y,z}$$

In general logarithmic form the formula for unification of time, harmonization energy of Bivacuum, electromagnetic and gravitational fields (48) turns to:

$$\mathbf{d}\ln\mathbf{t}_{x,y,z}^{i} = -\,\mathbf{d}\ln\mathbf{HaE}_{x,y,z}^{i} = -\,\mathbf{d}\ln\left(\omega_{HaE}^{i}\right)_{x,y,z} = -\,\mathbf{d}\ln\left(E_E + E_G\right)_{x,y,z}^{i} = \tag{48a}$$

$$= -\,\mathbf{d}\ln\left[\,(T_k)_{\parallel tr} + (\mathbf{T}_k)_{\perp tr}\right]_{x,y,z}^{i} = -\mathbf{d}\ln[\alpha m_C^+\mathbf{v}^2 + \beta m_C^+\mathbf{v}^2]_{x,y,z}^{i} =$$

$$= -\mathbf{d}\ln[m_C^+(v_{gr}^{ext})_{tr}^2]_{x,y,z} \tag{48b}$$

$$\mathbf{d}\ln\mathbf{t}_{x,y,z}^{i} = -\mathbf{d}\ln\left[\frac{\hbar^2}{m_C^+\mathbf{L}_C^2}(\alpha+\beta)\right]_{x,y,z} \cong -\mathbf{d}\ln\left[\frac{\alpha\hbar^2}{m_C^+\mathbf{L}_C^2}\right]_{x,y,z} \tag{48c}$$

where the curvature, characterizing the resulting rotational-translational momentum of elementary particle with actual mass m_C^+ and external group velocity $(\mathbf{v})$ for general case is:

$$\left[\mathbf{L}_C = \frac{\hbar}{m_C^+\mathbf{v}}\right]_{x,y,z} \tag{48d}$$

The expression for time itself, interrelating the measurable parameters like the density of electromagnetic and gravitational energy in given volume of space and the resulting translational kinetic energy of any closed system in the same space volume, leading from eq.48, is:

$$t_{x,y,z} = -\left[\frac{m_C^+(\mathbf{v}_{gr}^{ext})_{tr}^2}{d[m_C^+(\mathbf{v}_{gr}^{ext})_{tr}^2]/dt}\right]_{x,y,z} = -\left[\frac{d\,(E_E+E_G)}{d\,(E_E+E_G)\,/dt}\right]_{x,y,z} \tag{48e}$$

where the actual inertial mass of elementary particle: $m_C^+ = m_0[1-(\mathbf{v}/\mathbf{c})^2]^{-1/2}$

We can see from the formula above and eq. 48 that the acceleration of particles of closed system increasing its translational kinetic energy (for example, due to heating of matter in any phase or melting or boiling or any way of entropy increasing), accompanied by the positive values of

$$\left(d[m_C^+(v_{gr}^{ext})_{tr}^2]/dt\right)_{x,y,z}$$

and

$$\left(d\,(E_E+E_G)\,/dt\right)_{x,y,z}$$

means the negative time course (t). In other words, the pace of time is negative and time slowing down with the increasing of translational and the resulting velocity of particles $(\mathbf{v})$, composing the closed system under consideration.

For the other hand, the decreasing of particles velocity, related with temperature and entropy decreasing should be accompanied by the opposite temporal effect, i.e. positive pace of time in a system. In the absence of accelerations the pace of time is zero. These consequences of our time concept are in total accordance with consequences of special theory of relativity, confirmed experimentally.

In the gas phase, when the spatial distribution of the particles kinetic energy and changes are isotropic: $\left[m_C^+(\mathbf{v}_{gr}^{ext})_{tr}^2\right]_x = \left[m_C^+(\mathbf{v}_{gr}^{ext})_{tr}^2\right]_y = \left[m_C^+(\mathbf{v}_{gr}^{ext})_{tr}^2\right]_z$, the time and its pace is also spatially isotropic. However, in vortices of liquids and rotating solid cylinders or gyroscopes, when the linear velocity $(\mathbf{v}_{gr}^{ext})_{x,y}$ and kinetic energy of particles is growing up with distance from the rotation center, the time and its pace due centripetal acceleration $(\mathbf{v}^2/R)$ increasing will slow down. The asymmetry of anharmonic oscillation of particles in crystal cells, related with anisotropic spatial distribution of their kinetic energy also means the spatial asymmetry of pace of time in such closed sub-systems.

The quantized time interval: $t = n\,\tau_{C\rightleftharpoons W}^{\phi}$ is determined by the period of $[C \rightleftharpoons W]$ pulsation of elementary particles at Golden mean conditions, when their translational group velocity is tending to zero $\mathbf{v}_{tr} \to 0$ and:

$$\tau_{C\rightleftharpoons W}^{\phi} = \frac{1}{\nu_{C\rightleftharpoons W}^{\phi}} = \frac{h}{m_0 c^2} \tag{48f}$$

$$t = n\,\tau_{C\rightleftharpoons W}^{\phi}$$

For the closed system of N interacting coherent particles (i.e. in state of Bose condensation) at the permanent for this system time period ($t = \tau_{C\rightleftharpoons W} = 1/\nu_{C\rightleftharpoons W} = const$), when $\delta t = \delta\tau = 0$, the **Principle of least action**, leading from (47 and 45a,b), can be presented as:

$$\sum_k^N \left(\delta\, HaE^i\right)_k T = \frac{1}{\nu}\sum_k^N \delta\left(HaE^i_k\right) = 0 \quad at: \quad t = \tau_{C\rightleftharpoons W} = 1/\nu_{C\rightleftharpoons W} = const > 0 \tag{49}$$

$$or: \sum_k^N \delta\left(HaE^i_k\right) = \sum_k^N \delta[E_E + E_G]^i$$
$$= \sum_k \delta\left[\left|m_C^+ v^2_{\|tr}\right|^i + \left|m_C^+ v^2_{\perp tr}\right|^i\right] = 0 \tag{49a}$$

This means that at the permanent Harmonization energy of Bivacuum: $E^i_{HaE} = \hbar\left|\omega^i_{HaE}\right| = const$, the *variations of the electric and gravitational potentials should compensate each other*, as well, as corresponding longitudinal and transversal translational contributions to resulting kinetic energy of elementary particles:

$$\left[\sum_k^N \delta E^i_E = \sum_k^N -\ \delta E^i_G\right]_{x,y,z} \tag{49b}$$

$$\text{or:} \quad \left[\sum_k^N \delta\left|m_C^+ v^2_{\|tr}\right|^i = \sum_k^N -\ \delta\left|m_C^+ v^2_{\perp tr}\right|^i\right]_{x,y,z}$$

Such kind of electromagnetism - gravitation compensation phenomena, based on Least action, also may contribute to Bivacuum mediated interaction between Sender [S] and Receiver [R].

Except *Harmonization energy of Bivacuum* (HaE - 45a), we may introduce also the *Harmonization force of Bivacuum* (HaF), acting on internal and external dynamics of elementary particles and atoms, like time derivative of Harmonization energy momentum ($F^i_{HaF} = \partial p^i_{HaE}/\partial t)_{x,y,z}$, where from (44b) the momentum: $p^i_{HaE} = E^i_{HaE}/c = \hbar\left|\omega_{HaE}\right|^i/c = \hbar/L^i_{HaE}$

The *Harmonization force* of Bivacuum can expressed as:

$$HaF^i_{x,y,z} = \left(\partial p^i_{HaE}/\partial t\right)_{x,y,z} = \frac{\partial}{\partial t}\left[\left|m_C^+ - m_C^-\right|_{\|tr} c + \left|m_C^+ - m_C^-\right|_{\perp tr}\right]^i_{x,y,z} c = \tag{50}$$

$$= \frac{\partial}{\partial t}\left[\alpha/c\left|m_C^+ v^2\right|^i + \beta/c\left|m_C^+ v^2\right|^i\right]_{x,y,z} = \frac{\partial}{\partial t}\left[E^i_{el} + E^i_G\right]_{x,y,z} \tag{51}$$

This expression means, that HaF of Bivacuum, acting on matter, is a result of Bivacuum *feedback reaction*, determined by temporal change of electromagnetic and gravitational fields, radiated by matter.

The another possible way of HaF presentation, leads from HaE definition (45c): $HaE^i = h\nu^i_{HaE} = [E_E + E_G]^i = HaF^i\lambda^i_{HaE}$ and from (45a):

$$HaF^i_{x,y,z} = \frac{h\left(\nu^2_{HaE}\right)^i_{x,y,z}}{c} = \left(\frac{h\nu_{HaE}}{\lambda_{HaE}}\right)^i_{x,y,z} = \left(\frac{(\omega_{C\rightleftharpoons W} - \omega_0)^i}{2\pi c}\left[E^i_E + E^i_G\right]\right)_{x,y,z} \tag{52}$$

$$or: \quad HaF^i_{x,y,z} = \left(\frac{\left[E^i_E + E^i_G\right]^2}{hc}\right)^i_{x,y,z} = \left(\frac{[\alpha m^+_C v^2 + \beta m^+_C v^2]^2}{hc}\right)^i_{x,y,z} = \tag{52a}$$

where:

$$(\lambda_{HaE})_{x,y,z} = 2\pi c/(\omega_{C\rightleftharpoons W} - \omega_0)_{x,y,z} = hc/\left[E_E + E_G\right]^i_{x,y,z} \tag{53}$$

is the wave length of HaE field; $(\omega_{C\rightleftharpoons W} - \omega_0) = \omega_{HaE}$ is a frequency of Harmonization energy of Bivacuum, acting on dynamics of pulsing elementary particles; L_G and L_E are the curvatures, characterizing the gravitational and electromagnetic potentials (see eqs. 48d and 48e).

We get the important conclusion, based on (51a), that the bigger is electromagnetic and gravitational radiation of particle/matter, proportional to particles kinetic energy, the bigger is Harmonization force of Bivacuum, acting on this particles. This means that HaF tends to slow down the external dynamics of particles, increasing the probability of coherent Bose condensation of matter and shifting the properties of matter closer to those of Golden mean.

14 Quantum entanglement between coherent elementary particles. The Virtual Replica (VR) of matter in Bivacuum

For explanation of nonlocal quantum entanglement between two or more particles we will use the notion of massless *Virtual spin waves (VirSW)*, excited by the angular moments of cumulative virtual clouds (CVC) of each of sub-elementary particles in triplets $\langle[F^-_\uparrow \bowtie F^+_\downarrow] + F^\pm_\updownarrow\rangle$ in their [W] phase. The VirSW represents the oscillation of dynamic equilibrium between Bivacuum fermions of opposite spins: $[BVF^\uparrow \rightleftharpoons BVF^\downarrow]$, without changing their energy. We put forward a conjecture, that VirSW, excited by unpaired $F^\pm_\updownarrow\rangle$, are highly anisotropic, depending on orientation (polarization) of triplets in space, and able to form *nonlocal virtual wave-guides (VirWG)* with internal magnetic field and refraction index much higher than the external ones ($\mathbf{H}_{in} >> \mathbf{H}_{ext}$; $n_{in} > n_{ext}$). The experimental data are existing, indeed, pointing to change of vacuum refraction index in strong magnetic fields (Ginsburg, 1987).

The standing Virtual spin waves (VirSW), excited by $[C \rightleftharpoons W]$ pulsation of unpaired sub-elementary particle $F_{\updownarrow}^{\pm}\rangle$ of Sender [S] and Receiver [R] form virtual wave-guide $VirWG_{F_{\updownarrow}^{\pm}\rangle}$ between [S] and [R]:

$$[VirSW^{S}_{F_{\updownarrow}^{\pm}\rangle} \Leftrightarrow VirSW^{R}_{F_{\updownarrow}^{\pm}\rangle}] \rightarrow (VirWG)_{F_{\updownarrow}^{\pm}\rangle} \equiv (VirWG)_{F_{\updownarrow}^{\pm}\rangle}$$

The Virtual spin waves, excited by $[C \rightleftharpoons W]$ pulsation of pair $[F_{\uparrow}^{-} \bowtie F_{\downarrow}^{+}]$ of triplets $\langle[F_{\uparrow}^{-} \bowtie F_{\downarrow}^{+}] + F_{\updownarrow}^{\pm}\rangle$ can form a pair of $[VirWG^{\circlearrowright} \bowtie VirWG^{\circlearrowleft}]_{F_{\uparrow}^{-} \bowtie F_{\downarrow}^{+}}$, which participate in Virtual Replica of matter formation. However, the energy of virtual photons, channelled by this pair, almost totally compensate each other.

The $VirWG^{S,R}$, formed by two uncompensated $F_{\updownarrow}^{\pm}\rangle$ of [S] and [R] is most effective as a phase/information conductor between remote elementary particles of the same frequency and counterphase $[C \rightleftharpoons W]$ pulsation.

We suggest, that the real and virtual photons, as a part of CVC energy, can propagate via $VirWG^{S,R}$. The Faraday cage can't shield such kind of EM field, tunnelling throw Bivacuum virtual wave-guide.

The mechanism of VirWG between Sender [S] and Receiver [R], as $[VirSW^{S}_{F_{\updownarrow}^{\pm}\rangle} \Leftrightarrow VirSW^{R}_{F_{\updownarrow}^{\pm}\rangle}]$ and $[VirWG^{\circlearrowright} \bowtie VirWG^{\circlearrowleft}]^{S,R}_{F_{\uparrow}^{-} \bowtie F_{\downarrow}^{+}}$ formation is determined by realization of Principle of least action for system of N coherent elementary particles in form of (49a):

$$\sum_{k}^{N} \delta\, (HaE_{k}^{i})_{x,y,z} = \sum_{k}^{N} \delta [E_E + E_G]^{i}_{x,y,z} = \sum_{k}^{N} \delta \left[\left| m_C^{+} v_{\| tr}^{2} \right|^{i} + \left| m_C^{+} v_{\perp tr}^{2} \right|^{i} \right]_{x,y,z} = 0 \tag{54}$$

where: HaE_{k}^{i} is Harmonization energy of Bivacuum; $E_E = \left| m_C^{+} v_{\| tr}^{2} \right|^{i}$ and $E_G = \left| m_C^{+} v_{\perp tr}^{2} \right|^{i}$ are electromagnetic and gravitational potentials of elementary particles of certain trajectories in 3D space (x, y, z).

The *nonlocal* interaction via VirWG - channel between two coherent elementary particles may change a phase/information of 2nd particle after changing the phase (spin) of the 1st one, but not the energy, related to irreversible electromagnetic and gravitational contributions to energy of [C$\rightleftharpoons W$] pulsation of elementary particles $(E_{C \rightleftharpoons W} = m_C^{+} c^2 = \hbar \omega_{C \rightleftharpoons W})$. The existence of VirWG between particles and external modulation of their phase/spin could be the main element of natural quantum computers.

Let us consider $[C \rightleftharpoons W]$ pulsation of unpaired $\left(\mathbf{F}_{\updownarrow}^{\pm}\right)$ and quasisymmetric pair $[\mathbf{F}_{\uparrow}^{-} \bowtie \mathbf{F}_{\downarrow}^{+}]$ of any selected triplet (electron or positron), related to [emission $\rightleftharpoons$ absorption] of Cumulative Virtual Clouds ($CVC^{\pm}$) of sub-quantum particles of positive and negative energies:

$$\langle [\mathbf{F}_{\uparrow}^{-} \bowtie \mathbf{F}_{\downarrow}^{+}]_C + \left(\mathbf{F}_{\updownarrow}^{\pm}\right)_W \rangle^i \underset{-E_{CVC^\pm}}{\overset{+E_{CVC^\pm}}{<=====>}} \langle [\mathbf{F}_{\uparrow}^{-} \bowtie \mathbf{F}_{\downarrow}^{+}]_W + \left(\mathbf{F}_{\updownarrow}^{\pm}\right)_C \rangle^i \tag{55}$$

For unpaired sub-elementary particle of any generation ($i = e, \mu, \tau$):

$$\left(\mathbf{F}_{\updownarrow}^{\pm}\right)_W \underset{-E_{CVC^\pm}}{\overset{+E_{CVC^\pm}}{<=====>}} \left(\mathbf{F}_{\updownarrow}^{\pm}\right)_C \tag{56}$$

The corresponding cumulative virtual cloud energy ($E_{CVC^\pm}$):

$$E_{CVC^\pm} = E_W = E_C = E_0 + E_E + E_G = \left(m_C^+ \mathbf{v}_{rot}^2\right)^\phi + m_C^+ \mathbf{v}_{\|tr}^2 + m_C^+ \mathbf{v}_{\perp tr}^2 = \tag{57}$$
$$= m_0 \omega_0^2 L_0^2 + \alpha m_C^+ \mathbf{v}^2 + \beta m_C^+ \mathbf{v}^2 = m_C^+ \mathbf{v}^2$$

where the rest mass energy, corresponding to Golden mean conditions (Kaivarainen, 2002) is

$$E_0 = \left(m_C^+ \mathbf{v}_{rot}^2\right)^\phi = m_0 c^2 = m_0 \omega_0^2 L_0^2 \tag{57a}$$

and the resulting external group velocity squared ($\mathbf{v}^2$) is defined by (45d).

At the Golden mean (ϕ) conditions the resulting zero-point velocity of elementary particles is permanent and can be expressed as:

$$\mathbf{v}_0^2 = \mathbf{v}_{\min}^2 = \mathbf{c}^2\phi + \left(\mathbf{v}_{\|tr}^{\phi}\right)^2 + \left(\mathbf{v}_{\perp tr}^{\phi}\right)^2 = \mathbf{c}^2\phi + \mathbf{c}^2\alpha\phi + \mathbf{c}^2\beta\phi = \mathbf{c}^2\left(\phi + \alpha\phi + \beta\phi\right) \tag{58}$$

Only the part of energy of CVC$^\pm$ (57), determined by the rest mass of sub-elementary particle $E_0 = m_0 c^2 = E_S$, determined by its rotation/spinning, is totally reversible in the process of $[C \rightleftharpoons W]$ pulsation of sub-elementary particles.

The contributions to CVC$^\pm$ energy, related to electromagnetism and gravitation, are radiated irreversibly and lead to certain Bivacuum fermions dynamics perturbations, like small symmetry shifts between properties of rotors ($\mathbf{V}^+$) and antirotors ($\mathbf{V}^-$) of Bivacuum fermions and their nonzero external momentum in secondary Bivacuum. For $[C \rightleftharpoons W]$ pulsation of pair $[\mathbf{F}_{\uparrow}^{-} \bowtie \mathbf{F}_{\downarrow}^{+}]$ we have:

$$[\mathbf{F}_{\uparrow}^{-} \bowtie \mathbf{F}_{\downarrow}^{+}]_C \underset{-[E_{CVC^+} + E_{CVC^-}]}{\overset{[E_{CVC^+} + E_{CVC^-}]}{<==========>}} [\mathbf{F}_{\uparrow}^{-} \bowtie \mathbf{F}_{\downarrow}^{+}]_W \tag{59}$$

It is important to outline *three important consequences*, following from such consideration:

1. The *experimentally* registrable **energy** of $[C \rightleftharpoons W]$ pulsation of triplets, equal to that of unpaired sub-elementary particle $\left(\mathbf{F}_{\updownarrow}^{\pm}\right)$, is limited only by electromagnetic and gravitational contributions:

$$E_{\mathbf{F}_{\updownarrow}^{\pm}}^{C\rightleftharpoons W} = E_E + E_G = m_C^+ \mathbf{v}_{\|tr}^2 + m_C^+ \mathbf{v}_{\perp tr}^2 = \alpha m_C^+ \mathbf{v}^2 + \beta m_C^+ \mathbf{v}^2 \tag{60}$$

The energy of virtual spin waves, averaged during period of $[C \rightleftharpoons W]$ pulsation, is zero:

$$E_{VirSW} = \left(m_0\omega_0^2 L_0^2\right)_{C\to W} - \left(m_0\omega_0^2 L_0^2\right)_{W\to C} = 0 \tag{61}$$

due to the total reversibility of rotational (spin) contribution to CVC energy. Consequently, the nonlocal VirWG, formed by VirSW, is a carrier of the phase only. Possibly, in some cases, it may serve, as a *channel* conducting virtual photons, related with electromagnetic contribution of unpaired sub-elementary particle: $E_E = m_C^+ \mathbf{v}_{\|tr}^2 = \alpha m_C^+ \mathbf{v}^2$. In such a case the EM radiation of the electron should be highly anisotropic;

2. To keep a structure of triplets of sub-elementary particles $\langle [F_\uparrow^- \bowtie F_\downarrow^+] + F_\updownarrow^\pm \rangle$ stable and energetically symmetric, the changes of the energy of unpaired sub-elementary fermion: $\Delta E_{\mathbf{F}_\updownarrow^\pm}$, should be accompanied by similar changes in the absolute energy of each of other sub-elementary particle and antiparticle in pair $[\mathbf{F}_\uparrow^- \bowtie \mathbf{F}_\downarrow^+]$ of triplets:

$$\begin{aligned} \left|-\Delta E^{\mathbf{F}_\uparrow^-}_{\mathbf{F}_\uparrow^- \bowtie \mathbf{F}_\downarrow^+}\right| &= \left|\Delta E^{\mathbf{F}_\uparrow^+}_{\mathbf{F}_\uparrow^- \bowtie \mathbf{F}_\downarrow^+}\right| = \Delta E_{\mathbf{F}_\updownarrow^\pm} = \\ &= [\Delta(m_C^+ \mathbf{v}_{\|tr}^2) + \Delta(m_C^+ \mathbf{v}_{\perp tr}^2)]_{\mathbf{F}_\updownarrow^\pm} = [\Delta E_E + \Delta E_G]_{\mathbf{F}_\updownarrow^\pm} \end{aligned} \tag{62}$$

The energy increments of sub-elementary particle and sub-elementary antiparticles in pairs $[\mathbf{F}_\uparrow^- \bowtie \mathbf{F}_\downarrow^+]$ are almost equal and opposite by sign to each other. Consequently, their sum is always close to zero $\Delta E^{\mathbf{F}_\uparrow^+}_{\mathbf{F}_\uparrow^- \bowtie \mathbf{F}_\downarrow^+} + \left(-\Delta E^{\mathbf{F}_\uparrow^-}_{\mathbf{F}_\uparrow^- \bowtie \mathbf{F}_\downarrow^+}\right) \cong 0$ and they have small contribution to resulting energy of triplets, determined presumably by unpaired sub-elementary fermion;

3. *Virtual Replica (VR) of matter (phantom), as a result of Bivacuum perturbations.*

The interference between CVC$^+$ and CVC$^-$, emitted$\rightleftharpoons$absorbed by pairs $[\mathbf{F}_\uparrow^+ \bowtie \mathbf{F}_\downarrow^-]_{x,y}$:

$$\left(\Delta E^{\mathbf{F}_\uparrow^+}_{\mathbf{F}_\uparrow^- \bowtie \mathbf{F}_\downarrow^+}\right)_{x,y,z} = \Delta\left(m_0\omega_0^2 L_0^2 + \alpha m_C^+ \mathbf{v}^2 + \beta m_C^+ \mathbf{v}^2\right)_{x,y,z} \tag{63}$$

$$\left(-\Delta E^{\mathbf{F}_\uparrow^-}_{\mathbf{F}_\uparrow^- \bowtie \mathbf{F}_\downarrow^+}\right)_{x,y,z} = -\Delta\left(m_0\omega_0^2 L_0^2 + \alpha m_C^+ \mathbf{v}^2 + \beta m_C^+ \mathbf{v}^2\right)_{x,y,z} \tag{63a}$$

and fundamental virtual pressure waves (VPW$^-$ and VPW$^+$) of Bivacuum in the process of exchange interaction between elementary particles and Bivacuum in a course of $[C \rightleftharpoons W]$ pulsation of triplets is responsible for VR$^\pm$(phantom) formation. The fundamental frequency of Bivacuum, corresponding to that of Golden mean: $\omega_0 = m_0c^2/\hbar$ (Kaivarainen, 2002), is close to frequency of $[C \rightleftharpoons W]$ pulsation of sub-elementary particles/antiparticles in nonrelativistic conditions ($\mathbf{v} << \mathbf{c}$), when:

$$\omega_{C \rightleftharpoons W}^{\langle[\mathbf{F}_{\uparrow}^{-}\bowtie\mathbf{F}_{\downarrow}^{+}]+\mathbf{F}_{\updownarrow}^{\pm}\rangle} = \frac{1}{\hbar}\left(m_0 c^2 + \alpha m_C^{+}\mathbf{v}^2 + \beta m_C^{+}\mathbf{v}^2\right) \tag{64}$$

Anisotropic in general case perturbation of rotors ($\mathbf{V}^{+}$) and antirotors ($\mathbf{V}^{-}$) of Bivacuum dipoles ($\mathbf{BVF}^{\updownarrow} = \mathbf{V}^{+} \Updownarrow \mathbf{V}^{-}$) see eqs. (1-3) by electromagnetic and gravitational potentials of each of paired sub-elementary fermions $[\mathbf{F}_{\uparrow}^{-} \bowtie \mathbf{F}_{\downarrow}^{+}]$:

$$\left(E_{\mathbf{F}_{\uparrow}^{-}\bowtie\mathbf{F}_{\downarrow}^{+}}^{\mathbf{F}_{\uparrow}^{+}} \cong -E_{\mathbf{F}_{\uparrow}^{-}\bowtie\mathbf{F}_{\downarrow}^{+}}^{\mathbf{F}_{\uparrow}^{-}} = m_0\omega_0^2 L_0^2 + \alpha m_C^{+}\mathbf{v}^2 + \beta m_C^{+}\mathbf{v}^2 = m_0 c^2 + [E_E + E_G] = m_C^{+}\mathbf{v}^2\right)_{x,y,z} \tag{65}$$

is responsible for creation of *Virtual Replicas (VR)* of elementary, particles, atoms, molecules, condensed matter and its different forms, including living organisms, planets and stars (Kaivarainen, 2003b). The phase/informational contribution to VR is provided by Virtual spin waves $(\mathrm{VirSW})_{\mathbf{F}_{\uparrow}^{-}\bowtie\mathbf{F}_{\downarrow}^{+}}$, excited by the angel momentums of Cumulative virtual clouds (CVC) of [Wave] phase of pairs $[\mathbf{F}_{\uparrow}^{-} \bowtie \mathbf{F}_{\downarrow}^{+}]$. The properties of VR are dependent directly on properties of unpaired sub-elementary fermions $F_{\updownarrow}^{\pm}\rangle$, detectable in direct experiments, strictly interrelated with those of $[\mathbf{F}_{\uparrow}^{-} \bowtie \mathbf{F}_{\downarrow}^{+}]$, hidden for direct experimental evaluation.

The nonlocal long-range correlation between two coherent elementary particles or antiparticles: Sender [S] and Receiver [R] via Bivacuum are provided by massless and nonlocal standing Virtual spin waves (VirSW), activated by cumulative virtual clouds $CVC_S^{\pm\circlearrowleft}$ *and* $CVC_R^{\pm\circlearrowleft}$ with opposite *angular momentums/spins* ($S = \pm\frac{1}{2}\hbar$). These clouds are emitted by unpaired sub-elementary fermions of two triplets $\langle[F_{\uparrow}^{-} \bowtie F_{\downarrow}^{+}] + F_{\updownarrow}^{\pm}\rangle$ with *opposite spins*, as a result of $\mathbf{C} \rightarrow \mathbf{W}$ transition:

$$[\mathbf{F}_{\uparrow}^{\pm}]_S^{\mathbf{C}\to\mathbf{W}} \overset{BvSO_S}{\rightarrowtail} \mathbf{CVC}_S^{\pm\circlearrowleft} \underset{VirSW_S \rightleftharpoons VirSW_R}{\overset{K_{BVF^{\uparrow}\rightleftharpoons BVF^{\downarrow}}}{\Longleftrightarrow\Longleftrightarrow}} \mathbf{CVC}_R^{\pm\circlearrowleft} \overset{BvSO_R}{\leftarrowtail} [\mathbf{F}_{\downarrow}^{\pm}]_R^{\mathbf{C}\to\mathbf{W}} \tag{66}$$

Two Bivacuum fermions: $BVF^{\uparrow}$ and $BVF^{\downarrow}$ with *opposite spins* (66) have a properties of virtual Cooper pairs. Spin-spin exchange interaction between them may course the ***attraction*** in pairs $[BVF^{\uparrow} \bowtie BVF^{\downarrow}]$ and ***"contraction"*** of Bivacuum. The virtual *Pauli repulsion* (PR) pressure in Bivacuum between BVF of the same spin state $[BVF^{\uparrow} \bowtie BVF^{\uparrow}]$, can be induced by interaction between unpaired sub-elementary fermions of remote elementary particles with the ***same* spins** and phase of $[C \rightleftharpoons W]$ pulsation (67):

$$[\mathbf{F}_{\uparrow}^{\pm}]_S^{\mathbf{C}\to\mathbf{W}} \overset{BvSO_S}{\rightarrowtail} \mathbf{CVC}_S^{\pm\circlearrowleft} \underset{VirSW_S \bowtie VirSW_R}{\overset{K_{BVF^{\uparrow}\rightleftharpoons BVF^{\downarrow}}}{\Longleftrightarrow\Longleftrightarrow}} \mathbf{CVC}_R^{\pm\circlearrowleft} \overset{BvSO_R}{\leftarrowtail} [\mathbf{F}_{\uparrow}^{\pm}]_R^{\mathbf{C}\to\mathbf{W}} \tag{67}$$

where $BvSO_{S,R}$ means Bivacuum symmetry oscillation, accompanied the emission $\rightleftharpoons$ absorption of cumulative virtual cloud (CVC) of similar angular momentums/spins.

Such effect may be accompanied by Bivacuum ***"expansion"***.

The $[C \rightleftharpoons W]$ pulsation of uncompensated sub-elementary particle $\mathbf{F}_{\updownarrow}^{\pm}\rangle$ and quasisymmetric pair $[\mathbf{F}_{\uparrow}^{-} \bowtie \mathbf{F}_{\downarrow}^{+}]$ of each triplet of elementary particle $\langle[\mathbf{F}_{\uparrow}^{-} \bowtie \mathbf{F}_{\downarrow}^{+}] + \mathbf{F}_{\updownarrow}^{\pm}\rangle$ is accompanied by VirSW excitation with the same frequency:

$$\left[\omega_{C\rightleftharpoons W} = \omega_{VirSW} = (m_C^{+} - m_C^{-})\, c^2/\hbar\right] \tag{68}$$

$L = \hbar/P^{ext}$ is the actual de Broglie wave of particle; $P^{ext} = m_C^{+}\omega_C L^{+}$ is the external momentum of particle.

The anisotropic amplitude probability $(A_{C\rightleftharpoons W})_{x,y,z}$ of resonant exchange interaction between **two** particles: 'sender (S)' and 'receiver (R)' may be qualitatively described, using well known model of **damped harmonic oscillator** interacting with external alternating field:

$$[A_{C\rightleftharpoons W}]_{x,y,z} \sim \left[\frac{1}{(m_C^{+})_R}\frac{[HaF]_{x,y,z}}{\omega_R^2 - \omega_S^2 + \mathrm{Im}\ \gamma\omega_S}\right]_{x,y,z} \tag{69}$$

where: ω_R and ω_S are the frequencies of $C \rightleftharpoons W$ pulsation of sub-elementary particles of (S) and (R);

γ is a damping coefficient due to *decoherence effects*, generated by exchange interaction of CVC$^{\pm}$ of pairs $\left[\mathbf{F}_{\uparrow}^{-} \bowtie \mathbf{F}_{\downarrow}^{+}\right]$ of triplets $\langle[\mathbf{F}_{\uparrow}^{-} \bowtie \mathbf{F}_{\downarrow}^{+}] + \mathbf{F}_{\updownarrow}^{\pm}\rangle$ with local fluctuations of Bivacuum; $\left(m_C^{+}\right)_R$ is the actual mass of particle (R); $[HaF]_{x,y,z}$ is a spatially anisotropic *Harmonization force of Bivacuum (52a)*.

The influence of Harmonization force of Bivacuum with fundamental Golden mean frequency $\omega_0 = m_0 c^2/\hbar$ stimulates the synchronization of (S) and (R) pulsations, i.e. $\omega_R \rightarrow \omega_S \rightarrow \omega_0$.

Consequently, the effectiveness of nonlocal interaction between two or more distant elementary particles is dependent on synchronization of their $[C \rightleftharpoons W]$ pulsations under the influence of Harmonization force of Bivacuum and quality of virtual wave-guide (VirWG)^{S,R} between Sender and Receiver, formed by standing virtual spin-waves $[VirSW^{S}_{F_{\updownarrow}^{\pm}\rangle} \Leftrightarrow VirSW^{R}_{F_{\updownarrow}^{\pm}\rangle}]$.

The nonlocal cosmophysical flicker noise can be responsible for so-called macroscopic fluctuations (Shnoll, 2001), resulting, in accordance to our Unified model, from beats/interference between harmonics of different components of Virtual Replicas (VR) of the Solar system, including Earth, Moon and Sun and VR of detectors. Like the Schumann resonance, this may enhance the correlation of quantum dynamics and entanglement between Sender and Receiver (detector) due to feedback reaction [Matter $\rightleftharpoons$ Virtual Replica of this Matter].

The mechanism, proposed, may explain the theoretical (Einstein, et all. 1935; Cramer, 1986) and experimental evidence in proof of nonlocal interaction between coherent elementary particles (Aspect and Gragier, 1983), atoms and even between remote coherent clusters of molecules in form of mesoscopic Bose condensation (mBC) (Kaivarainen, 2001a; 2003b).

Our Unified model predicts, that the same mechanism may provide the distant quantum entanglement between macroscopic systems, including biological ones, if $[C \rightleftharpoons W]$ *pulsations of their particles are 'tuned' to each other by the above mentioned mechanisms and they have close spatial polarization (orientation) and symmetry of their Virtual replicas (Kaivarainen, 2003 a,b).*

15 References

Aspect A. and Gragier P. Experiments on Einstein-Podolsky-Rosen type correlations with pairs of visible photons. In: *Quantum theory of measurement.* Eds. Wheeler J.A., Zurek W.H. Princeton University Press, (1983).

Dirac P. Book: *The principles of quantum mechanics.* Claredon press, Oxford, 1958.

Bohm D. and Hiley B.J. *The Undivided Universe. An ontological interpretation of quantum mechanics.* Routledge. London, New York, (1993).

Cramer J. G., The transactional interpretation of quantum mechanics. *Rev. Morden Physics*, 58, 647-688, (1986).

Einstein A., Podolsky B. and Rosen N. *Phys. Rev.* 47, 777-787, (1935).

Jin D.Z., Dubin D.H. E. Characteristics of two-dimensional turbulence that self-organizes into vortex crystals. *Phys. Rev. Lett.*, 84(7), 1443-1447, (2000).

Kaivarainen A. Book: Hierarchic concept of matter and field. *Water, biosystems and elementary particles.* New York, NY, pp. 482, ISBN 0-9642557-0-7 (1995).

Kaivarainen A., Hierarchic theory of condensed matter and its computerized application to water and ice. In the Archives of Los-Alamos: *http://arXiv.org/abs/physics/0102086*, (2001).

Kaivarainen A., Hierarchic theory of matter, general for liquids and solids: ice, water and phase transitions. *American Institute of Physics Conference Proceedings*, **573**, 181-200, (2001a).

Kaivarainen A. Bivacuum, sub-elementary particles and dynamic model of corpuscle-wave duality. CASYS: International Journal of Computing Anticipatory Systems. **10**, 121-137, (2001b).

Kaivarainen A. Unified model of Bivacuum, particles duality, electromagnetism, gravitation & time. The superflows energy of asymmetric Bivacuum. In the archives of Los-Alamos *http://arXiv.org/abs/physics/0207027*, (2002).

Kaivarainen A. New Hierarchic theory of water and its role in biosystems. The quantum Psi problem. *Proceedings of the International conference: "Energy and informational transfer in biological systems: How physics could enrich biological understanding".* F. Musumeci, L.S. Brishik, M.W. Ho (editors), World Scientific, pp. 82-147, (2003a).

Kaivarainen A. *New Hierarchic theory of water and its role in biosystems. Bivacuum mediated time effects, electromagnetic, gravitational & mental interactions.* *http://www.karelia.ru/~alexk/new_articles/index.html and in the Electronic library of Temporology Institute (Moscow University): http://www.chronos.msu.ru/relectropublications.html* (2003b).

Krasnoholovets V. On the nature of spin, inertia and gravity of a moving canonical particle. *Indian journal of theoretical physics*, 48, no.2, pp. 97-132, 2000.

Schecter D. A. and Dubin D. Vortex motion driven by background vorticity gradient. *Phys. Rev. Lett.*, 83 (11), 2191-2193, (1999).

In: Frontiers in Quantum Physics Research ISBN 1-59454-002-2
Editor: F. Columbus and V. Krasnoholovets, pp. 129-142

On Angular Momentum Operator in Quantum Field Theory

Bozhidar Z. Iliev[*†‡]

Abstract

Relations between two definitions of the (total) angular momentum operator, as a generator of rotations and in the Lagrangian formalism, are explored in quantum field theory. Generally, these definitions result in different angular momentum operators, which are suitable for different purposes in the theory. From the spin and orbital angular momentum operators (in the Lagrangian formalism) are extracted additive terms which are conserved operators and whose sum is the total angular momentum operator.

Key-Words:
Quantum field theory, Angular momentum operators
Angular momentum operator in quantum field theory
Spin and orbital angular momentum operators, Generators of rotations

1. Introduction

Two different definitions of the total angular momentum operator in quantum field theory are in current usage. The one defines it as a conserved operator arising via the Noether's theorem for rotation-invariant Lagrangians; we call the arising operator the canonical (or physical) angular momentum operator, or simply the angular momentum operator. The other one defines the angular momentum operator as a generator of the representation of rotations in the Minkowski spacetime on the space of operators acting on the Hilbert space of some system of quantum fields; we call the so-arising operator the rotational (or mathematical) angular momentum operator. As we shall see, this second operator is defined up to a constant second-rank tensor, which allows its identification with the physical angular momentum operator on some subset of the Hilbert space of states of a quantum system; as a rule, that subset is a proper subset.

[*]Laboratory of Mathematical Modeling in Physics, Institute for Nuclear Research and Nuclear Energy, Bulgarian Academy of Sciences, Boul. Tzarigradsko chaussée 72, 1784 Sofia, Bulgaria
[†]E-mail address: bozho@inrne.bas.bg
[‡]URL: http://theo.inrne.bas.bg/~bozho/

The present paper is similar to [1] and can be regarded as its continuation.

The lay-out of the work is as follows.

In Sect. 2 is reviewed the notion of angular momentum operator in the Lagrangian formalism. In Sect. 3 is considered the problem of conservation of spin and orbital angular momentum operators. From these operators are extracted additive parts, which are conserved operators and whose sum is the (total) angular momentum operator. Sect. 4 contains a brief review of the angular momentum operator as a generator of rotations. In Sect. 5 are discussed different commutation relations involving the canonical or rotational angular momentum operators. In Sect. 6 is shown that on some set the canonical and rotational angular momentum operators can coincide, but, generally, these are different operators. The basic results of the work are summarized in Sect. 7.

In what follows, we suppose that there is given a system of quantum fields, described via field operators $\varphi_i(x)$, $i = 1, \dots, n \in \mathbb{N}$, $x \in M$ over the 4-dimensional Minkowski spacetime M endowed with standard Lorentzian metric tensor $\eta_{\mu\nu}$ with signature $(+\,-\,-\,-)$.[1] The system's Hilbert space of states is denoted by $\mathcal{F}$ and all considerations are in Heisenberg picture of motion if the opposite is not stated explicitly. The Greek indices $\mu, \nu, \dots$ run from 0 to $3 = \dim M - 1$ and the Einstein's summation convention is assumed over indices repeated on different levels. The coordinates of a point $x \in M$ are denoted by x^μ, $\boldsymbol{x} := (x^1, x^2, x^3)$, $\mathrm{d}^3\boldsymbol{x} := \mathrm{d}x^1\,\mathrm{d}x^2\,\mathrm{d}x^3$, and the derivative with respect to x^μ is $\frac{\partial}{\partial x^\mu} =: \partial_\mu$. The imaginary unit is denoted by i and $\hbar$ and c stand for the Planck's constant (divided by 2π) and the velocity of light in vacuum, respectively.

2. The canonical angular momentum

Most of the material in this section is standard and can be found, for instance, in [2, 3, 4, 5].

Suppose, a system of quantum fields, represented by field operators $\varphi_i(x)\colon \mathcal{F} \to \mathcal{F}$, $i = 1, \dots, N$, is described by a Lagrangian $\mathcal{L} = \mathcal{L}(x) = \mathcal{L}(\varphi_i(x), \partial_\mu\varphi_i(x))$ depending on the fields and their first partial derivatives. Let us introduce the quantities

$$\pi^{i\mu} := \pi^{i\mu}(x) := \frac{\partial\,\mathcal{L}}{\partial(\partial_\mu\varphi_i(x))}, \tag{2.1}$$

called sometimes generalized momenta. As pointed in [6], here the derivatives with respect to $\partial_\mu\varphi_i(x)$, as well as with respect to other generally non-commuting arguments, should be considered as *mappings from some subspace ω of the operator*

[1] The quantum fields should be regarded as operator-valued distributions (acting on a relevant space of test functions) in the rigorous mathematical setting of Lagrangian quantum field theory. This approach will be considered elsewhere.

space $\{\mathcal{F} \to \mathcal{F}\}$ *over* $\mathcal{F}$ *on* $\{\mathcal{F} \to \mathcal{F}\}$, i.e. $\pi^{i\mu} \colon \omega \to \{\mathcal{F} \to \mathcal{F}\}$.[2]

The system's *(canonical) energy-momentum (tensorial) operator* is

$$\mathcal{T}^{\mu\nu} := \mathcal{T}^{\mu\nu}(x) := \sum_i \pi^{i\mu}(x)\big(\partial^\nu \varphi_i(x)\big) - \eta^{\mu\nu}\,\mathcal{L}(x) \tag{2.2}$$

and satisfies the continuity equation

$$\partial_\mu \mathcal{T}^{\mu\nu} = 0, \tag{2.3}$$

as a result of which the (canonical, dynamical, Noetherian) *momentum operator*

$$\mathcal{P}_\mu := \frac{1}{c}\int_{x^0=\text{const}} \mathcal{T}_{0\mu}(x)\,\mathrm{d}^3\boldsymbol{x} \tag{2.4}$$

is a conserved operator, i.e. $\frac{\mathrm{d}\mathcal{P}_\mu}{\mathrm{d}x^0} = 0$ and hence $\partial_\nu\,\mathcal{P}_\mu = 0$.

Suppose under a 4-rotation $x^\mu \mapsto x'^\mu = x^\mu + \varepsilon^{\mu\nu}x_\nu$, with $x_\nu := \eta_{\nu\mu}x^\mu$ and antisymmetric real parameters $\varepsilon^{\mu\nu} = -\varepsilon^{\nu\mu}$, the field operators transform as $\varphi_i(x) \mapsto \varphi'_i(x') := \varphi_i(x) + \sum_{\mu<\nu} I^j_{i\mu\nu}\varphi_j(x)\varepsilon^{\mu\nu} + \cdots$, where the dots stand for second and higher order terms in $\varepsilon^{\mu\nu}$ and the numbers $I^j_{i\mu\nu} = -I^j_{i\nu\mu}$ characterize the behaviour of the field operators under rotations.

The *total angular momentum density operator* of a system of quantum fields $\varphi_i(x)$ is

$$\mathcal{M}^\lambda_{\mu\nu}(x) = \mathcal{L}^\lambda_{\mu\nu}(x) + \mathcal{S}^\lambda_{\mu\nu}(x) \tag{2.5}$$

where

$$\mathcal{L}^\lambda_{\mu\nu}(x) := x_\mu\,\mathcal{T}^\lambda{}_\nu(x) - x_\nu\,\mathcal{T}^\lambda{}_\mu(x) \tag{2.6}$$

$$\mathcal{S}^\lambda_{\mu\nu}(x) := \sum_{i,j} \pi^{i\lambda}(x)\big(\varphi_j(x)\big) I^j_{i\mu\nu} \tag{2.7}$$

are respectively the *orbital and spin angular momentum density operators.* As a result of the continuity equation

$$\partial_\lambda\,\mathcal{M}^\lambda_{\mu\nu}(x) = 0, \tag{2.8}$$

the *(total) angular momentum operator*

$$\mathcal{M}_{\mu\nu} = \mathcal{L}_{\mu\nu}(x^0) + \mathcal{S}_{\mu\nu}(x^0) \tag{2.9}$$

---

[2] However, in some simple cases, the derivatives with respect to non-commuting variables may be computed by following the rules of the analysis of commuting variables by preserving the order of all operators; if this is the case, then $\pi^{i\mu}$ may be regarded as a mapping $\mathcal{F} \to \mathcal{F}$ and one can simply write, e.g., $\pi^{i\mu} \circ \varphi_j$ instead of $\pi^{i\mu}(\varphi_j)$. It should be emphasized, the treatment of derivatives like (2.1) as mappings $\omega \to \{\mathcal{F} \to \mathcal{F}\}$ combined with Schwinger action principle solves completely the problem for operator ordering in the conserved quantities [6, sec. 4].

where

$$\mathcal{L}_{\mu\nu}(x^0) := \frac{1}{c}\int_{x^0=\text{const}} \mathcal{L}^0_{\mu\nu}(x)\,\mathrm{d}^3\boldsymbol{x} \tag{2.10}$$

$$\mathcal{S}_{\mu\nu}(x^0) := \frac{1}{c}\int_{x^0=\text{const}} \mathcal{S}^0_{\mu\nu}(x)\,\mathrm{d}^3\boldsymbol{x} \tag{2.11}$$

are respectively the *orbital and spin angular momentum operators*, is a conserved quantity, i.e.

$$\frac{\mathrm{d}\,\mathcal{M}_{\mu\nu}}{\mathrm{d}x^0} = 0. \tag{2.12}$$

Notice, in the general case, the operators (2.10) and (2.11) are not conserved (see Sect. 3 below).

3. Conservation laws

Since from (2.3) and (2.5)–(2.8) follow the equations

$$\partial_\lambda \mathcal{L}^\lambda_{\mu\nu} = \mathcal{T}_{\mu\nu} - \mathcal{T}_{\nu\mu} \tag{3.1}$$

$$\partial_\lambda \mathcal{S}^\lambda_{\mu\nu} + \partial_\lambda \mathcal{L}^\lambda_{\mu\nu} = 0, \tag{3.2}$$

in the general case, when the (canonical) energy-momentum tensor $\mathcal{T}_{\mu\nu}$ is non-symmetric,[3] the spin and orbital angular momentum operators are not conserved. However, from the operators (2.10) and (2.11) can be extracted additive conserved ones, which are, in fact, the invariants characteristics of the spin and orbital angular properties of quantum systems.

For the purpose, define the antisymmetric operators

$$t_{\mu\nu}(x) := \int_{x^0_0}^{x^0} \big(\mathcal{T}_{\mu\nu}(x) - \mathcal{T}_{\nu\mu}(x)\big)\,\mathrm{d}x^0 = -t_{\nu\mu} \tag{3.3}$$

with x^0_0 being some arbitrarily fixed instant of the time coordinate x^0. Let us put

$${}^0\mathcal{S}_{\mu\nu} := \frac{1}{c}\int_{x^0=\text{const}} \{\,\mathcal{S}^0_{\mu\nu}(x) + t_{\mu\nu}(x)\}\,\mathrm{d}^3\boldsymbol{x} = \mathcal{S}_{\mu\nu}(x_0) + \frac{1}{c}\int_{x^0=\text{const}} t_{\mu\nu}(x)\,\mathrm{d}^3\boldsymbol{x} \tag{3.4a}$$

$${}^0\mathcal{L}_{\mu\nu} := \frac{1}{c}\int_{x^0=\text{const}} \{\,\mathcal{L}^0_{\mu\nu}(x) - t_{\mu\nu}(x)\}\,\mathrm{d}^3\boldsymbol{x} = \mathcal{L}_{\mu\nu}(x_0) - \frac{1}{c}\int_{x^0=\text{const}} t_{\mu\nu}(x)\,\mathrm{d}^3\boldsymbol{x}. \tag{3.4b}$$

[3] By adding to (2.2) a full divergence term, one can form a symmetric energy-momentum tensor; for details, see [7, sec. 2 and the references therein]. However, this does not change anything in our conclusions as expressions, like the r.h.s. of (3.1) with $\mathcal{T}_{\mu\nu}$ defined by (2.2), remain the same if one works with the new symmetric tensor.

Since

$$t_{\mu\nu}(x) = 0 \qquad \text{for } \mathcal{T}_{\mu\nu} = \mathcal{T}_{\nu\mu}, \tag{3.5}$$

we have

$${}^0\mathcal{S}_{\mu\nu} = \mathcal{S}_{\mu\nu}(x_0) \quad {}^0\mathcal{L}_{\mu\nu} = \mathcal{L}_{\mu\nu}(x_0) \qquad \text{for } \mathcal{T}_{\mu\nu} = \mathcal{T}_{\nu\mu}. \tag{3.6}$$

Applying (3.1)–(3.3), we get

$$\frac{\mathrm{d}}{\mathrm{d}x^0}\,{}^0\mathcal{L}_{\mu\nu} = \frac{1}{c}\int_{x^0=\text{const}} \mathrm{d}^3\boldsymbol{x}\{\partial_0\,\mathcal{L}^0_{\mu\nu}(x) - \partial_0 t_{\mu\nu}(x)\} = -\frac{1}{c}\int_{x^0=\text{const}} \mathrm{d}^3\boldsymbol{x}\sum_{a=1}^{3}\partial_a\,\mathcal{L}^a_{\mu\nu}(x)$$

$$\begin{aligned}\frac{\mathrm{d}}{\mathrm{d}x^0}\,{}^0\mathcal{S}_{\mu\nu} &= \frac{1}{c}\int_{x^0=\text{const}} \mathrm{d}^3\boldsymbol{x}\{\partial_0\,\mathcal{S}^0_{\mu\nu}(x) + \partial_0 t_{\mu\nu}(x)\}\\ &= \frac{1}{c}\int_{x^0=\text{const}} \mathrm{d}^3\boldsymbol{x}\Big\{-\sum_{a=1}^{3}\partial_a\,\mathcal{S}^a_{\mu\nu}(x) - \partial_\lambda\,\mathcal{L}^\lambda_{\mu\nu}(x) + \mathcal{T}_{\mu\nu}(x) - \mathcal{T}_{\nu\mu}(x)\Big\}\\ &= -\frac{1}{c}\int_{x^0=\text{const}} \mathrm{d}^3\boldsymbol{x}\sum_{a=1}^{3}\partial_a\,\mathcal{S}^a_{\mu\nu}(x).\end{aligned}$$

Let us suppose that the field operators tend to zero sufficiently fast at spacial infinity and $\mathcal{L}^a_{\mu\nu}(x)$, $\mathcal{S}^a_{\mu\nu}(x) \to 0$ when x tends to spacial infinity. Then, from the last equalities, we derive the conservation laws

$$\frac{\mathrm{d}}{\mathrm{d}x^0}\,{}^0\mathcal{L}_{\mu\nu} = 0 \qquad \frac{\mathrm{d}}{\mathrm{d}x^0}\,{}^0\mathcal{S}_{\mu\nu} = 0. \tag{3.7}$$

Thus, the operators (3.4) are conserved. Besides, due to equations (2.9) and (3.4), their sum is exactly the angular momentum operator,

$$\mathcal{M}_{\mu\nu} = {}^0\mathcal{L}_{\mu\nu} + {}^0\mathcal{S}_{\mu\nu}. \tag{3.8}$$

Moreover, if one starts from the definitions (3.3), (3.4), (2.1)–(2.7), and (2.9)–(2.11), one can prove (3.7) via a direct calculation (involving the field equations) the validity of (3.7) and, consequently, the conservation law (2.12) becomes a corollary of the ones for ${}^0\mathcal{L}_{\mu\nu}$ and ${}^0\mathcal{S}_{\mu\nu}$.

Since the operator ${}^0\mathcal{S}_{\mu\nu}$ characterizes entirely internal properties of the considered system of quantum fields, it is suitable to be called its *spin (or spin charge) operator*. Similar name, the *orbital operator*, is more or less applicable for ${}^0\mathcal{L}_{\mu\nu}$ too. Particular examples of these quantities will be presented elsewhere.

The above considerations are, evidently, true in the case of classical Lagrangian formalism of commuting variables too.[4]

---

[4] The only essential change in this case is that expressions like $\pi^{i\mu}(\varphi_j)$ should be replaced with $\pi^{i\mu}\varphi_j$, where an ordinary multiplication between functions is understood.

4. The generators of rotations

Besides (2.5), there is a second definition of the total angular momentum operator, which defines it as a generator of the representation of rotational subgroup of Poincaré group on the space of operators acting on the Hilbert space $\mathcal{F}$ of the fields $\varphi_i(x)$.[5] The so-arising operator will be referred as the *rotational* (or mathematical) *angular momentum operator* and will be denoted by $\mathcal{M}^{\mathrm{r}}_{\mu\nu}$. It is defined as follows. If $x \mapsto x'$, with $x'^{\lambda} = x^{\lambda} + \varepsilon^{\lambda\nu} x_{\nu} = x^{\lambda} + \sum_{\mu<\nu} \varepsilon^{\mu\nu}(\delta^{\lambda}_{\mu} x_{\nu} - \delta^{\lambda}_{\nu} x_{\mu})$, $\varepsilon^{\mu\nu} = -\varepsilon^{\nu\mu}$, is a rotation of the Minkowski spacetime, then it induces the transformation

$$\mathcal{A}(x) \mapsto \mathcal{A}(x') =: \mathrm{e}^{\frac{1}{\mathrm{i}\hbar}\frac{1}{2}\varepsilon^{\mu\nu}\mathcal{M}^{\mathrm{r}}_{\mu\nu}} \circ \mathcal{A}(x) \circ \mathrm{e}^{-\frac{1}{\mathrm{i}\hbar}\frac{1}{2}\varepsilon^{\mu\nu}\mathcal{M}^{\mathrm{r}}_{\mu\nu}}, \tag{4.1}$$

where $\circ$ denotes composition of mappings, $\mathcal{A}(x)\colon \mathcal{F} \to \mathcal{F}$ is a linear operator and $\mathcal{M}^{\mathrm{r}}_{\mu\nu}\colon \mathcal{F} \to \mathcal{F}$ are some operators. In differential form, equation (4.1) is equivalent to

$$[\mathcal{A}(x), \mathcal{M}^{\mathrm{r}}_{\mu\nu}]_{\_} = \mathrm{i}\hbar\Big(x_{\mu}\frac{\partial \mathcal{A}(x)}{\partial x^{\nu}} - x_{\nu}\frac{\partial \mathcal{A}(x)}{\partial x^{\mu}}\Big), \tag{4.2}$$

where $[\mathcal{A}, \mathcal{B}]_{\_} := \mathcal{A} \circ \mathcal{B} - \mathcal{B} \circ \mathcal{A}$ is the commutator of $\mathcal{A}, \mathcal{B}\colon \mathcal{F} \to \mathcal{F}$.

However, the behaviour of the field operators $\varphi_i(x)$ under rotations is, generally, more complicated than (4.1), viz. [2, 3]

$$\varphi_i(x) \mapsto \sum_j (S^{-1})_i^{\,j}(\varepsilon)\varphi_j(x') = \mathrm{e}^{\frac{1}{\mathrm{i}\hbar}\frac{1}{2}\varepsilon^{\mu\nu}\mathcal{M}^{\mathrm{r}}_{\mu\nu}} \circ \varphi_i(x) \circ \mathrm{e}^{-\frac{1}{\mathrm{i}\hbar}\frac{1}{2}\varepsilon^{\mu\nu}\mathcal{M}^{\mathrm{r}}_{\mu\nu}}, \tag{4.3}$$

with $S = [S_i^{\,j}(\varepsilon)]$ being a depending on $\varepsilon = [\varepsilon^{\mu\nu}]$ non-degenerate matrix such that $S|_{\varepsilon^{\mu\nu}=0}$ is the identity matrix of corresponding size. The appearance of a, generally, non-unit matrix S in (4.3) is due to the fact that under the set of field operators $\{\varphi_i(x)\}$ is understood the collection of all of the *components* $\varphi_i(x)$ of the fields forming a given system of quantum fields. This means that if, say, the field operators $\varphi_{i_1}(x), \dots, \varphi_{i_n}(x)$ for some indices $i_1, \dots, i_n$, $n \in \mathbb{N}$, represent a particular quantum field, they are components of an operator vector (a vector-valued operator) ϕ with respect to some basis $\{f_{i_1}, \dots, f_{i_n}\}$ of operator vector space to which ϕ belongs, $\phi(x) = \varphi_{i_1}(x) f_{i_1} + \dots + \varphi_{i_n}(x) f_{i_n}$. Under rotations, $\phi(x)$ transforms according to (4.1), but its components $\varphi_{i_1}(x), \dots, \varphi_{i_n}(x)$ generally do not; they transform in conformity with (4.3) because a change $x \mapsto x'$ is supposed to induce a linear change $f_{i_\alpha} \mapsto f'_{i_\alpha} = \sum_{\beta=1}^{n} \big((S^{\phi})^{-1}\big)_{i_\alpha}^{\;\;i_\beta}(\varepsilon) f_{i_\beta}$, $\alpha = 1, \dots, n$, with a non-degenerate matrix $S^{\phi}(\varepsilon) := [{S^{\phi}}_{i_\alpha}^{\;\;i_\beta}(\varepsilon)]$ [6] and $\phi(x') = \varphi_{i_1}(x') f'_{i_1} + \dots + \varphi_{i_n}(x') f'_{i_n}$.

[5] In axiomatic quantum field theory, this is practically the only definition of (total) angular momentum; see, e.g., [5, sec. 3.1.2], [8, p. 146] and [9, sec. 7.1]. This definition can also be found the (text)books on Lagrangian/canonical quantum field theory; see, for instance, [4, sec. 2.1], [5, sec. 3.1.2], [2, § 68], and [3, sec. 9.4].

[6] The matrix $S(\varepsilon)$ in (4.3) is a direct sum of the matrices $S^{\phi}(\varepsilon)$ for the independent fields ϕ forming the system under consideration.

Often (4.3) is written in a differential form as [2, eq. (11.73)]

$$[\varphi_i(x), \mathcal{M}^{\mathrm{r}}_{\mu\nu}]_- = \mathrm{i}\hbar\Big(x_\mu \frac{\partial \varphi_i(x)}{\partial x^\nu} - x_\nu \frac{\partial \varphi_i(x)}{\partial x^\mu}\Big) + \mathrm{i}\hbar \sum_j I^j_{i\mu\nu} \varphi_j(x), \tag{4.4}$$

where $I^j_{i\mu\nu} = -I^j_{i\nu\mu}$ are defined by $I^j_{i\mu\nu} := \frac{\partial S_i^{\;j}(\varepsilon)}{\partial \varepsilon^{\mu\nu}}\Big|_{\varepsilon=0}$ for $\mu < \nu$, i.e. $S_i^{\;j}(\varepsilon) = \delta^j_i + \frac{1}{2} I^j_{i\mu\nu}\varepsilon^{\mu\nu} + \cdots$ with δ^j_i being the Kronecker delta-symbol and the dots stay for higher order terms in $\varepsilon^{\mu\nu}$.

There is a simple relation between $\mathcal{M}^{\mathrm{r}}_{\mu\nu}$ and the rotation operator on $\mathcal{F}$. Let $\mathcal{M}^{\mathrm{QM}}_{\mu\nu}$, where QM stands for Quantum Mechanics (see below), denotes the Hermitian generator of rotations in $\mathcal{F}$, i.e. if $\mathcal{X}(x) \in \mathcal{F}$, then

$$\mathcal{X}(x) \mapsto \mathcal{X}(x') = \mathrm{e}^{\frac{1}{\mathrm{i}\hbar}\frac{1}{2}\varepsilon^{\mu\nu}\mathcal{M}^{\mathrm{QM}}_{\mu\nu}}(\mathcal{X}(x)) \tag{4.5}$$

with $x'^\mu = x^\mu + \varepsilon^{\mu\nu} x_\nu$. Explicitly, we have (see [5], [8, eq. (6.5)] or [9, eq. (7.14)])

$$\mathcal{M}^{\mathrm{QM}}_{\mu\nu} = \mathrm{i}\hbar(x_\mu \partial_\nu - x_\nu \partial_\mu), \tag{4.6}$$

which is exactly the orbital angular momentum operator in quantum mechanics if one restricts μ and ν to the range 1, 2, 3, forms the corresponding to (4.6) 3-vector operator (see, e.g., [10]) and identifies $\mathcal{F}$ with the Hilbert space of states of quantum mechanics[7]. The equalities

$$\mathcal{A}(x') = \mathrm{e}^{\frac{1}{\mathrm{i}\hbar}\frac{1}{2}\varepsilon^{\mu\nu}\mathcal{M}^{\mathrm{QM}}_{\mu\nu}} \circ \mathcal{A}(x) \circ \mathrm{e}^{-\frac{1}{\mathrm{i}\hbar}\frac{1}{2}\varepsilon^{\mu\nu}\mathcal{M}^{\mathrm{QM}}_{\mu\nu}} \tag{4.7}$$

$$\sum_j (S^{-1})_i^{\;j}(\varepsilon)\varphi_j(x') = \mathrm{e}^{\frac{1}{\mathrm{i}\hbar}\frac{1}{2}\varepsilon^{\mu\nu}\mathcal{M}^{\mathrm{QM}}_{\mu\nu}} \circ \varphi_i(x) \circ \mathrm{e}^{-\frac{1}{\mathrm{i}\hbar}\frac{1}{2}\varepsilon^{\mu\nu}\mathcal{M}^{\mathrm{QM}}_{\mu\nu}} \tag{4.8}$$

are simple corollaries of (4.5) and in differential form read

$$[\mathcal{A}(x), \mathcal{M}^{\mathrm{QM}}_{\mu\nu}]_- = \mathrm{i}\hbar\Big(x_\mu \frac{\partial \mathcal{A}(x)}{\partial x^\nu} - x_\nu \frac{\partial \mathcal{A}(x)}{\partial x^\mu}\Big), \tag{4.9}$$

$$[\varphi_i(x), \mathcal{M}^{\mathrm{QM}}_{\mu\nu}]_- = \mathrm{i}\hbar\Big(x_\mu \frac{\partial \varphi_i(x)}{\partial x^\nu} - x_\nu \frac{\partial \varphi_i(x)}{\partial x^\mu}\Big) + \mathrm{i}\hbar \sum_j I^j_{i\mu\nu}\varphi_j(x). \tag{4.10}$$

Comparing the last equations with (4.2) and (4.4), we find

$$[\mathcal{A}(x), \mathcal{M}^{\mathrm{QM}}_{\mu\nu} - \mathcal{M}^{\mathrm{r}}_{\mu\nu}]_- = 0 \tag{4.11a}$$

$$[\varphi_i(x), \mathcal{M}^{\mathrm{QM}}_{\mu\nu} - \mathcal{M}^{\mathrm{r}}_{\mu\nu}]_- = 0. \tag{4.11b}$$

[7] Since $\mathcal{F}$ is different from the Hilbert space of states of quantum mechanics, the superscript QM in $\mathcal{M}^{\mathrm{QM}}_{\mu\nu}$ only reminds the analogy between the r.h.s. of (4.6) and a similar operator in quantum mechanics.

If we admit (4.11a) (or (4.2)) to hold for *every* $\mathcal{A}(x)\colon \mathcal{F} \to \mathcal{F}$, the Schur's lemma[8] implies

$$\mathcal{M}^{\mathrm{r}}_{\mu\nu} = \mathcal{M}^{\mathrm{QM}}_{\mu\nu} + m_{\mu\nu}\,\mathsf{id}_{\mathcal{F}} = \mathrm{i}\hbar(x_\mu \partial_\nu - x_\nu \partial_\mu) + m_{\mu\nu}\,\mathsf{id}_{\mathcal{F}}, \tag{4.12}$$

where $\mathsf{id}_{\mathcal{F}}$ is the identity mapping of $\mathcal{F}$ and $m_{\mu\nu}$ are real numbers with dimension of angular momentum and forming the covariant components of some tensor of second rank.

One should be aware of the fact that the notation $\mathcal{M}^{\mathrm{QM}}_{\mu\nu}$ for the generator of rotations on $\mathcal{F}$ only emphasizes on the analogy with a similar operator in quantum mechanics (see also equation (4.6)); however, these two operators are completely different as they act on different spaces, the Hilbert space of states of quantum field theory and quantum mechanics respectively, which cannot be identified. For that reason, we cannot say that (4.12) with $m_{\mu\nu} = 0$ implies that the angular momentum of a system in quantum field theory and in quantum mechanics are equal up to a sign.

5. Discussion

The problem for coincidence of the both definitions of (total) angular momentum operator, the canonical and as generator of rotations, is a natural one and its positive answer is, more or less, implicitly assumed in the literature [4, 5]. However, these definitions originate from different approaches to quantum field theory: the canonical is due to the Lagrangian formalism [3, 14, 5], while the another one finds its natural place in axiomatic quantum field theory [8, 9].

As a condition weaker than

$$\mathcal{M}_{\mu\nu} = \mathcal{M}^{\mathrm{r}}_{\mu\nu}, \tag{5.1}$$

the relation (4.4), or its integral version (4.3), is assumed with $\mathcal{M}_{\mu\nu}$ for $\mathcal{M}^{\mathrm{r}}_{\mu\nu}$, i.e.

$$[\varphi_i(x), \mathcal{M}_{\mu\nu}]_{\_} = \mathrm{i}\hbar\Big(x_\mu \frac{\partial\varphi_i(x)}{\partial x^\nu} - x_\nu \frac{\partial\varphi_i(x)}{\partial x^\mu}\Big) + \mathrm{i}\hbar \sum_j I^j_{i\mu\nu}\varphi_j(x). \tag{5.2}$$

For instance, in [2, § 68] or in [3, sections 9.3 and 9.4], the last equation is proved under some explicitly written conditions concerning the transformation properties of the field operators and state vectors under Poincaré transformations. But, whatever the conditions leading to (5.2) are, they and (5.2) are external to the Lagrangian formalism by means of which the canonical angular momentum operator $\mathcal{M}_{\mu\nu}$ is defined. Usually, equation (5.2) is considered as one of the stipulations ensuring the relativistic covariance of the Lagrangian formalism and the theory is restricted to Lagrangians for which (5.2) is a consequence of the field equations (and, possibly, some restrictions on the formalism).[9]

[8] See, e.g, [11, appendix II], [12, sec. 8.2], [13, ch. 5, sec. 3].

[9] See [2], especially the comments on this item in section 68 of that book.

It should be noted, the general equation (4.2) with $\mathcal{M}_{\mu\nu}$ for $\mathcal{M}^{\mathrm{r}}_{\mu\nu}$ and arbitrary operator $\mathcal{A}(x)$, i.e.

$$[\mathcal{A}(x), \mathcal{M}_{\mu\nu}]_{\_} = \mathrm{i}\hbar\Big(x_\mu \frac{\partial \mathcal{A}(x)}{\partial x^\nu} - x_\nu \frac{\partial \mathcal{A}(x)}{\partial x^\mu}\Big), \tag{5.3}$$

cannot be valid; a simple counter example is provided by $\mathcal{A}(x) = \mathcal{P}_\mu$, with $\mathcal{P}_\mu$ being the canonical momentum operator of the considered system of quantum fields, for which the r.h.s. of (5.3) vanishes, as $\partial_\lambda \mathcal{P}_\mu = 0$, but $[\mathcal{M}_{\mu\nu}, \mathcal{P}_\lambda]_{\_} = \mathrm{i}\hbar(\eta_{\lambda\mu}\mathcal{P}_\nu - \eta_{\lambda\nu}\mathcal{P}_\mu)$ (see, e.g., [4, eq. (2-83)] or [8, eq. (6.1)]). However, if it happens (5.3) to hold for operators $\mathcal{A}(x)$ that form an irreducible unitary representation of some group, then, combining (5.3) and (4.2) and applying the Schur lemma, we get

$$\mathcal{M}_{\mu\nu} = \mathcal{M}^{\mathrm{r}}_{\mu\nu} + n_{\mu\nu}\,\mathsf{id}_{\mathcal{F}} = \mathcal{M}^{\mathrm{QM}}_{\mu\nu} + (m_{\mu\nu} + n_{\mu\nu})\,\mathsf{id}_{\mathcal{F}}, \tag{5.4}$$

where (4.12) was used and $n_{\mu\nu}$ are constant covariant components of some second-rank tensor.

Defining the operator

$$\mathcal{P}^{\mathrm{QM}}_\mu := \mathrm{i}\hbar\frac{\partial}{\partial x^\mu}, \tag{5.5}$$

which is the 4-dimensional analogue of the momentum operator in quantum mechanics, we see that it and $\mathcal{M}^{\mathrm{QM}}_{\mu\nu}$ (see (4.6)) satisfy the following relations

$$[\mathcal{P}^{\mathrm{QM}}_\mu, \mathcal{P}^{\mathrm{QM}}_\nu]_{\_} = 0 \tag{5.6a}$$
$$[\mathcal{M}^{\mathrm{QM}}_{\mu\nu}, \mathcal{P}^{\mathrm{QM}}_\lambda]_{\_} = -\mathrm{i}\hbar(\eta_{\lambda\mu}\mathcal{P}^{\mathrm{QM}}_\nu - \eta_{\lambda\nu}\mathcal{P}^{\mathrm{QM}}_\mu) \tag{5.6b}$$
$$[\mathcal{M}^{\mathrm{QM}}_{\mu\nu}, \mathcal{M}^{\mathrm{QM}}_{\kappa\lambda}]_{\_} = -\mathrm{i}\hbar\big(\eta_{\mu\kappa}\mathcal{M}^{\mathrm{QM}}_{\nu\lambda} + \eta_{\kappa\nu}\mathcal{M}^{\mathrm{QM}}_{\lambda\mu} + \eta_{\nu\lambda}\mathcal{M}^{\mathrm{QM}}_{\mu\kappa} + \eta_{\lambda\mu}\mathcal{M}^{\mathrm{QM}}_{\kappa\nu}\big), \tag{5.6c}$$

which characterize the Lie algebra of the Poincaré group (see [8, sec. 6.1], [9, sec. 7.1], and [15]). From (5.6) it is easily seen, the operators (4.12) and

$$\mathcal{P}^{\mathrm{t}}_\mu = -\mathcal{P}^{\mathrm{QM}}_\mu + p_\mu\,\mathsf{id}_{\mathcal{F}}, \tag{5.7}$$

where p_μ are constant covariant components of a 4-vector, satisfy the Lie algebra of the Poincaré group (see [8, sec. 6.1], [9, sec. 7.1] and [4, pp. 76–78]), i.e.

$$[\mathcal{P}^{\mathrm{t}}_\mu, \mathcal{P}^{\mathrm{t}}_\nu]_{\_} = 0 \tag{5.8a}$$
$$[\mathcal{M}^{\mathrm{r}}_{\mu\nu}, \mathcal{P}^{\mathrm{t}}_\lambda]_{\_} = \mathrm{i}\hbar(\eta_{\lambda\mu}\mathcal{P}^{\mathrm{t}}_\nu - \eta_{\lambda\nu}\mathcal{P}^{\mathrm{t}}_\mu) \tag{5.8b}$$
$$[\mathcal{M}^{\mathrm{r}}_{\mu\nu}, \mathcal{M}^{\mathrm{r}}_{\kappa\lambda}]_{\_} = \mathrm{i}\hbar\big(\eta_{\mu\kappa}\mathcal{M}^{\mathrm{r}}_{\nu\lambda} + \eta_{\kappa\nu}\mathcal{M}^{\mathrm{r}}_{\lambda\mu} + \eta_{\nu\lambda}\mathcal{M}^{\mathrm{r}}_{\mu\kappa} + \eta_{\lambda\mu}\mathcal{M}^{\mathrm{r}}_{\kappa\nu}\big) \tag{5.8c}$$

if and only if

$$p_\mu = 0 \quad m_{\mu\nu} = 0. \tag{5.9}$$

Thus, the relations (5.8) remove completely the arbitrariness in the operators $\mathcal{P}^{\mathrm{t}}_\mu$ and $\mathcal{M}^{\mathrm{r}}_{\mu\nu}$. The equations (5.8) are often [4] assumed to hold for the canonical momentum and angular momentum operators,

$$[\mathcal{P}_\mu, \mathcal{P}_\nu]_- = 0 \tag{5.10a}$$

$$[\mathcal{M}_{\mu\nu}, \mathcal{P}_\lambda]_- = \mathrm{i}\hbar(\eta_{\lambda\mu}\mathcal{P}_\nu - \eta_{\lambda\nu}\mathcal{P}_\mu) \tag{5.10b}$$

$$[\mathcal{M}_{\mu\nu}, \mathcal{M}_{\kappa\lambda}]_- = \mathrm{i}\hbar(\eta_{\mu\kappa}\mathcal{M}_{\nu\lambda} + \eta_{\kappa\nu}\mathcal{M}_{\lambda\mu} + \eta_{\nu\lambda}\mathcal{M}_{\mu\kappa} + \eta_{\lambda\mu}\mathcal{M}_{\kappa\nu}). \tag{5.10c}$$

However, these equations, as well as (5.2), are external to the Lagrangian formalism and, consequently, their validity should be checked for any particular Lagrangian.[10]

6. Inferences

Regardless of the fact that the equation (5.2) holds in most cases, it is quite different from the similar to it relation (4.4). Indeed, the relation (4.4) is an identity with respect to the field operators $\varphi_i(x)$, while (5.2) can be considered as an equation with respect to them. Thus, (5.2) can be considered as equations of motion relative to the field operators (known as (part of) the Heisenberg equations/relations), but (4.4) are identically valid with respect to these operators. Consequently, from this position, the possible equality (5.1) is unacceptable because it will entail (5.2) as an identity regardless of the Lagrangian one starts off.

Let $\mathcal{X} \in \mathcal{F}$ be a state vector of the considered system of quantum fields. Since we work in Heisenberg picture, it is a constant vector and, consequently, we have

$$\mathcal{M}^{\mathrm{QM}}_{\mu\nu}(\mathcal{X}) = 0 \quad \mathrm{e}^{\frac{1}{\mathrm{i}\hbar}\frac{1}{2}\varepsilon^{\mu\nu}\mathcal{M}^{\mathrm{QM}}_{\mu\nu}}(x) = 0 \quad \mathcal{P}^{\mathrm{QM}}_\mu(\mathcal{X}) = 0. \tag{6.1}$$

Then (4.12) implies

$$\mathcal{M}^{\mathrm{r}}_{\mu\nu}(\mathcal{X}) = m_{\mu\nu}\mathcal{X}. \tag{6.2}$$

As we intend to interpret $\mathcal{M}^{\mathrm{r}}_{\mu\nu}$ as system's total angular momentum operator, the last equation entails the *interpretation of* $m_{\mu\nu}$ *as components of the total 4-angular momentum of the system.* To justify this interpretation, one should assume $\mathcal{X}$ to be also an eigenvector of the canonical angular momentum operator with the same eigenvalues, i.e.

$$\mathcal{M}_{\mu\nu}(\mathcal{X}) = m_{\mu\nu}\mathcal{X}. \tag{6.3}$$

From here two conclusions can be made. On one hand, the equality (5.4) may be valid only for $n_{\mu\nu} = 0$, but, as we said earlier, this equation cannot hold in the general case. On other hand, equations (6.2) and (6.3) imply

$$\mathcal{M}_{\mu\nu}\big|_{\mathcal{D}_m} = \mathcal{M}^{\mathrm{r}}_{\mu\nu}\big|_{\mathcal{D}_m} \tag{6.4}$$

[10] Elsewhere we shall demonstrate that, if (5.10a) and (5.10c) hold, then the equation (5.10b) is *not* valid for a free spin 0, $\frac{1}{2}$ and 1 fields; more precisely, for these fields, it is true with an opposite sign of its r.h.s., i.e. with $-\mathrm{i}\hbar$ instead of $\mathrm{i}\hbar$.

where

$$\mathcal{D}_m := \{\mathcal{X} \in \mathcal{F} : \mathcal{M}_{\mu\nu}(\mathcal{X}) = m_{\mu\nu}\mathcal{X}\}. \tag{6.5}$$

Generally, the relation (6.4) is weaker than (5.1) and implies it if a basis of $\mathcal{F}$ can be formed from vectors in $\mathcal{D}_m$.

An alternative to (6.4) and (5.1) are the equalities

$$[\varphi_i(x), \mathcal{M}_{\mu\nu}]_{\_} = [\varphi_i(x), \mathcal{M}^{\mathrm{r}}_{\mu\nu}]_{\_} = \mathrm{i}\hbar\Big(x_\mu \frac{\partial\varphi_i(x)}{\partial x^\nu} - x_\nu \frac{\partial\varphi_i(x)}{\partial x^\mu}\Big) + \mathrm{i}\hbar \sum_j I^j_{i\mu\nu}\varphi_j(x), \tag{6.6}$$

which do not imply (5.1). The above discussion also shows that the equality $\mathcal{M}_{\mu\nu} = \mathcal{M}^{\mathrm{QM}}_{\mu\nu}$ can be valid only for states with vanishing total angular momentum.

In Sect. 5, we mentioned that the operators $\mathcal{M}_{\mu\nu}$ and $\mathcal{M}^{\mathrm{r}}_{\mu\nu}$ may satisfy the relations (5.8) or (5.9), respectively. However, in view of (5.9) and (6.2)–(6.4), the equations (5.8) may be valid only when applied to states with vanishing angular momentum (and momentum — see [1]). Therefore the relations (5.8) are, generally, unacceptable. However, in the Lagrangian formalism, the relations (5.10) may or may not hold, depending on the particular Lagrangian employed.

7. Conclusion

The following results should be mentioned as major ones of this paper:

(i) The generator $\mathcal{M}^{\mathrm{QM}}_{\mu\nu}$ (of the representation) of rotations in system's Hilbert space of states is *not* the (total) angular momentum operator in quantum field theory, but it is closely related to a kind of such operator (see (4.12)).

(ii) The rotational angular momentum operator $\mathcal{M}^{\mathrm{r}}_{\mu\nu}$ is a generator of (the representation of) rotations in the space of operators acting on system's Hilbert space of states. It depends on a second-rank tensor $m_{\mu\nu}$ with constant (relative to Poincaré group) components.

(iii) The canonical total angular momentum operator $\mathcal{M}_{\mu\nu}$ is, generally, different from the rotational one. But, if one identifies the tensor $m_{\mu\nu}$ with the total angular momentum of the system under consideration, the restrictions of $\mathcal{M}_{\mu\nu}$ and $\mathcal{M}^{\mathrm{r}}_{\mu\nu}$ to the set (6.5) coincide.

(iv) An operator $\mathcal{M}^{\mathrm{r}}_{\mu\nu}$, satisfying the commutation relations (5.8) of the Lie algebra of the Poincaré group, describes a system with vanishing angular momentum. The operator $\mathcal{M}_{\mu\nu}$ may or may not satisfy (5.10), depending on the Lagrangian describing the system explored.

(v) When commutators with field operators are concerned, the operators $\mathcal{M}_{\mu\nu}$ and $\mathcal{M}^{\mathrm{r}}_{\mu\nu}$ may be regarded as interchangeable ones (see (6.6)). However, the relations (4.4) are identities, while (5.2) are equations relative to the field operators and their validity depends on the Lagrangian employed.

(vi) The spin and orbital angular momentum operators (in the Lagrangian formalism) contain additive terms which are conserved operators and their sum is the total angular momentum operator. (This result is completely valid in the classical Lagrangian formalism too, when functions, not operators, are involved.)

As it is noted in [2, § 68], the quantum field theory must be such that the (canonical) angular momentum operator $\mathcal{M}_{\mu\nu}$, given in Heisenberg picture via (2.9), must satisfy the Heisenberg relations/equations (5.2). This puts some restrictions on the arbitrariness of the (canonical) energy-momentum tensorial operator $\mathcal{T}_{\mu\nu}$ and spin angular momentum density operator $\mathcal{S}^{\lambda}_{\mu\nu}$ (see (2.6), (2.10) and (2.11)), obtained, via the (first) Noether theorem, from the system's Lagrangian. Consequently, this puts some, quite general, restrictions on the possible Lagrangians describing systems of quantum fields.

If (5.2) holds, then, evidently, its r.h.s. is a sum of two parts, the first related to the orbital angular momentum and the second one — to the spin angular momentum. This observation suggests the idea of spliting the total angular momentum as a sum

$$\mathcal{M}_{\mu\nu} = \mathcal{M}^{\text{or}}_{\mu\nu}(x) + \mathcal{M}^{\text{sp}}_{\mu\nu}(x), \tag{7.1}$$

where $\mathcal{M}^{\text{or}}_{\mu\nu}(x)$ and $\mathcal{M}^{\text{sp}}_{\mu\nu}(x)$ characterize the 'pure' orbital and spin, respectively, properties of the system, and

$$[\varphi_i(x), \mathcal{M}^{\text{or}}_{\mu\nu}(x)]_{\_} = \mathrm{i}\hbar\Big(x_\mu\frac{\partial}{\partial x^\nu} - x_\nu\frac{\partial}{\partial x^\mu}\Big)\varphi_i(x) \tag{7.2}$$

$$[\varphi_i(x), \mathcal{M}^{\text{sp}}_{\mu\nu}(x)]_{\_} = \mathrm{i}\hbar I^{j}_{i\mu\nu}\varphi_j(x) \tag{7.3}$$

from where (5.2) follows. If we accept the Heisenberg relation for the momentum operator $\mathcal{P}_\mu$ of the system, i.e. [2, 3, 4, 5]

$$[\varphi_i(x), \mathcal{P}_\mu]_{\_} = \mathrm{i}\hbar\frac{\partial}{\partial x^\mu}\varphi_i(x), \tag{7.4}$$

then we can set

$$\mathcal{M}^{\text{or}}_{\mu\nu}(x) = x_\mu\mathcal{P}_\nu - x_\nu\mathcal{P}_\mu \tag{7.5}$$

$$\mathcal{M}^{\text{sp}}_{\mu\nu}(x) = \mathcal{M}_{\mu\nu} - \big(x_\mu\mathcal{P}_\nu - x_\nu\mathcal{P}_\mu\big). \tag{7.6}$$

Such a splitting can be justified by the explicit form of $\mathcal{M}_{\mu\nu}$ for free fields in a kind of 4-dimensional analogue of the Schrödinger picture of motion, which will be considered elsewhere. The operators (7.5) and (7.6), similarly to (2.6) and (2.7), are non-conserved quantities. The physical sense of the operator (7.5) is that it represents the angular momentum of the system due to its movement as a whole. Respectively, the operator (7.6) describes the system's angular momentum as a result of its internal movement and/or structure. Elsewhere we shall present an explicit splitting, like (7.1), for free fields in which the operators $\mathcal{M}^{\text{or}}_{\mu\nu}(x)$ and $\mathcal{M}^{\text{sp}}_{\mu\nu}(x)$ will be conserved ones and will represent the pure orbital and spin, respectively, angular momentum of the system considered.

References

[1] Bozhidar Z. Iliev. On momentum operator in quantum field theory. http://www.arXiv.org e-Print archive, E-print No. hep-th/0206008, 2002. See chapter 6 in this book.

[2] J. D. Bjorken and S. D. Drell. *Relativistic quantum fields*, volume 2. McGraw-Hill Book Company, New York, 1965. Russian translation: Nauka, Moscow, 1978.

[3] N. N. Bogolyubov and D. V. Shirkov. *Introduction to the theory of quantized fields.* Nauka, Moscow, third edition, 1976. In Russian. English translation: Wiley, New York, 1980.

[4] Paul Roman. *Introduction to quantum field theory.* John Wiley&Sons, Inc., New York-London-Sydney-Toronto, 1969.

[5] C. Itzykson and J.-B. Zuber. *Quantum field theory.* McGraw-Hill Book Company, New York, 1980. Russian translation (in two volumes): Mir, Moscow, 1984.

[6] Bozhidar Z. Iliev. On operator differentiation in the action principle in quantum field theory. In "Trends in Complex Analysis, Differential Geometry and Mathematical Physics", *Proceedings of the 6th International Workshop on Complex Structures and Vector Fields*, 3-6 September 2002, St Knstantin resort (near Varna), Bulgaria, Editors Stancho Dimiev and Kouei Sekigava, World Scientific, New Jersey-London-Singapore-Hong Kong, 2003. pp. 76–107. http://www.arXiv.org e-Print archive, E-print No. hep-th/0204003, April 2002.

[7] W. Pauli. Relativistic field theories of elementary particles. *Rev. Mod. Phys.*, 13:203–232, 1941. Russian translation in [16, pp. 372–423].

[8] N. N. Bogolubov, A. A. Logunov, and I. T. Todorov. *Introduction to axiomatic quantum field theory.* W. A. Benjamin, Inc., London, 1975. Translation from Russian: Nauka, Moscow, 1969.

[9] N. N. Bogolubov, A. A. Logunov, A. I. Oksak, and I. T. Todorov. *General principles of quantum field theory.* Nauka, Moscow, 1987. In Russian. English translation: Kluwer Academic Publishers, Dordrecht, 1989.

[10] A. M. L. Messiah. *Quantum mechanics*, volume I. North Holland, Amsterdam, 1961. Russian translation: Nauka, Moscow, 1978.

[11] Yu. B. Rumer and A. I. Fet. *Group theory and quantized fields.* Nauka, Moscow, 1977. In Russian.

[12] A. A. Kirillov. *Elements of the theory of representations.* Springer, Berlin, 1976. Translation from Russian (second ed., Nauka, Moscow, 1978).

[13] Asim Barut and Ryszard Roczka. *Theory of group representations and applications.* PWN — Polish Scientific Publishers, Waszawa, 1977. Russian translation: Mir, Moscow, 1980.

[14] J. D. Bjorken and S. D. Drell. *Relativistic quantum mechanics*, volume 1 and 2. McGraw-Hill Book Company, New York, 1964, 1965. Russian translation: Nauka, Moscow, 1978.

[15] Pierre Ramond. *Field theory: a modern primer*, volume 51 of *Frontiers in physics.* Reading, MA Benjamin-Cummings, London-Amsterdam-Don Mills, Ontario-Sidney-Tokio, 1 edition, 1981. 2nd rev. print, Frontiers in physics vol. 74, Adison Wesley Publ. Co., Redwood city, CA, 1989; Russian translation from the first ed.: Moscow, Mir 1984.

[16] Wolfgang Pauli. *Works on quantum theory. Articles 1928–1958.* Nauka Publ., Moscow, 1977. In Russian.

In: Frontiers in Quantum Physics Research ISBN 1-59454-002-2
Editor: F. Columbus and V. Krasnoholovets, pp. 143-156

On Momentum Operator in Quantum Field Theory

Bozhidar Z. Iliev*†‡

Abstract

The interrelations between the two definitions of momentum operator, via the canonical energy-momentum tensorial operator and as translation operator (on the operator space), are studied in quantum field theory. These definitions give rise to similar but, generally, different momentum operators, each of them having its own place in the theory. Some speculations on the relations between quantum field theory and quantum mechanics are presented.

Key-Words:
Quantum field theory, Momentum operator
Momentum operator in Lagrangian quantum field theory
Generators of translations

1. Introduction

Two definitions of a momentum operator exist in quantum field theory. The first one defines it as a conserved operator arising via the Noether's theorem for translation invariant Lagrangians; we call the arising operator the canonical (or physical) momentum operator, or simply the momentum operator. The second definition defines the momentum operator as a generator of the representation of translations in the Minkowski spacetime on the space of operators acting on the Hilbert space of some system of quantum fields; we call the so-arising operator the translation (or mathematical) momentum operator. As we shall see, this second operator is defined up to a constant 4-vector which allows its identification with the physical momentum operator on some subset of the Hilbert space of states of a quantum system; as a rule, that subset is a proper subset.

The lay-out of the work is as follows.

*Laboratory of Mathematical Modeling in Physics, Institute for Nuclear Research and Nuclear Energy, Bulgarian Academy of Sciences, Boul. Tzarigradsko chaussée 72, 1784 Sofia, Bulgaria
†E-mail address: bozho@inrne.bas.bg
‡URL: http://theo.inrne.bas.bg/~bozho/

In Sect. 2 the rigorous formulation of the definitions mentioned above are presented and a relation is derived between the (physical) momentum operator and the translation operator acting on system's Hilbert space; the last operator being similar to the momentum operator used in quantum mechanics. In Sect. 3 an analysis of these definitions is done and some relations between the different momentum operators are established. Certain conclusions from these results are made in Sect. 4. In particular, the possible equivalence between physical and mathematical momentum operators is discussed and the inference is made that, generally, these are different operators. The results obtained are summarized in Sect. 5. It also contains some speculations on the links between quantum field theory and (non-relativistic and relativistic) quantum mechanics.

In what follows, we suppose that there is given a system of quantum fields, described via field operators $\varphi_i(x)$, $i = 1, \dots, n \in \mathbb{N}$, $x \in M$ over the 4-dimensional Minkowski spacetime M endowed with standard Lorentzian metric tensor $\eta_{\mu\nu}$ with signature $(+ - - -)$.[1] The system's Hilbert space of states is denoted by $\mathcal{F}$ and all considerations are in Heisenberg picture of motion if the opposite is not stated explicitly. The Greek indices $\mu, \nu, \dots$ run from 0 to $3 = \dim M - 1$ and the Einstein's summation convention is assumed over indices repeating on different levels. The coordinates of a point $x \in M$ are denoted by x^μ, $\boldsymbol{x} := (x^1, x^2, x^3)$ and the derivative with respect to x^μ is $\frac{\partial}{\partial x^\mu} =: \partial_\mu$. The imaginary unit is denoted by i and $\hbar$ and c stand for the Planck's constant (divided by 2π) and the velocity of light in vacuum, respectively.

2. The two momentum operators

There are two quite similar understandings of the energy-momentum vectorial operator, called simply (4-)momentum operator and denoted by $\mathcal{P}_\mu$, in (translation invariant) quantum field theory.

The first definition of momentum operator defines it, in the Lagrangian quantum field theory, through the canonical energy-momentum tensorial operator $\mathcal{T}_{\mu\nu}$ [1, eq. (2-41)] of a system of quantum fields $\varphi_i(x)$, i.e. [2]

$$\mathcal{P}_\mu := \frac{1}{c} \int \mathcal{T}^0{}_\mu(x)\,\mathrm{d}^3\boldsymbol{x}. \tag{2.1}$$

Here the integration is over the equal-time surface $x^0 = ct = \text{const}$ with c being the velocity of light, x is a point in the Minkowski spacetime M of special relativity endowed with a metric tensor $\eta_{\mu\nu}$, $[\eta_{\mu\nu}] = \operatorname{diag}(1, -1, -1, -1)$ by means of which are raised/lowered the Greek indices. We call the so-defined operator $\mathcal{P}_\mu$ the canonical

[1] The quantum fields should be regarded as operator-valued distributions (acting on a relevant space of test functions) in the rigorous mathematical setting of Lagrangian quantum field theory. This approach will be considered elsewhere.

[2] See, e.g., [2, 1, 3, 4].

(or physical, dynamical, Noetherian) momentum operator or simply the *momentum operator*. The operator (2.1) is a conserved quantity [2, 1, 3], i.e. $\frac{\mathrm{d}\,\mathcal{P}_\mu}{\mathrm{d}x^0} = 0$. So, since $\mathcal{P}_\mu$ does not depend on x^a for $a = 1, 2, 3$, it is valid the equality

$$\frac{\partial\,\mathcal{P}_\mu}{\partial x^\nu} = 0. \tag{2.2}$$

The second definition of the momentum operator identifies it with the generator of representation of the translation subgroup of the Poincaré group on the space of operators acting on the Hilbert space $\mathcal{F}$ of some system of quantum fields $\varphi_i(x)$.[3] The so-arising operator will be referred as the *translation* (or mathematical) *momentum operator* and will be denoted by $\mathcal{P}^{\mathrm{t}}_\mu$. It is defined as follows. Let a and x be points in Minkowski spacetime M and $\mathcal{A}(x)\colon \mathcal{F} \to \mathcal{F}$ be a field operator $\varphi_i(x)$ or an operator corresponding to some dynamical variable $\mathbf{A}$. The translation $x \mapsto x+a$ in M entails

$$\mathcal{A}(x) \mapsto \mathcal{A}(x+a) = \mathrm{e}^{-\frac{1}{\mathrm{i}\hbar} a^\mu\,\mathcal{P}^{\mathrm{t}}_\mu} \circ \mathcal{A}(x) \circ \mathrm{e}^{\frac{1}{\mathrm{i}\hbar} a^\mu\,\mathcal{P}^{\mathrm{t}}_\mu}, \tag{2.3}$$

where $\hbar$ is the Planck constant (divided by 2π), which, in a differential form, can be rewritten as

$$\mathrm{i}\hbar \frac{\partial}{\partial x^\mu} \mathcal{A}(x) = [\mathcal{A}(x), \mathcal{P}^{\mathrm{t}}_\mu]_{\_} \tag{2.3$'$}$$

where $[\mathcal{A}, \mathcal{B}]_{\_} := \mathcal{A} \circ \mathcal{B} - \mathcal{B} \circ \mathcal{A}$ is the commutator of $\mathcal{A}$, $\mathcal{B}\colon \mathcal{F} \to \mathcal{F}$ with $\circ$ denoting the composition of mappings.

There is a simple relation between $\mathcal{P}^{\mathrm{t}}_\mu$ and the translation operator on $\mathcal{F}$. Let $\mathcal{P}^{\mathrm{QM}}_\mu$, where the superscript QM stands for Quantum Mechanics (see below and Sect. 5), denotes the generator of translation operator on system's Hilbert space $\mathcal{F}$, i.e. $\mathcal{P}^{\mathrm{QM}}_\mu$ is the Hermitian generator of the mapping $\mathcal{X}(x) \mapsto \mathcal{X}(x+a)$ for any points $x, a \in M$. Explicitly, we have

$$\mathcal{X}(x) \mapsto \mathcal{X}(x+a) = \mathrm{e}^{\frac{1}{\mathrm{i}\hbar} a^\mu\,\mathcal{P}^{\mathrm{QM}}_\mu} (\mathcal{X}(x)). \tag{2.4}$$

The equality

$$\mathcal{A}(x+a) = \mathrm{e}^{\frac{1}{\mathrm{i}\hbar} a^\mu\,\mathcal{P}^{\mathrm{QM}}_\mu} \circ \mathcal{A}(x) \circ \mathrm{e}^{-\frac{1}{\mathrm{i}\hbar} a^\mu\,\mathcal{P}^{\mathrm{QM}}_\mu} \tag{2.5}$$

is a simple corollary of (2.4): $\mathrm{e}^{+\cdots} \circ \mathcal{A}(x) \circ \mathrm{e}^{-\cdots}(\mathcal{X}(x)) = \mathrm{e}^{+\cdots}(\mathcal{A}(x)(\mathcal{X}(x-a))) = \mathcal{A}(x+a)(\mathcal{X}(x-a+a)) = \mathcal{A}(x+a)(\mathcal{X}(x))$. Similarly to (??$'$), the differential form of (2.5) is

$$\mathrm{i}\hbar \frac{\partial}{\partial x^\mu} \mathcal{A}(x) = -[\mathcal{A}(x), \mathcal{P}^{\mathrm{QM}}_\mu]_{\_}. \tag{2.5$'$}$$

[3] See, for instance, [5, p. 146], [6, sec. 7.1], [1, sec. 2.1.1 (iiia)], [4, sec. 3.1.2], and [7, § 68].

Subtracting (??′) from (??′), we find

$$[\mathcal{A}(x), \mathcal{P}_\mu^{\mathrm{QM}} + \mathcal{P}_\mu^{\mathrm{t}}]_- = 0. \tag{2.6}$$

Consequently, if $\mathcal{A}$ is *arbitrary*, this equality, by virtue of Schur's lemma[4] implies

$$\mathcal{P}_\mu^{\mathrm{t}} = -\mathcal{P}_\mu^{\mathrm{QM}} + p_\mu \,\mathsf{id}_{\mathcal{F}} \tag{2.7}$$

where $\mathsf{id}_{\mathcal{F}}$ is the identity mapping of $\mathcal{F}$ and p_μ are constant (real — see Sect. 4) numbers, with dimension of 4-momentum, representing the covariant components of some vector p.

Notice (cf. [2, subsec. 9.3], if $\langle\cdot|\cdot\rangle\colon \mathcal{F} \to \mathcal{F}$ is the Hermitian scaler product of $\mathcal{F}$, the definition (2.3) is chosen so that $\langle \mathcal{X}(x)|\,\mathcal{A}(x+a)(\mathcal{Y}(x))\rangle = \langle \mathcal{X}(x+a)|\,\mathcal{A}(x)(\mathcal{Y}(x+a))\rangle$, while in a case of (2.5) it is fulfilled $\langle \mathcal{X}(x+a)|\,\mathcal{A}(x+a)(\mathcal{Y}(x+a))\rangle = \langle \mathcal{X}(x)|\,\mathcal{A}(x)(\mathcal{Y}(x))\rangle$ for any $\mathcal{X}(x), \mathcal{Y}(x) \in \mathcal{F}$.

One should be aware of the fact that the notation $\mathcal{P}_\mu^{\mathrm{QM}}$ for the generator of translations on $\mathcal{F}$ only emphasizes on the analogy with a similar operator in quantum mechanics (see also equation (3.3) below); however, these two operators are completely different as they act on different spaces, the Hilbert space of states of quantum field theory and quantum mechanics respectively, which cannot be identified. For that reason, we cannot say that (2.7) with $p_\mu = 0$ implies, for instance, that if the energy of a system is positive in quantum field theory, then it is negative in quantum mechanics and *vice versa*.

3. Discussion

At this point a problem arises: are the operators $\mathcal{P}_\mu^{\mathrm{t}}$ and $\mathcal{P}_\mu$ identical? That is, can we write $\mathcal{P}_\mu^{\mathrm{t}} = \mathcal{P}_\mu$ and identify this operator with system's 'true' momentum operator, or, more generally, what is the connection between $\mathcal{P}_\mu^{\mathrm{t}}$ and $\mathcal{P}_\mu$, if any? The author of these lines fails to find the answer, as well as the explicit formulation, of that problem in the literature; the only exception being [7, § 68] where in the discussion following eq. (11.71) in *loc. cit.*, it is mentioned of a possible problem that may arise if the canonical momentum operator, in our terminology, does not satisfy equations like (3.5) below.

Consider two approaches to the above problem.

In [1, subsec. 2.1, p. 70] is *implicitly* assumed a connection between $\mathcal{P}_\mu^{\mathrm{t}}$ and $\mathcal{P}_\mu$,

[4] See, e.g, [8, appendix II], [9, sec. 8.2], [10, ch. 5, sec. 3].

by, in fact, postulating that (??′) should be valid for $\mathcal{P}_\mu$,[5] i.e.

$$\mathrm{i}\hbar \frac{\partial \mathcal{A}(x)}{\partial x^\mu} = [\mathcal{A}(x),\, \mathcal{P}_\mu]_{\_} \tag{3.1}$$

for any operator $\mathcal{A}(x)$. Combining this with (??′), we get

$$[\mathcal{A}(x),\, \mathcal{P}_\mu - \mathcal{P}^{\mathrm{t}}_\mu]_{\_} = 0$$

for *every* operator $\mathcal{A}\colon \mathcal{F} \to \mathcal{F}$. Hereof the Schur's lemma implies

$$\mathcal{P}_\mu = \mathcal{P}^{\mathrm{t}}_\mu + q_\mu \,\mathsf{id}_{\mathcal{F}} \tag{3.2}$$

for some 4-vector field q with *constant* covariant components q_μ.

In [2, §§ 9,10], we see a mixture of $\mathcal{P}_\mu$ and $\mathcal{P}^{\mathrm{t}}_\mu$, denoted there by the single symbol P^n, and only with some effort, from the context, one can find out whether the authors have in mind $\mathcal{P}_\mu$ or $\mathcal{P}^{\mathrm{t}}_\mu$. It is known, the generators of translations in a vector space $\mathcal{F}$ are the partial derivative operators ∂_μ on $\mathcal{F}$, so that the explicit form of $\mathcal{P}^{\mathrm{QM}}_\mu$ is[6]

$$\mathcal{P}^{\mathrm{QM}}_\mu = \mathrm{i}\hbar \partial^{\mathcal{F}}_\mu, \tag{3.3}$$

where, if $\mathcal{X}\colon x \mapsto \mathcal{X}(x) \in \mathcal{F}$, $\partial^{\mathcal{F}}_\mu\colon \mathcal{X} \mapsto \partial^{\mathcal{F}}_\mu(\mathcal{X})\colon x \mapsto \frac{\partial \mathcal{X}}{\partial x^\mu}\big|_x$. This equality is practically derived in [2, subsec. 9.3] where a remark is made that the *r.h.s. of* (3.3) *cannot serve as an energy-momentum vectorial operator* as its application to any state vector, which is constant in the Heisenberg picture, is identically equal to zero. The conclusion is that the momentum operator must be something else, specified in the *loc. cit.* as our translation momentum operator $\mathcal{P}^{\mathrm{QM}}_\mu$ until subsection 10.2 in *loc. cit.*, where as it is taken our physical momentum operator $\mathcal{P}_\mu$. However, the final results in [2] are correct as, in fact, in *loc. cit.* only the relations (cf. (??′))

$$\mathrm{i}\hbar \frac{\partial}{\partial x^\mu} \varphi_i(x) = [\varphi_i(x),\, \mathcal{P}^{\mathrm{t}}_\mu]_{\_}, \tag{3.4}$$

[5] In fact, in [1, subsec. 2.1] is actually *proved* equation (3.5) below from which follows (3.1) for $\mathcal{A}(x)$ of a type of polynomial or convergent power series in the field operators $\varphi_i(x)$. Actually the equality (3.1) cannot hold for arbitrary operator $\mathcal{A}(x)$. A simple counterexample is provided by any quantum system possessing a non-vanishing momentum and angular momentum operators $\mathcal{P}_\mu$ and $\mathcal{M}_{\mu\nu}$, respectively. Indeed, on one hand, we have $\partial_\lambda \mathcal{M}_{\mu\nu} = 0$ as $\mathcal{M}_{\mu\nu}$ is a conserved quantity [1, 2, 7], and, on other hand, $[\mathcal{M}_{\mu\nu},\, \mathcal{P}_\lambda]_{\_} = \mathrm{i}\hbar(\eta_{\lambda\mu}\mathcal{P}_\nu - \eta_{\lambda\nu}\mathcal{P}_\mu)$ (see, e.g., [1, p. 77]) and, consequently $\mathrm{i}\hbar \frac{\partial \mathcal{M}_{\mu\nu}}{\partial x^\lambda} = 0 \neq [\mathcal{M}_{\mu\nu},\, \mathcal{P}_\lambda]_{\_}$.

[6] Proof: Expanding $\mathcal{X}(x+a) = \sum_\sigma \mathcal{X}^\sigma e_\sigma$, where $\{e_\sigma\}$ is an independent of x basis of $\mathcal{F}$, into a Taylor's series around the point x, we get

$$\mathcal{X}(x+a) = \sum_\sigma \Big(\sum_{n=0}^{\infty} \frac{1}{n!} a^{\mu_1} \cdots a^{\mu_n} \frac{\partial^n \mathcal{X}^\sigma}{\partial x^{\mu_1} \ldots \partial x^{\mu_n}}\Big|_x \Big) e_\sigma = \sum_\sigma \Big(\sum_{n=0}^{\infty} \frac{1}{n!} (a^{\mu_1}\partial_{\mu_1}) \cdots (a^{\mu_n}\partial_{\mu_n}) \mathcal{X}^\sigma \Big)\Big|_x e_\sigma$$

$$= \sum_\sigma \Big(\sum_{n=0}^{\infty} \frac{1}{n!} [(a^\mu \partial_\mu)^n \mathcal{X}^\sigma]\Big|_x \Big) e_\sigma = \sum_{n=0}^{\infty} \frac{1}{n!} [(a^\mu \partial^{\mathcal{F}}_\mu)^n \mathcal{X}]\Big|_x = (\mathrm{e}^{a^\mu \partial^{\mathcal{F}}_\mu}(\mathcal{X}))(x),$$

that is $\mathrm{e}^{a^\mu \partial^{\mathcal{F}}_\mu}\colon \mathcal{X} \mapsto \mathrm{e}^{a^\mu \partial^{\mathcal{F}}_\mu}(\mathcal{X})\colon x \mapsto \mathcal{X}(x+a)$. Using some freedom of the notation, one usually writes $\frac{\partial}{\partial x^\mu}$ for $\partial^{\mathcal{F}}_\mu$ and the last result is written in the form (2.4).

where $\varphi_i(x)$ is any field operator, are employed (see, e.g., subsections 9.3–10.1 in *loc. cit.*). Further, in [2, subsec. 10.2], the authors assume (3.4) to hold with $\mathcal{P}_\mu$ for $\mathcal{P}_\mu^{\mathrm{t}}$,[7] i.e. (cf. (3.1))

$$\mathrm{i}\hbar \frac{\partial}{\partial x^\mu} \varphi_i(x) = [\varphi_i(x), \mathcal{P}_\mu]_{\_}. \tag{3.5}$$

From these equalities, know as *Heisenberg relations or Heisenberg equations of motion* for the field operators, they derive the (anti)commutation relations for the frequency parts of the field operators as well as for creation and annihilation operators (for free fields). Consequently, the theory is so built that in it both relations, (3.5) and (3.4), hold.

Remark 3.1. The relations (3.5) are external for the Lagrangian formalism and their validity depends on the particular Lagrangian employed. Usually [7, § 68] only Lagrangians for which (3.5) holds are used.

Suppose $\mathcal{A}(x)$ is function of the field operators and their partial derivatives (of finite order). If $\mathcal{A}(x)$ does not depend explicitly on x as a separate argument and it is a polynomial or convergent power series in the field operators and their derivatives, then (3.5) implies (3.1) for such functions $\mathcal{A}(x)$.[8] If we assume that such operators $\mathcal{A}(x)$ can form an irreducible unitary represention of some group, then, equation (3.2) follows from (3.1) and (??′) and the Schur's lemma.

Similar is the situation in other works devoted to the grounds of quantum field theory. E.g., in [4, § 3.1.2] both momentum operators are identified, while in [11, sec. 3.1] the momentum operator is identified with the Hermitian generator of a unitary representation of the translation subgroup of the Poincaré group. However, as we demonstrated, the generators of translations on the operator space are defined up to a multiples of the identity mapping of the system's Hilbert space. This arbitrariness seems not be used until now by implicitly setting the mentioned multiplier to zero.

4. Inferences

In Sect. 3 we saw that the translation momentum operator $\mathcal{P}_\mu^{\mathrm{t}}$ of a system of quantum fields is given by the r.h.s. of (2.7),

$$\mathcal{P}_\mu^{\mathrm{t}} = -\mathcal{P}_\mu^{\mathrm{QM}} + p_\mu \,\mathsf{id}_{\mathcal{F}}, \tag{4.1}$$

[7] This is a theorem; see, e.g., [1, p. 70]. The commutation relations must be compatible with (3.5) [7, § 68].

[8] To prove this, one has to calculate the commutators of every term in the expansion of $\mathcal{A}(x)$ with $\mathcal{P}_\mu$ by using the identity $[a \circ b, c]_{\_} \equiv [a, c]_{\_} \circ b + a \circ [b, c]_{\_}$ for any operators a, b, and c, to express the remaining commutators through (3.5), and, at the end, to sum all of the terms obtained. If $\mathcal{A}(x)$ depends on derivatives of $\varphi_i(x)$, in the proof one should take into account the equality $[\partial_\mu(\varphi_i), \mathcal{P}_\nu]_{\_} = \partial_\mu[\varphi_i, \mathcal{P}_\nu]_{\_} = \mathrm{i}\hbar\partial_\mu \circ \partial_\nu(\varphi_i)$, which is a corollary of $\partial_\mu \mathcal{P}_\nu \equiv 0$.

which, in view of (3.3), may be rewritten as

$$\mathcal{P}_\mu^{\mathrm{t}} = -\mathrm{i}\hbar\frac{\partial}{\partial x^\mu} + p_\mu \,\mathsf{id}_{\mathcal{F}}, \tag{4.2}$$

where p_μ are the covariant components of a 4-vector.[9]

It, obviously, satisfies the relation[10]

$$\mathrm{i}\hbar\frac{\partial\,\mathcal{A}(x)}{\partial x^\mu} = [\mathcal{A}(x),\, \mathcal{P}_\mu^{\mathrm{t}}]_{\_} \tag{4.3}$$

for any $\mathcal{A}(x)\colon \mathcal{F} \to \mathcal{F}$. A little below (see the discussion after (4.6)), the *possible* equality

$$\mathcal{P}_\mu^{\mathrm{t}} = \mathcal{P}_\mu \tag{4.4}$$

will be explored. If $\mathcal{A}(x)$ is (observable or not) operator constructed form the field operators and their partial derivatives, one can assumed (4.4) to hold when results involving $\mathcal{P}_\mu^{\mathrm{t}}$ are extracted from (4.3). However, if we want to look on (4.3) as (implicit Heisenberg) equations of motion, as it is done often in the literature, the equality (4.4) is unacceptable since (4.3) entails (3.1) as an identity. Consequently, when the equations of motion have to be considered, one should work with (3.1), not with (4.3). In particular, the *Heisenberg equations of motion for the field operators* φ_i *are* (3.5), not (3.4) or (4.3) with $\mathcal{A}(x) = \varphi_i(x)$ and $\mathcal{P}_\mu^{\mathrm{t}}$ given by (4.1).

Assume now $\mathcal{X} \in \mathcal{F}$ is a state vector of a system of quantum fields. It is a constant vector as we work in the Heisenberg picture, i.e. we have the equivalent equations

$$\mathrm{e}^{\frac{1}{\mathrm{i}\hbar}a^\mu \mathcal{P}_\mu^{\mathrm{QM}}}\mathcal{X} = \mathcal{X} \qquad \mathcal{P}_\mu^{\mathrm{QM}}(\mathcal{X}) = 0 \qquad \frac{\partial\,\mathcal{X}}{\partial x^\mu} = 0. \tag{4.5}$$

Then (4.1) and (3.2) imply respectively

$$\mathcal{P}_\mu^{\mathrm{t}}(\mathcal{X}) = p_\mu\,\mathcal{X} \tag{4.6}$$

$$\mathcal{P}_\mu(\mathcal{X}) = (p_\mu + q_\mu)\,\mathcal{X}. \tag{4.7}$$

So, any state vector is an eigenvector for $\mathcal{P}_\mu^{\mathrm{t}}$ with eigenvalue p_μ. As we would like to interpret $\mathcal{P}_\mu^{\mathrm{t}}$ a (total) 4-momentum operator of a system, (4.6) entails that p_μ *should be considered as components of the total 4-energy-momentum* vector p of the system under consideration. Notice, the 4-vector field p_μ generally depends on the state at which the system is, in particular $p_\mu = 0$ corresponds to its vacuum state(s). Of course, the proposed interpretation of p_μ is physically sensible if p_μ

[9] Since the Lorentz/Poincaré transformations, employed in quantum field theory, are linear transformations with *constant* (in spacetime) coefficients, the assertion that a vector has constant components is a covariant one.

[10] In fact, the r.h.s. of (4.1) is the general solution of (4.3) with respect to $\mathcal{P}_\mu^{\mathrm{t}}$ if (4.3) holds for every $\mathcal{A}(x)\colon \mathcal{F} \to \mathcal{F}$.

are *real* which we assume from now on.[11] This interpretation of the numbers p_μ allows their identification with the eigenvalues $p_\mu + q_\mu$ of the canonical momentum operator $\mathcal{P}_\mu$ when it acts on state vectors, viz. we should have

$$q_\mu = 0 \tag{4.8}$$

or

$$\mathcal{P}_\mu(\mathcal{X}) = p_\mu \mathcal{X} \tag{4.9}$$

for any state vector $\mathcal{X}$ describing system's state with total 4-momentum vector p_μ. Consequently, from (3.2), (4.1), and (4.8), we see that

$$\mathcal{P}_\mu|_{\mathcal{D}_p} = (-\mathcal{P}_\mu^{\text{QM}} + p_\mu \,\mathsf{id}_{\mathcal{F}})|_{\mathcal{D}_p} \qquad (= \mathcal{P}_\mu^{\text{t}}|_{\mathcal{D}_p}), \tag{4.10}$$

where

$$\mathcal{D}_p := \{\mathcal{X} \in \mathcal{F} : \mathcal{P}_\mu(\mathcal{X}) = p_\mu \mathcal{X}\}, \tag{4.11}$$

which, as it is easily seen, is a simple consequence of the conservation of the energy-momentum for a closed (translation invariant) system.[12] If a base of $\mathcal{F}$ can be formed from vectors in $\mathcal{D}_p$, from (4.10) the equality (4.4) will follow.[13] But, generally, we can only assert that

$$[\varphi_i(x), \mathcal{P}_\mu^{\text{t}}]_{\_} = [\varphi_i(x), \mathcal{P}_\mu]_{\_} = \mathrm{i}\hbar \frac{\partial \varphi_i(x)}{\partial x^\mu} \qquad (= -[\varphi_i(x), \mathcal{P}_\mu^{\text{QM}}]_{\_}) \tag{4.12}$$

which does *not* imply (4.4). The above discussion also reveals that the equality $\mathcal{P}_\mu^{\text{t}}(\mathcal{X}) = -\mathcal{P}_\mu^{\text{QM}}(\mathcal{X})$ can be valid only for states with zero 4-momentum, $p_\mu = 0$, i.e. only for the vacuum state(s).

It should be mentioned, the equality (4.4) entails (3.1) for arbitrary operator $\mathcal{A}(x)$, which, as pointed in footnote 5, leads to contradictions.

From (4.6) or directly from the explicit relation (4.1), we derive

$$\mathrm{e}^{-\frac{1}{\mathrm{i}\hbar} a^\mu \mathcal{P}_\mu^{\text{t}}}(\mathcal{X}) = \mathrm{e}^{\mathrm{i}(\frac{1}{\hbar} a^\mu p_\mu)} \mathcal{X}. \tag{4.13}$$

[11] Since $\partial/\partial x^\mu$ is anti-Hermitian operator, the assumption that p_μ are real, which is equivalent to the Hermiticity of $p_\mu \,\mathsf{id}_{\mathcal{F}}$, is tantamount to the Hermiticity of $\mathcal{P}_\mu^{\text{t}}$, $(\mathcal{P}_\mu^{\text{t}})^\dagger = \mathcal{P}_\mu^{\text{t}}$, and, consequently, to the unitarity of $\exp(\pm\frac{1}{\mathrm{i}\hbar} a^\mu \mathcal{P}_\mu^{\text{t}})$, as one expects it to be. (Recall, until now nobody has put the Hermiticity of the momentum operator under question for any on of its definitions.)

[12] It is worth noting, similar considerations in quantum mechanics give rise to the Schrödinger equation. Indeed, defining the 'mathematical' energy by $\mathcal{E}^{\text{m}} := \mathrm{i}\hbar \frac{\partial}{\partial t}$, $x^0 =: ct$, and the 'canonical' one by $\mathcal{E}^{\text{c}} = \mathcal{H}$, $\mathcal{H}$ being the system's Hamiltonian, we see that the equation $\mathcal{E}^{\text{c}}(\psi) = \mathcal{E}^{\text{m}}(\psi)$ is identical with the Schrödinger equation for a wavefunction ψ. If the system is closed, the common eigenvalues of $\mathcal{E}^{\text{c}}$ and $\mathcal{E}^{\text{m}}$ represent the (stationary) energy levels of the system under consideration.

[13] In Sect. 5 it will be proved that, generally, this is not the case; see, e.g., (5.8) and the sentence after it.

Hence, the action of the unitary operators $U(a,\mathbf{1}) := \mathrm{e}^{-\frac{1}{\mathrm{i}\hbar} a^\mu \mathcal{P}^{\mathrm{t}}_\mu}$, which form a unitary representation of the translation subgroup of the Poincaré group, on state vectors reduce to a simple multiplication with a phase factor. This means that the vectors $\mathcal{X}$ and $U(a,\mathbf{1})(\mathcal{X})$ describe one and the same state of the system as the state vectors, in general, are defined up to a phase factor.

As we see, the situation with $\mathcal{P}^{\mathrm{t}}_\mu$ is completely different from the one with $\mathcal{P}^{\mathrm{QM}}_\mu$ for which $\mathcal{P}^{\mathrm{QM}}_\mu(\mathcal{X}) \equiv 0$, i.e. if one takes $\mathcal{P}^{\mathrm{QM}}_\mu$ as a 'true' momentum operator, any state will be characterized by identically vanishing 4-momentum vector.

The relation

$$\mathcal{A}(x+a) = \mathrm{e}^{-\frac{1}{\mathrm{i}\hbar} a^\mu \mathcal{P}^{\mathrm{t}}_\mu} \circ \mathcal{A}(x) \circ \mathrm{e}^{+\frac{1}{\mathrm{i}\hbar} a^\mu \mathcal{P}^{\mathrm{t}}_\mu} \tag{4.14}$$

is a corollary of (4.2) and (2.3) as $p_\mu \,\mathsf{id}_{\mathcal{F}}$, $p_\mu = \text{const}$, commutes with all operators on $\mathcal{F}$. Hence, $\mathcal{P}^{\mathrm{t}}_\mu$, as given by (4.1), is a generator of (the representation of) the translations on the operators on $\mathcal{F}$. But, in view of (4.13), $\mathcal{P}_\mu$ is *not* a generator of (the representation of) the translations on the vectors in $\mathcal{F}$. As a result of (4.10), the same conclusions are valid and with respect to the canonical momentum operator $\mathcal{P}_\mu$ on the domain $\mathcal{D}_p$, defined via (4.11), for any given p.

Ending this section, we want to note that the components of momentum operator(s) commute. In fact, equation (3.1) with $\mathcal{A}(x) = \mathcal{P}_\nu$ implies

$$[\mathcal{P}_\mu, \mathcal{P}_\nu]_{\_} = 0 \tag{4.15}$$

due to (2.2).[14] Similar equality for the translation momentum operator $\mathcal{P}^{\mathrm{t}}_\mu$, i.e.

$$[\mathcal{P}^{\mathrm{t}}_\mu, \mathcal{P}^{\mathrm{t}}_\nu]_{\_} = 0 \tag{4.16}$$

is a direct consequence of (4.2) due to $\partial_\nu p_\mu \equiv 0$ (by definition p_μ are constant (real) numbers).

5. Conclusion

The main results of our previous exposition can be formulated as follows.

(i) The generator $\mathcal{P}^{\mathrm{QM}}_\mu$ of (the representation of) the translations in system's Hilbert space is *not* the momentum operator in quantum field theory. However, there is a close connection between both operators (see equation (4.1)).

(ii) The translation momentum operator $\mathcal{P}^{\mathrm{t}}_\mu$ of a quantum system is a generator of (the representation of) the translations in the space of operators acting on system's Hilbert space. It depends on a 4-vector p with constant (relative to Poincaré group) components.

---

[14] This proof of (4.15) is not quite rigorous as, in view of (2.1), $\mathcal{P}_\mu$ is not a function, but an operator-valued functional of the field operators $\varphi_i(x)$. Rigorously (4.15) is a corollary of (2.1), the conservation law (2.2), and the equal-time (anti-)commutation relations [1, 3]. For other proof, see, e.g., [1, p. 76].

(iii) The (canonical/physical) momentum operator $\mathcal{P}_\mu$ is, generally, different from the translation momentum operator. However, the restrictions of $\mathcal{P}_\mu$ and $\mathcal{P}^{\mathrm{t}}_\mu$ in the set (4.11) coincide due to the identification of the vector p with the vector of eigenvalues of $\mathcal{P}_\mu$.

(iv) When commutators with field operators or functions of them are concerned, the operators $\mathcal{P}_\mu$ and $\mathcal{P}^{\mathrm{t}}_\mu$ may be considered as interchangeable ones (see (4.12)). However, equalities, like (**??**′), in particular (3.4), are identities, while ones, like (3.5), are equations relative to the field operators and their validity depends on the particular Lagrangian from which $\mathcal{P}_\mu$ is constructed.

As it is noted in [7, § 68], the quantum field theory must be such that the (canonical) momentum operator $\mathcal{P}_\mu$, given in Heisenberg picture via (2.1), must satisfy the Heisenberg relations/equations (3.5). This puts some restrictions on the arbitrariness of the (canonical) energy-momentum tensorial operator $\mathcal{T}^{\mu\nu}$ entering into (2.1) and obtained, via the (first) Noether theorem, from the system's Lagrangian. Consequently, at the end, this puts some, quite general, restrictions on the possible Lagrangians describing systems of quantum fields.

Our analysis of the momentum operator in quantum field theory can be transferred *mutatis mutandis* on the angular momentum operator and similar ones arising via the (first) Noether theorem from Lagrangians invariant under some spacetime continuous symmetries.

Since in the description of the dynamics of a quantum system enters only the (physical) momentum operator $\mathcal{P}_\mu$, we share the opinion that it is more important than the mathematical momentum operator $\mathcal{P}^{\mathrm{t}}_\mu$; the latter one playing an auxiliary role mainly in the derivation of Heisenberg equations of motion or the transformation properties of quantum fields.

Now we would like to look on the non-relativistic and relativistic quantum mechanics from the view-point of the above results. Since these theories are, usually, formulated in the Schrödinger picture of motion, we shall, first of all, 'translate' the momentum operator into it. Besides, as in quantum field theory only Hamiltonians, which do not explicitly depend on the spacetime coordinates are considered, we shall suppose the system's Hamiltonian $\mathcal{H} = c\mathcal{P}_0$ to be of such a type in Heisenberg picture of motion.

The transition from Heisenberg picture to Schrödinger one is performed via the mappings

$$\mathcal{X} \mapsto {}^{\mathrm{S}}\mathcal{X} = \mathrm{e}^{\frac{1}{\mathrm{i}\hbar}(t-t_0)\mathcal{H}}(\mathcal{X}) \tag{5.1}$$

$$\mathcal{A}(x) \mapsto {}^{S}\mathcal{A}(x) = \mathrm{e}^{\frac{1}{\mathrm{i}\hbar}(t-t_0)\mathcal{H}} \circ \mathcal{A}(x) \circ \mathrm{e}^{-\frac{1}{\mathrm{i}\hbar}(t-t_0)\mathcal{H}}, \tag{5.2}$$

where $\mathcal{X} \in \mathcal{F}$ and $\mathcal{A}(x)\colon \mathcal{F} \to \mathcal{F}$ are arbitrary, t is the time (coordinate) and t_0 is arbitrarily fixed instant of time. In particular, (5.2) with $\mathcal{A}(x) = \mathcal{P}^{\mathrm{t}}_\mu$, gives (see (4.2) and recall that $x^0 = ct$, c being the velocity of light in vacuum)

$$c\,{}^{\mathrm{S}}\mathcal{P}_0^{\mathrm{t}} = c\,\mathcal{P}_0^{\mathrm{t}} + c\,\mathcal{P}_0 = c\,\mathcal{P}_0^{\mathrm{t}} + \mathcal{H} = -\mathrm{i}\hbar\frac{\partial}{\partial t} + e\,\mathsf{id}_{\mathcal{F}} + \mathcal{H} \tag{5.3a}$$

$${}^{\mathrm{S}}\mathcal{P}_a^{\mathrm{t}} = \mathcal{P}_a^{\mathrm{t}} = -\,\mathcal{P}_a^{\mathrm{QM}} + p_a\,\mathsf{id}_{\mathcal{F}} = -\mathrm{i}\hbar\frac{\partial}{\partial x^a} + p_a\,\mathsf{id}_{\mathcal{F}} \qquad a = 1,2,3. \tag{5.3b}$$

Here $p_0 = \frac{1}{c}e$ is an eigenvalue of $\mathcal{P}_0 = \frac{1}{c}\mathcal{H} = \frac{1}{c}\,{}^{\mathrm{S}}\mathcal{H} = {}^{\mathrm{S}}\mathcal{P}_0$.

Let ${}^{\mathrm{S}}\mathcal{X}_e$ be a state vector corresponding to a state with fixed energy e, i.e.

$${}^{\mathrm{S}}\mathcal{H}({}^{\mathrm{S}}\mathcal{X}_e) = \mathcal{H}({}^{\mathrm{S}}\mathcal{X}_e) = e\,{}^{\mathrm{S}}\mathcal{X}_e \quad \mathcal{H}(\mathcal{X}_e) = e\,\mathcal{X}_e. \tag{5.4}$$

A straightforward calculation gives

$$c\,\mathcal{P}_0^{\mathrm{t}}({}^{\mathrm{S}}\mathcal{X}_e) = -\mathrm{i}\hbar\frac{\partial\,{}^{\mathrm{S}}\mathcal{X}_e}{\partial t} + e\,{}^{\mathrm{S}}\mathcal{X}_e = 0 \tag{5.5}$$

which, in view of (5.4), is a version of the Schrödinger equation

$$\mathrm{i}\hbar\frac{\partial\,{}^{\mathrm{S}}\mathcal{X}_e}{\partial t} = {}^{\mathrm{S}}\mathcal{H}({}^{\mathrm{S}}\mathcal{X}_e). \tag{5.6}$$

Besides, equations (5.3a) and (5.5) imply

$$c\,{}^{\mathrm{S}}\mathcal{P}_0^{\mathrm{t}}({}^{\mathrm{S}}\mathcal{X}_e) = {}^{\mathrm{S}}\mathcal{H}({}^{\mathrm{S}}\mathcal{X}_e) = \mathcal{H}({}^{\mathrm{S}}\mathcal{X}_e) = c\,\mathcal{P}_0({}^{\mathrm{S}}\mathcal{X}_e) = e\,{}^{\mathrm{S}}\mathcal{X}_e. \tag{5.7}$$

Therefore, if $\mathcal{X}_e \neq 0$, then (see (5.5))

$$\mathcal{P}_0^{\mathrm{t}}({}^{\mathrm{S}}\mathcal{X}_e) = 0 \neq \frac{e}{c}\,{}^{\mathrm{S}}\mathcal{X}_e = \mathcal{P}_0({}^{\mathrm{S}}\mathcal{X}_e). \tag{5.8}$$

Hence, the non-zero state vectors, representing states with fixed and non-vanishing energy in the Schrödinger picture, can serve as example of vectors on which the equality (4.4) (in Heisenberg picture) *cannot* hold. However, it must be emphasized on the fact that a vector ${}^{\mathrm{S}}\mathcal{X}_e$ with ${}^{\mathrm{S}}\mathcal{X}_e \neq 0$ does not represent a physically realizable state in Heisenberg picture as it is a time-depending vector, contrary to the physically realizable ones which, by definition, are constant in this picture of motion.

Consider states with fixed 3-momentum vector $\boldsymbol{p} = (p^1, p^2, p^3) = -(p_1, p_2, p_3)$.[15] In 3-dimensional notation, the equation (5.3b) reads

$${}^{\mathrm{S}}\boldsymbol{\mathcal{P}}^{\mathrm{t}} = \boldsymbol{\mathcal{P}}^{\mathrm{t}} = -\,{}^{\mathrm{S}}\boldsymbol{\mathcal{P}}^{\mathrm{QM}} + \boldsymbol{p}\,\mathsf{id}_{\mathcal{F}} = \mathrm{i}\hbar\boldsymbol{\nabla} + \boldsymbol{p}\,\mathsf{id}_{\mathcal{F}} \tag{5.9}$$

---

[15] The Lorentz metric is suppose to be of signature $(+\,-\,-\,-)$. Here and below the boldface symbols denote 3-dimensional vectors formed from the corresponding 4-vectors.

with $\boldsymbol{\nabla} := \bigl(\frac{\partial}{\partial x^1}, \frac{\partial}{\partial x^2}, \frac{\partial}{\partial x^3}\bigr)$ and $\boldsymbol{p}$ being an eigenvector of $\boldsymbol{\mathcal{P}}$. Suppose $\mathcal{X}_{\boldsymbol{p}}$ is an eigenvector of $\boldsymbol{\mathcal{P}}$ with $\boldsymbol{p}$ as eigenvalue, viz.

$$\boldsymbol{\mathcal{P}}(\mathcal{X}_{\boldsymbol{p}}) = \boldsymbol{p}\,\mathcal{X}_{\boldsymbol{p}} \qquad {}^{\mathrm{S}}\boldsymbol{\mathcal{P}}({}^{\mathrm{S}}\mathcal{X}_{\boldsymbol{p}}) = \boldsymbol{p}\,{}^{\mathrm{S}}\mathcal{X}_{\boldsymbol{p}}. \tag{5.10}$$

Let us assume, as in non-relativistic quantum mechanics, that the vector ${}^{\mathrm{S}}\mathcal{X}_{\boldsymbol{p}}$ to be also an eigenvector of $\boldsymbol{\mathcal{P}}^{\mathrm{QM}} = -\mathrm{i}\hbar\boldsymbol{\nabla} = {}^{\mathrm{S}}\boldsymbol{\mathcal{P}}^{\mathrm{QM}}$ with the same eigenvalues, i.e.

$${}^{\mathrm{S}}\boldsymbol{\mathcal{P}}^{\mathrm{QM}}({}^{\mathrm{S}}\mathcal{X}_{\boldsymbol{p}}) = \boldsymbol{p}\,{}^{\mathrm{S}}\mathcal{X}_{\boldsymbol{p}} \tag{5.11}$$

or, equivalently, ${}^{\mathrm{S}}\mathcal{X}_{\boldsymbol{p}} = \mathrm{e}^{-\frac{1}{\mathrm{i}\hbar}(\boldsymbol{x}-\boldsymbol{x}_0)\cdot\boldsymbol{p}}({}^{\mathrm{S}}\mathcal{X}_{\boldsymbol{p}}|_{x_0})$, equation (5.9) yields

$${}^{\mathrm{S}}\boldsymbol{\mathcal{P}}^{\mathrm{t}}({}^{\mathrm{S}}\mathcal{X}_{\boldsymbol{p}}) = \boldsymbol{\mathcal{P}}^{\mathrm{t}}({}^{\mathrm{S}}\mathcal{X}_{\boldsymbol{p}}) \equiv 0 \tag{5.12}$$

which in the Heisenberg picture reads

$$\boldsymbol{\mathcal{P}}^{\mathrm{t}}(\mathcal{X}_{\boldsymbol{p}}) = 0. \tag{5.13}$$

By virtue of (5.10), the last equality means that

$$\boldsymbol{p} = 0 \tag{5.14}$$

if $\mathcal{X}_{\boldsymbol{p}} \neq 0$. Consequently, from the view-point of quantum field theory, the quantum mechanics describes systems with zero 3-momentum. This unpleasant conclusion is not rigorous. It simply shows that (5.11) is not compatible with other axioms of quantum field theory or, said differently, the quantum mechanics and quantum field theories rest on different, not completely compatible postulates and, in particular, their Hilbert spaces of states are not identical. Since we plan to give a satisfactory solution of the problem of comparison of the grounds of and interrelations between these theories elsewhere, below are presented non-rigorous, possibly intuitive and naive, conclusions from the above-written material.

The non-relativistic quantum mechanics can be obtained from quantum field theory in Schrödinger picture by extracting from the latter theory only the Schrödinger equation (5.6) and ignoring all other aspects of it. Besides, this equation is generalized in a sense that it is assumed to hold for arbitrary, generally non-closed or translation non-invariant, systems, i.e.

$$\mathrm{i}\hbar\frac{\partial\,{}^{\mathrm{S}}\mathcal{X}}{\partial t} = {}^{\mathrm{S}}\mathcal{H}(t,\boldsymbol{x})({}^{\mathrm{S}}\mathcal{X}). \tag{5.15}$$

The explicit form of the Hamiltonian ${}^{\mathrm{S}}\mathcal{H}(t,\boldsymbol{x})$, as well as of other operators, representing dynamical variables, are almost 'put by hands' by postulating them; the only guiding principle being the compatibility with some classical analogues, if such exist for a given variable or system. In particular, it happens that the operator representing the 3-momentum vector is exactly $\boldsymbol{\mathcal{P}}^{\mathrm{QM}} = -\mathrm{i}\hbar\boldsymbol{\nabla}$, i.e. $\mathcal{P}_a^{\mathrm{QM}} = -(\boldsymbol{\mathcal{P}}^{\mathrm{QM}})^a =$

$i\hbar\frac{\partial}{\partial x^a}$, $a = 1, 2, 3$. It should also be noted, the replacement of the field equations of Lagrangian quantum field theory with the Schrödinger equation (5.15) entails, among other thing, a change of the Hilbert space of states of this theory with the one of non-relativistic quantum mechanics.

Now a few words about relativistic quantum mechanics are in order. The situation in that case is similar to the non-relativistic one with the only difference that the Hamiltonian operator ${}^{\mathrm{S}}\mathcal{H}(t, \boldsymbol{x})$ should be consistent with special relativity, not with classical mechanics. For instance, for a point particle with rest mass m, it must be such that

$$({}^{\mathrm{S}}\mathcal{H}(t, \boldsymbol{x}))^2 = c^2(\boldsymbol{\mathcal{P}}^{\mathrm{QM}})^2 + m^2c^4\,\mathsf{id}_{\mathcal{F}} \tag{5.16}$$

where $(\boldsymbol{\mathcal{P}}^{\mathrm{QM}})^2$ is the square of the 3-dimensional part of the operator (3.3). From this relation, under some additional assumptions, the whole relativistic quantum mechanics can be derived, as it is done, e.g., in [12].

References

[1] Paul Roman. *Introduction to quantum field theory.* John Wiley&Sons, Inc., New York-London-Sydney-Toronto, 1969.

[2] N. N. Bogolyubov and D. V. Shirkov. *Introduction to the theory of quantized fields.* Nauka, Moscow, third edition, 1976. In Russian. English translation: Wiley, New York, 1980.

[3] J. D. Bjorken and S. D. Drell. *Relativistic quantum mechanics*, volume 1 and 2. McGraw-Hill Book Company, New York, 1964, 1965. Russian translation: Nauka, Moscow, 1978.

[4] C. Itzykson and J.-B. Zuber. *Quantum field theory.* McGraw-Hill Book Company, New York, 1980. Russian translation (in two volumes): Mir, Moscow, 1984.

[5] N. N. Bogolubov, A. A. Logunov, and I. T. Todorov. *Introduction to axiomatic quantum field theory.* W. A. Benjamin, Inc., London, 1975. Translation from Russian: Nauka, Moscow, 1969.

[6] N. N. Bogolubov, A. A. Logunov, A. I. Oksak, and I. T. Todorov. *General principles of quantum field theory.* Nauka, Moscow, 1987. In Russian. English translation: Kluwer Academic Publishers, Dordrecht, 1989.

[7] J. D. Bjorken and S. D. Drell. *Relativistic quantum fields*, volume 2. McGraw-Hill Book Company, New York, 1965. Russian translation: Nauka, Moscow, 1978.

[8] Yu. B. Rumer and A. I. Fet. *Group theory and quantized fields.* Nauka, Moscow, 1977. In Russian.

[9] A. A. Kirillov. *Elements of the theory of representations.* Springer, Berlin, 1976. Translation from Russian (second ed., Nauka, Moscow, 1978).

[10] Asim Barut and Ryszard Roczka. *Theory of group representations and applications.* PWN — Polish Scientific Publishers, Waszawa, 1977. Russian translation: Mir, Moscow, 1980.

[11] R. F. Streater and A. S. Wightman. *PCT, spin and statistics and all that.* W. A. Benjamin, Inc., New York-Amsterdam, 1964. Russian translation: Nauka, Moscow, 1966.

[12] J. D. Bjorken and S. D. Drell. *Relativistic quantum mechanics*, volume 1. McGraw-Hill Book Company, New York, 1964. Russian translation: Nauka, Moscow, 1978.

In: Frontiers in Quantum Physics Research ISBN 1-59454-002-2
Editor: F. Columbus and V. Krasnoholovets, pp. 157-166

On Conserved Operator Quantities in Quantum Field Theory

Bozhidar Z. Iliev*†‡

Abstract

Conserved operator quantities in quantum field theory can be defined via the Noether theorem in the Lagrangian formalism and as generators of some transformations. These definitions lead to generally different conserved operators which are suitable for different purposes. Some relations involving conserved operators are analyzed.

Key-Words:
Quantum field theory, Conserved operators in quantum field theory
Noether theorem, Noetherian (dynamical) conserved operators
Generators of symmetry transformations

1. Introduction

There are two approaches for introduction of conserved operator quantities in quantum field theory. The first one is based on the Lagrangian formalism and defines them via the first Noether theorem as conserved operators corresponding to smooth transformations leaving invariant the action integral of an investigated system; these are the canonical conserved operators. The second set of conserved operators consists of generators of some transformations of state vectors (and observables). Since these operators are of pure mathematical origin, we call them mathematical conserved quantities (operators). The present paper is devoted to a discussion of some relations between the mentioned two kinds of conserved quantities in quantum field theory. It is pointed that the two types of conserved operators are generally different and may coincide on some subspace of the system's Hilbert space of states.

The present work generalizes part of the results of [1, 2] and may be considered as a continuation of these papers.

*Laboratory of Mathematical Modeling in Physics, Institute for Nuclear Research and Nuclear Energy, Bulgarian Academy of Sciences, Boul. Tzarigradsko chaussée 72, 1784 Sofia, Bulgaria
†E-mail address: bozho@inrne.bas.bg
‡URL: http://theo.inrne.bas.bg/~bozho/

In what follows, we suppose that there is given a system of quantum fields, described via field operators $\varphi_i(x)$, $i = 1, \dots, n \in \mathbb{N}$, $x \in M$ over the 4-dimensional Minkowski spacetime M endowed with standard Lorentzian metric tensor $\eta_{\mu\nu}$ with signature $(+ - - -)$.[1] The system's Hilbert space of states is denoted by $\mathcal{F}$ and all considerations are in Heisenberg picture of motion if the opposite is not stated explicitly. The Greek indices $\mu, \nu, \dots$ run from 0 to $3 = \dim M - 1$ and the Einstein's summation convention is assumed over indices repeated on different levels. The coordinates of a point $x \in M$ are denoted by x^μ, $\boldsymbol{x} := (x^1, x^2, x^3)$, $\mathrm{d}^3\boldsymbol{x} := \mathrm{d}x^1\,\mathrm{d}x^2\,\mathrm{d}x^3$, and the derivative with respect to x^μ is $\frac{\partial}{\partial x^\mu} =: \partial_\mu$. The imaginary unit is denoted by i and $\hbar$ and c stand for the Planck's constant (divided by 2π) and the velocity of light in vacuum, respectively.

2. Canonical conserved quantities

Suppose a system of *classical* fields $\varphi_i(x)$, $i = 1, \dots, n \in \mathbb{N}$, over the Minkowski spacetime M, $x \in M$, is described via a Lagrangian L depending on them and their first partial derivatives $\partial_\mu\varphi_i(x) = \frac{\partial\varphi_i(x)}{\partial x^\mu}$, $\{x^\mu\}$ being the (local) coordinates of $x \in M$, i.e. $L = L(\varphi_j(x), \partial_\nu\varphi_i(x))$. Here and henceforth the Greek indices $\mu, \nu, \dots$ run from 0 to $\dim M - 1 = 3$ and the Latin indices $i, j, \dots$ run from 1 to some integer n. The equations of motion for $\varphi_i(x)$, known as the *Euler-Lagrange equations*, are[2] $\frac{\partial L}{\partial\varphi_i(x)} - \frac{\partial}{\partial x^\mu}\Big(\frac{\partial L}{\partial(\partial_\mu\varphi_i(x))}\Big) = 0$ and are derived from the variational principle of stationary action, known as the *action principle* (see, e.g, [3, § 1], [4, § 67], [5, pp. 19–20]).

The (first) Noether theorem [3, § 2] says that, if the action's variation is invariant under C^1 transformations

$$
\begin{aligned}
&x \mapsto x^\omega = x^\omega(x) \quad x^\omega|_{\omega=\mathbf{0}} = x \qquad \omega = (\omega^{(1)}, \dots, \omega^{(s)}) \\
&\varphi_i(x) \mapsto \varphi_i^\omega(x^\omega) \quad \varphi_i^\omega(x^\omega)|_{\omega=\mathbf{0}} = \varphi_i(x)
\end{aligned}
\tag{2.1}
$$

depending on $s \in \mathbb{N}$ independent real parameters $\omega^{(1)}, \dots, \omega^{(s)}$, then the quantities ('Noether currents')

$$
\theta^\mu_{(\alpha)}(x) := -\pi^{i\mu}\Big\{\frac{\partial\varphi_i^\omega(x^\omega)}{\partial\omega^{(\alpha)}}\Big|_{\omega=0} - (\partial_\nu\varphi_i(x))\frac{\partial x^{\omega\,\nu}}{\partial\omega^{(\alpha)}}\Big|_{\omega=0}\Big\} - L(x)\frac{\partial x^{\omega\,\mu}}{\partial\omega^{(\alpha)}}\Big|_{\omega=0},
\tag{2.2}
$$

where $\alpha = 1, \dots, s$ and

$$
\pi^{i\mu} := \frac{\partial L}{\partial(\partial_\mu\varphi_i(x))},
\tag{2.3}
$$

[1] The quantum fields should be regarded as operator-valued distributions (acting on a relevant space of test functions) in the rigorous mathematical setting of Lagrangian quantum field theory. This approach will be considered elsewhere.

[2] In this paper the Einstein's summation convention over indices appearing twice on different levels is assumed over the whole range of their values.

are conserved in a sense that their divergences vanish, viz.

$$\partial_\mu \theta^\mu_{(\alpha)}(x) = 0. \tag{2.4}$$

Respectively, the quantities

$$C_{(\alpha)}(x) := \frac{1}{c} \int \theta^0_{(\alpha)} \, \mathrm{d}^3 \boldsymbol{x}, \tag{2.5}$$

which in fact may depend only on x^0, are conserved in a sense that

$$\frac{\partial C_{(\alpha)}(x)}{\partial x^0} = 0 \tag{2.6}$$

and hence $\partial_\mu C_{(\alpha)} = 0$. The functions (constants) $C_{(\alpha)}$ are called *canonical (Noetherian, dynamical) conserved quantities* corresponding to the symmetry transformations (2.1) of the system considered.

Let us turn now our attention to a system of *quantum* fields represented by *field operators* $\varphi_i(x)\colon \mathcal{F} \to \mathcal{F}$, $i = 1, \dots, n \in \mathbb{N}$, acting on the system's Hilbert space $\mathcal{F}$ of states and described via a Lagrangian $\mathcal{L} = \mathcal{L}(x) = \mathcal{L}(\varphi_i(x), \partial_\mu \varphi_j(x))$. Suppose the system's action integral is invariant under the C^1 transformations (2.1). As a consequence of that supposition, one may expect the *operators* (2.2), with $\pi^{i\mu}$ defined via (2.3), to be conserved, i.e. the equations (2.4) to be valid. However, at this point two problems arise: (i) what is the meaning of the derivatives in (2.3) as $\partial_\mu \varphi_i(x)$ is *operator*, not a classical function? and (ii) in what order one should write the operators compositions in (2.2), e.g. shall we write $\pi^{i\mu} \circ \partial_\nu \varphi_i(x)$ or $\partial_\nu \varphi_i(x) \circ \pi^{i\mu}$? Usually [4, 3, 6] these problems are solved by (implicitly) adding to the theory additional assumptions concerning the operator ordering in (2.2) and meaning of derivatives with respect to operator-valued arguments.[3] In the work [7] we demonstrated that there is only one problem connected with a suitable definition of derivatives relative to operator-valued arguments and all other results follow directly from the (Schwinger's) action principle. The main point is that such derivatives are mappings form (a subset of) the space $\{\mathcal{F} \to \mathcal{F}\}$ of operators on $\mathcal{F}$ into $\{\mathcal{F} \to \mathcal{F}\}$ rather than operators $\mathcal{F} \to \mathcal{F}$. In particular, we have

$$\pi^{i\mu}(x) := \frac{\partial L}{\partial(\partial_\mu \varphi_i(x))} \colon \{\mathcal{F} \to \mathcal{F}\} \to \{\mathcal{F} \to \mathcal{F}\}. \tag{2.7}$$

For details and the rigorous definition of a derivative (of polynomial or convergent power series) relative to operator-valued argument, the reader is referred to [7].

[3] E.g., derivatives like the ones in (2.3) are calculated according to the rules of classical analysis of commuting variables by preserving the relative order of all terms in the Lagrangian. As pointed in [7], this rule corresponds to field variations proportional to the identity mapping $\mathsf{id}_\mathcal{F}$ of $\mathcal{F}$.

Accepting (2.7), we can write the quantum field analogue of (2.2), i.e. the 'Noether's current operators", as

$$\theta^{\mu}_{(\alpha)}(x) := -\sum_i \pi^{i\mu}(x)\Big(\frac{\partial\varphi_i^{\omega}(x^{\omega})}{\partial\omega^{(\alpha)}}\Big|_{\omega=0}\Big)$$

$$+\sum_{i,\nu}\pi^{i\mu}(x)\big(\partial_\nu\varphi_i(x)\big)\frac{\partial x^{\omega\,\nu}}{\partial\omega^{(\alpha)}}\Big|_{\omega=0} - L(x)\frac{\partial x^{\omega\,\mu}}{\partial\omega^{(\alpha)}}\Big|_{\omega=0}, \tag{2.8}$$

which immediately leads to the conservation laws (2.4) and (2.6). The quantities (2.5), with $\theta^{\mu}_{(\alpha)}$ given by (2.8), are called the *canonical (Noetherian, dynamical) conserved operators* corresponding to the symmetry transformations (2.1).

We end this section by the remark that the momentum, (total) angular momentum, and charge conserved operators are generated respectively by the transformation:

$$x \mapsto x + b \qquad \varphi_i(x) \mapsto \varphi_i(x) \tag{2.9a}$$

$$x^{\varkappa} \mapsto x^{\varkappa} + \varepsilon^{\varkappa\nu}x_\nu \qquad \varphi_i(x) \mapsto \varphi_i(x) + \frac{1}{2}I^{j}_{i\mu\nu}\varepsilon^{\mu\nu}\varphi_j(x) + \cdots \tag{2.9b}$$

$$x \mapsto x \qquad \varphi_i(x) \mapsto \mathrm{e}^{\frac{q}{i\hbar c}\lambda}\varphi_i(x), \tag{2.9c}$$

where $b \in M$, $\varepsilon^{\mu\nu} = -\varepsilon^{\nu\mu} \in \mathbb{R}$, and $\lambda \in \mathbb{R}$ are the parameters of the corresponding transformations, x_μ are the covariant coordinates of $x \in M$, the numbers $I^{j}_{i\mu\nu} = -I^{j}_{i\nu\mu}$ characterize the behaviour of the field operators under rotations, and the dots stand for higher order terms in $\varepsilon^{\mu\nu}$.

3. On observer dependence of state vectors and observables

Let two observers O and O' investigate one and the same system of quantum fields. The quantities relative to O' will be denoted as those relative to O by adding a prime to their kernel symbols. The transition $O \mapsto O'$ implies the changes

$$x \mapsto x' = L(x) \tag{3.1}$$

(of the coordinates) of a spacetime point $x = (x^0, x^1, x^2, x^3) \in M$ and

$$\mathcal{X}(x) \mapsto \mathcal{X}'(x') = \Lambda(\mathcal{X}(x)) \tag{3.2}$$

of a state vector $\mathcal{X}(x) \in \mathcal{F}$ of system of quantum fields $\varphi_i(x)$.[4] Requiring preservation of the scalar products in $\mathcal{F}$ under the change $O \mapsto O'$, which physically

[4] It is inessential for the following whether L (Λ) is an element (of a representation) of the Poincaré group or not; the former case is realized when O and O' are inertial observers.

corresponds to preservation of probability amplitudes, we see that Λ is a *unitary* operator,

$$\Lambda^{-1} = \Lambda^{\dagger} \tag{3.3}$$

where the dagger $\dagger$ denotes Hermitian conjugation (i.e., in mathematical terms, $\Lambda^{\dagger}$ is the adjoint to Λ operator).

Let $\mathbb{A}$ be a dynamical variable and $\mathcal{A}(x)\colon \mathcal{F} \to \mathcal{F}$ be the corresponding to it observable. The change $O \mapsto O'$ entails $\mathcal{A}(x) \mapsto \mathcal{A}(L(x))$. Supposing preservation of the mean (expectation) values (and the matrix elements of $\mathbb{A}$ (or $\mathcal{A}(x)$)) in states with finite norm under the change $O \mapsto O'$, we get

$$\mathcal{A}(L(x)) = (\Lambda^{\dagger})^{-1} \circ \mathcal{A}(x) \circ \Lambda^{-1} = \Lambda \circ \mathcal{A}(x) \circ \Lambda^{-1}. \tag{3.4}$$

As explained in [2, sect. 4] or in [4, 3], the field operators $\varphi_i(x)$ undergo more complicated change when one passes from O to O':

$$\varphi_i(x) \mapsto \sum_j (S^{-1})_i^{\,j}(L)\varphi_j(x) = \Lambda \circ \varphi_i(x) \circ \Lambda^{-1} \tag{3.5}$$

where the depending on L matrix $S = S(L) = [(S^{-1})_i^{\,j}(L)]$ characterizes the transformation properties of any particular field (e.g. scale or vector one) under $O \mapsto O'$ and is such that $S(L)|_{L=\mathsf{id}_M}$ is the identity matrix of relevant size.

4. Transformations with Hermitian generators

Let $\omega^1, \dots, \omega^s$, $s \in \mathbb{N}$, be real independent parameters and $\omega := (\omega^1, \dots, \omega^s) \in \mathbb{R}^s$. Suppose the changes (3.1) and (3.2) depend on ω and

$$\begin{aligned} & x \mapsto x' = L^{\omega}(x) = x^{\omega}(x) \quad x^{\omega}(x)|_{\omega=0} = x \\ & \Lambda = \Lambda^{\omega} = \exp\Big\{ \frac{\eta}{\mathrm{i}\hbar} \sum_{\alpha=1}^{s} \omega^{\alpha} \mathcal{J}_{\alpha}^{\mathrm{m}} \Big\}, \end{aligned} \tag{4.1}$$

where the operators $\mathcal{J}_{\alpha}^{\mathrm{m}}\colon \mathcal{F} \to \mathcal{F}$ are Hermitian,

$$(\mathcal{J}_{\alpha}^{\mathrm{m}})^{\dagger} = \mathcal{J}_{\alpha}^{\mathrm{m}}, \tag{4.2}$$

which ensures the validity of (3.3), and the particular choice of the constant $\eta \in \mathbb{R} \setminus \{0\}$ depends on what physical interpretation of $\mathcal{J}_{\alpha}^{\mathrm{m}}$ one intends to get.

Differentiating (3.4) and (3.5) with respect to ω^{α} and setting $\omega = 0$, we rewrite them in differential form respectively as

$$\eta[\mathcal{A}(x), \mathcal{J}_{\alpha}^{\mathrm{m}}]_{\_} = -\mathrm{i}\hbar \frac{\partial \mathcal{A}(x)}{\partial x^{\mu}} \frac{\partial x^{\omega\,\mu}}{\partial \omega^{\alpha}}\Big|_{\omega=0} \tag{4.3}$$

$$\eta[\varphi_i(x), \mathcal{J}_{\alpha}^{\mathrm{m}}]_{\_} = \mathrm{i}\hbar \sum_j I_{i\alpha}^{j} \varphi_j(x) - \mathrm{i}\hbar \frac{\partial \varphi_i(x)}{\partial x^{\mu}} \frac{\partial x^{\omega\,\mu}}{\partial \omega^{\alpha}}\Big|_{\omega=0} \tag{4.4}$$

where $I^j_{i\alpha} := \frac{\partial S_i^{\ j}(L^\omega)}{\partial \omega^\alpha}\Big|_{\omega=0}$, i.e. $S_i^{\ j}(L^\omega) = \delta^j_i + \sum_\alpha I^j_{i\alpha}\omega^\alpha + \cdots$ with δ^j_i being the Kronecker deltas and the dots denoting higher order terms in ω, and $[\mathcal{A}, \mathcal{B}]_{\_} := \mathcal{A}\circ\mathcal{B} - \mathcal{B}\circ\mathcal{A}$ is the commutator of operators $\mathcal{A}, \mathcal{B}\colon \mathcal{F}\to\mathcal{F}$.

In particular, to describe the quantum analogue of the transformations (2.9), in (4.1) we have to make respectively the replacements:

$$\omega^\alpha \mapsto b^\mu \qquad x^\omega \mapsto x^b = x + b \qquad \eta \mapsto -1 \quad \mathcal{J}^{\mathrm{m}}_\alpha \mapsto \mathcal{P}^{\mathrm{t}}_\mu \tag{4.5a}$$

$$\omega^\alpha \mapsto \varepsilon^{\mu\nu} \quad (\mu<\nu) \quad x^{\omega\,\mu} \mapsto x^{\varepsilon\,\mu} = x^\mu + \varepsilon^{\mu\nu} x_\nu \quad \eta \mapsto +1 \quad \mathcal{J}^{\mathrm{m}}_\alpha \mapsto \mathcal{M}^{\mathrm{r}}_{\mu\nu} \tag{4.5b}$$

$$\omega^\alpha \mapsto \lambda \qquad x^\omega \mapsto x^\lambda = x \qquad \eta \mapsto \frac{q}{c} \quad \mathcal{J}^{\mathrm{m}}_\alpha \mapsto \mathcal{Q}^{\mathrm{p}}, \tag{4.5c}$$

so that $\frac{\partial x^{\omega\,\varkappa}}{\partial \omega^\alpha}\big|_{\omega=0}$ reduces to $\delta^\varkappa_\mu$, $(\delta^\varkappa_\mu x_\nu - \delta^\varkappa_\nu x_\mu)$, and $0\in\mathbb{R}$, respectively. The operators $\mathcal{P}^{\mathrm{t}}_\mu$, $\mathcal{M}^{\mathrm{r}}_{\mu\nu}$, and $\mathcal{Q}^{\mathrm{p}}$ are the translation (mathematical) momentum operator, total rotational (mathematical) angular momentum operator, and constant phase transformation (mathematical) charge operator, respectively. In these cases, the equations (4.4) are known as the Heisenberg equations/relations for the operators mentioned [5, 3, 4] . For that reason, it is convenient to call (4.4) *Heisenberg equations/relations* (for the operators $\mathcal{J}^{\mathrm{m}}_\alpha$) in the general case.

The transformations (3.1) and (3.5), defined by the choice (4.1), are the *quantum* observer-transformation version of (2.1). For that reason, one can expect the (spacetime constant) operators $\mathcal{J}^{\mathrm{m}}_\alpha$ to play, in some sense, a role similar to the conserved operators (2.8); we shall call $\mathcal{J}^{\mathrm{m}}_\alpha$ *mathematical conserved operators* corresponding to the transformations (3.1) and (3.5) under the choices (4.1).

Suppose there exist operators $\mathcal{J}^{\mathrm{QM}}_\alpha$, where QM stands for quantum mechanics[5], generating the change $\mathcal{X}(x)\mapsto\mathcal{X}(x')$, i.e. such that ($|\eta|$ is the absolute value of η)

$$\mathcal{X}(x)\mapsto\mathcal{X}(x') = \Lambda^{\mathrm{QM}}(\mathcal{X}(x)) := \exp\Big\{\frac{|\eta|}{\mathrm{i}\hbar}\sum_{\alpha=1}^{s}\omega^\alpha\mathcal{J}^{\mathrm{QM}}_\alpha\Big\}(\mathcal{X}(x)). \tag{4.6}$$

Note that $\mathcal{J}^{\mathrm{QM}}_\alpha$ (as well as $\mathcal{J}^{\mathrm{m}}_\alpha$) may depend on x; for instance, the changes $x\mapsto x^\omega$ defined via (4.5a)–(4.5c) entail (4.6) with respectively ($\mathsf{id}_\mathcal{F}$ is the identity mapping of $\mathcal{F}$)

$$\mathcal{J}^{\mathrm{QM}}_\alpha \mapsto \mathcal{P}^{\mathrm{QM}}_\mu = \mathrm{i}\hbar\partial_\mu \tag{4.7a}$$

$$\mathcal{J}^{\mathrm{QM}}_\alpha \mapsto \mathcal{M}^{\mathrm{QM}}_{\mu\nu} = \mathrm{i}\hbar(x_\mu\partial_\nu - x_\nu\partial_\mu) \tag{4.7b}$$

$$\mathcal{J}^{\mathrm{QM}}_\alpha \mapsto \mathcal{Q}^{\mathrm{QM}}_\mu = \mathrm{e}^{\frac{q}{\mathrm{i}\hbar c}\lambda}\,\mathsf{id}_\mathcal{F}. \tag{4.7c}$$

[5] This notation reminds only some analogy with quantum mechanics. If one identifies $\mathcal{F}$ with the Hilbert space of this theory and makes some other assumptions, (part of) the generators $\mathcal{J}^{\mathrm{QM}}_\alpha$ will coincide with similar objects in quantum mechanics. However, as the Hilbert spaces of quantum field theory and quantum mechanics are different, the corresponding operators in these theories cannot be identified. See similar remarks in [1, 2] concerning the momentum and angular momentum operators, respectively.

The transformation (4.6) implies the changes

$$\mathcal{A}(x) \mapsto \mathcal{A}(x') = \Lambda^{\text{QM}} \circ \mathcal{A}(x) \circ (\Lambda^{\text{QM}})^{-1}. \tag{4.8}$$

$$\varphi_i(x) \mapsto \sum_j (S^{-1})_i^{\,j}(L)\varphi_j(x') = \Lambda^{\text{QM}} \circ \varphi_i(x) \circ (\Lambda^{\text{QM}})^{-1} \tag{4.9}$$

which, in differential form, entail

$$|\eta|[\mathcal{A}(x), \mathcal{J}_\alpha^{\text{QM}}]_{\_} = -\mathrm{i}\hbar \frac{\partial \mathcal{A}(x)}{\partial x^\mu} \frac{\partial x^{\omega\,\mu}}{\partial \omega^\alpha}\Big|_{\omega=0} \tag{4.10}$$

$$|\eta|[\varphi_i(x), \mathcal{J}_\alpha^{\text{QM}}]_{\_} = \mathrm{i}\hbar \sum_j I_{i\alpha}^{j} \varphi_j(x) - \mathrm{i}\hbar \frac{\partial \varphi_i(x)}{\partial x^\mu} \frac{\partial x^{\omega\,\mu}}{\partial \omega^\alpha}\Big|_{\omega=0} \tag{4.11}$$

Comparing these equations with (4.3) and (4.4), we find

$$[\mathcal{A}(x), \mathcal{J}_\alpha^{\text{m}} - \operatorname{sign}\eta\, \mathcal{J}_\alpha^{\text{QM}}]_{\_} = 0 \tag{4.12a}$$

$$[\varphi_i(x), \mathcal{J}_\alpha^{\text{m}} - \operatorname{sign}\eta\, \mathcal{J}_\alpha^{\text{QM}}]_{\_} = 0 \tag{4.12b}$$

where $\operatorname{sign}\eta := \eta/|\eta| \in \{-1. +1\}$ is the sign of $\eta \in \mathbb{R} \setminus \{0\}$. If we admit (4.12a) to hold for *every* $\mathcal{A}(x)\colon \mathcal{F} \to \mathcal{F}$, the Schur's lemma[6] implies

$$\mathcal{J}_\alpha^{\text{m}} = \operatorname{sign}\eta\, \mathcal{J}_\alpha^{\text{QM}} + j_\alpha \,\mathsf{id}_{\mathcal{F}}, \tag{4.13}$$

where j_α are real numbers (with the same dimension as the eigenvalues of $\mathcal{J}_\alpha^{\text{m}}$).

5. Discussion

Following the opinion established in the literature[7], the identification

$$C_{(\alpha)} = \mathcal{J}_\alpha^{\text{m}} \tag{5.1}$$

may seem 'natural' *prima facie* but, generally, it is unacceptable as its l.h.s. comes out from the Lagrangian formalism (via (2.8) and (2.5)), while its r.h.s. originates from pure mathematical (geometrical) considerations and is suitable for the axiomatic quantum field theory [11, 12].

As an equality weaker than (5.1), the Heisenberg relations (4.4) with $C_{(\alpha)}$ for $\mathcal{J}_\alpha^{\text{m}}$ can be assumed:

$$\eta[\varphi_i(x), C_{(\alpha)}]_{\_} = \mathrm{i}\hbar \sum_j I_{i\alpha}^{j} \varphi_j(x) - \mathrm{i}\hbar \frac{\partial \varphi_i(x)}{\partial x^\mu} \frac{\partial x^{\omega\,\mu}}{\partial \omega^\alpha}\Big|_{\omega=0}. \tag{5.2}$$

However, these equations as well as (5.1) are external to the Lagrangian formalism by means of which the canonical conserved operators are defined. As discussed in [4,

[6] See, e.g, [8, appendix II], [9, sec. 8.2], [10, ch. 5, sec. 3].

[7] See also the papers [1, 2] in which the momentum and angular momentum are analyzed.

§ 68] on particular examples, the validity of the equations (5.2) should be checked for any particular Lagrangian and they express (in the sense explained in *loc. cit.*) the relativistic covariance of the Lagrangian quantum field theory.

Generally the equation (4.3) with $C_{(\alpha)}$ for $\mathcal{J}_\alpha^{\mathrm{m}}$, viz.

$$\eta[\mathcal{A}(x), C_{(\alpha)}]_{\_} = -\mathrm{i}\hbar \frac{\partial \mathcal{A}(x)}{\partial x^\mu} \frac{\partial x^{\omega\,\mu}}{\partial \omega^\alpha}\Big|_{\omega=0}, \tag{5.3}$$

cannot hold; a counterexample being the choice of $\mathcal{A}(x)$ and $C_{(\alpha)}$ as the momentum and angular momentum operators (or *vice versa*). If (5.3) happens to be valid for operators $\mathcal{A}(x)$ forming an irreducible representation of some group, then, by virtue of (5.3) and (4.3), the Schur's lemma implies

$$C_{(\alpha)} = \operatorname{sign}\eta\, \mathcal{J}_\alpha^{\mathrm{m}} + i_\alpha \,\mathsf{id}_{\mathcal{F}} = \operatorname{sign}\eta\, \mathcal{J}_\alpha^{\mathrm{QM}} + (i_\alpha + j_\alpha)\,\mathsf{id}_{\mathcal{F}} \tag{5.4}$$

for some real numbers i_α (see also (4.13)).

Let a vector $\mathcal{X} \in \mathcal{F}$ represents a state of the system of quantum fields considered. It is a spacetime-constant vector as we are working in Heisenberg picture of motion. Consequently, we have $\mathcal{X}(x) = \mathcal{X}(x')$ which, when combined with (4.6), entails

$$\mathcal{J}_\alpha^{\mathrm{QM}}(\mathcal{X}) = 0. \tag{5.5}$$

So, applying (4.13) to $\mathcal{X}$, we get

$$\mathcal{J}_\alpha^{\mathrm{m}}(\mathcal{X}) = j_\alpha \mathcal{X}. \tag{5.6}$$

If one intends to interpret $\mathcal{J}_\alpha^{\mathrm{m}}$ as the conserved canonical operators $C_{(\alpha)}$ (see the possible equality (5.1)), then one should interpret j_α as the mean (expectation) value of $C_{(\alpha)}$, which will be the case if

$$C_{(\alpha)}(\mathcal{X}) = j_\alpha \mathcal{X}. \tag{5.7}$$

(Notice, (5.7) and (5.4) are compatible iff $i_\alpha = 0$.) The equations (5.6) and (5.7) imply

$$C_{(\alpha)}|_{\mathcal{D}_j} = \mathcal{J}_\alpha^{\mathrm{m}}|_{\mathcal{D}_j}. \tag{5.8}$$

where

$$\mathcal{D}_j := \{\mathcal{X} \in \mathcal{F} : C_{(\alpha)}(\mathcal{X}) = j_\alpha \mathcal{X}\}. \tag{5.9}$$

Generally the set $\mathcal{D}_j$ is a proper subset of $\mathcal{F}$ and hence (5.8) is weaker than (5.1); if a basis of $\mathcal{F}$ can be formed from vectors in $\mathcal{D}_j$, then (5.8) and (5.1) will be equivalent. But, in the general case, equations (5.2) and (4.4) lead only to

$$[\varphi_i(x), C_{(\alpha)}]_{\_} = [\varphi_i(x), \mathcal{J}_\alpha^{\mathrm{m}}]_{\_} \qquad \Big(= \frac{1}{\eta}\mathrm{i}\hbar \sum_j I_{i\alpha}^j \varphi_j(x) - \frac{1}{\eta}\mathrm{i}\hbar \frac{\partial \varphi_i(x)}{\partial x^\mu} \frac{\partial x^{\omega\,\mu}}{\partial \omega^\alpha}\Big|_{\omega=0}\Big), \tag{5.10}$$

but not to (5.1).

Ending this section, we note that the equality $C_{(\alpha)} = \mathcal{J}_\alpha^{\mathrm{QM}}$ is unacceptable as, in view of (5.5), it leads to identically vanishing eigenvalues of $C_{(\alpha)}$.

6. Conclusion

In this work we have analyzed two types of conserved operator quantities in quantum field theory, viz. the ones arising from the (first) Noether theorem in the framework of Lagrangian formalism and conserved operators having pure mathematical origin as generators of some transformations (and having natural place in the axiomatic approach). These operators are generally different and their equality is a problem which is external to the Lagrangian formalism and may be considered as possible subsidiary restrictions to it. However, using the arbitrariness (4.13) in the mathematical conserved operators, both types of conserved operators can be chosen to coincide on the set (5.9). As weaker conditions additionally imposed on the Lagrangian formalism, one can require the equality (5.10) between the commutators of the field operators and conserved operators. As it is known [4], the Heisenberg relations (5.2) are equations relative to the field operators, while (4.4) are identities with respect to them.

References

[1] Bozhidar Z. Iliev. On momentum operator in quantum field theory. http://www.arXiv.org e-Print archive, E-print No. hep-th/0206008, 2002. See chapter 6 in this book.

[2] Bozhidar Z. Iliev. On angular momentum operator in quantum field theory. http://www.arXiv.org e-Print archive, E-print No. hep-th/0211153, 2002. See chapter 5 in this book.

[3] N. N. Bogolyubov and D. V. Shirkov. *Introduction to the theory of quantized fields.* Nauka, Moscow, third edition, 1976. In Russian. English translation: Wiley, New York, 1980.

[4] J. D. Bjorken and S. D. Drell. *Relativistic quantum fields*, volume 2. McGraw-Hill Book Company, New York, 1965. Russian translation: Nauka, Moscow, 1978.

[5] Paul Roman. *Introduction to quantum field theory.* John Wiley&Sons, Inc., New York-London-Sydney-Toronto, 1969.

[6] C. Itzykson and J.-B. Zuber. *Quantum field theory.* McGraw-Hill Book Company, New York, 1980. Russian translation (in two volumes): Mir, Moscow, 1984.

[7] Bozhidar Z. Iliev. On the action principle in quantum field theory. In "Trends in Complex Analysis, Differential Geometry and Mathematical Physics", *Proceedings of the 6th International Workshop on Complex Structures and Vector*

Fields, 3-6 September 2002, St Knstantin resort (near Varna), Bulgaria, Editors Stancho Dimiev and Kouei Sekigava, World Scientific, New Jersey-London-Singapore-Hong Kong, 2003. pp. 76–107. http://www.arXiv.org e-Print archive, E-print No. hep-th/0204003, April 2002.

[8] Yu. B. Rumer and A. I. Fet. *Group theory and quantized fields.* Nauka, Moscow, 1977. In Russian.

[9] A. A. Kirillov. *Elements of the theory of representations.* Springer, Berlin, 1976. Translation from Russian (second ed., Nauka, Moscow, 1978).

[10] Asim Barut and Ryszard Roczka. *Theory of group representations and applications.* PWN — Polish Scientific Publishers, Waszawa, 1977. Russian translation: Mir, Moscow, 1980.

[11] N. N. Bogolubov, A. A. Logunov, and I. T. Todorov. *Introduction to axiomatic quantum field theory.* W. A. Benjamin, Inc., London, 1975. Translation from Russian: Nauka, Moscow, 1969.

[12] N. N. Bogolubov, A. A. Logunov, A. I. Oksak, and I. T. Todorov. *General principlės of quantum field theory.* Nauka, Moscow, 1987. In Russian. English translation: Kluwer Academic Publishers, Dordrecht, 1989.

In: Frontiers in Quantum Physics Research ISBN 1-59454-002-2
Editor: F. Columbus and V. Krasnoholovets, pp. 167-184

Continuous Radiation Emitted in the $(n;t)-$, $(n;\alpha)-$ and (n,p)-Reactions in Atomic Systems

Alexei M. Frolov

Department of Chemistry, Queen's University, Kingston, Ontario, Canada K7L 3N6

Abstract

The emission of continuous radiation in the $(n;t)-$, $(n;\alpha)-$ and $(n;p)-$nuclear reactions in light atoms and ions is discussed. The angular distributions of the emitted radiation and its spectrum are considered in detail. It is shown that the total radiation $I(t)$ rapidly decreases with time $I(t) \sim t^{-4}$ at large t. It is shown that the radiation emitted during the $(n;t)-$nuclear reactions in the rapidly moving ions and atoms can be used to study the electron-electron position correlations in these systems. The approach developed in our present study can also be applied to describe the emission of bremsstrahlung during the spontaneous and neutron-stimulated nuclear fission in heavy atoms. In particular, the circular motion of the relativistic lithium ions is discussed.

PACS number(s): 32.90.+a and 41.60.-m

1.1 Introduction

Presently, we consider radiation emitted during the nuclear $(n;t)-$ reactions in some light atoms and ions. Our main interest is related to the $(n, {}^6Li; t, \alpha)$ and $(n, {}^3He; t, p)-$reactions

$$n + {}^6_3Li = {}^4_2He + {}^3_1H + 4.785 MeV \quad , \tag{1}$$
$$n + {}^3_2He = {}^1_1H + {}^3_1H + 0.764 MeV \quad , \tag{2}$$

which are extensively used in many thermonuclear applications, including thermonuclear explosive devices [1]. In particular, these two reactions are used to increase the tritium-deuterium ratio and decrease the high-temperature bremsstrahlung loss I_B from thermonuclear fuel (in general, $I_B \sim \overline{Q^2} \cdot \overline{Q} \approx \overline{Q}^3$, where $\overline{Q}$ and $\overline{Q^2}$ are the mean atomic charge and quadratic charge, respectively [2]). This simplifies significantly the following thermonuclear burn-up in highly-compressed thermonuclear fuel (${}^6LiD$). Furthermore, these two $(n;t)-$reactions, Eqs.(1) - (2), allow one to avoid the two $(d, {}^3He; {}^4He, p)-$ and $(d, {}^6Li; 2{}^4He)$ -reactions in thermonuclear fuel which have extremely high thermal gain (18.354 MeV and 22.375 MeV, respectively), but cannot be easily started and controlled at $T \approx 4 - 10$ keV and $\rho \leq 1000$ $g \cdot cm^{-3}$. In fact, in this study we shall not discuss thermonuclear applications of these two reactions. Instead, our present main interest is related to the radiation emitted in the $(n;t)-$reactions in atoms and ions, and first of all, in the $(n; {}^6Li; {}^4He, t)-$reaction, Eq.(1).

In general, the nuclear $(n;t)-$reactions take place in few electron atoms and ions, rather than in bare nuclei. This means that such reactions are followed by the separation of electrons from the incidental (or maternal) atom. Moreover, for the $(n,{}^6Li;t,\alpha)$−reaction the fast tritium nucleus and α−particle are the sources of time-varying electromagnetic fields, which will accelerate the atomic electrons. In turn, the accelerated electrons will emit electromagnetic radiation. Our main interest is the properties of radiation emitted during the nuclear $(n;t)$−reactions in atoms and ions. Note that the incident [3] and final atomic states, in the case of Eqs.(1) - (2), are formally unbound since the velocities of the nuclear fragments significantly exceed the velocities of atomic electrons. On the other hand, the velocities of two nuclear fragments are significantly smaller than speed of light (see below). This essentially means that in this study we are dealing with non-relativistic free-free emission (or bremsstrahlung, for short). Our main goal is to develop a complete theory of non-relativistic bremsstrahlung in the $(n;t)$−reactions in atomic systems. In this study a number of realistic models are discussed. Such models can be used to analyze many essential features of the continuous radiation emitted in the $(n;t)$−reactions in atomic (and molecular) systems. Presently, we apply both classical and full-scale quantum-mechanical approaches. These two approaches are discussed in the third and fourth sections, respectively. The second section contains a brief description of the $(n;t)$−reactions in atoms and ions. The fifth section deals with the $(n;t)$−reactions in atoms and ions which are moving in some external magnetic fields. Concluding remarks can be found in the last section.

1.2 Neutron-tritium reactions in one-electron ions

Our present discussion begins with consideration of the nuclear $(n;t)$−reactions in one-electron, hydrogen-like ions. The kinetic energies T_i (MeV) and velocities v_i $(cm \cdot sec^{-1})$ of the two charged nuclear fragments formed in the (n,t)−reactions Eqs.(1) - (2) with slow (or thermal) neutrons ($v_n \approx 0$ and $T_n \approx 0$) can be evaluated as follows

$$T_i \approx \frac{M_j \Delta E}{M_1 + M_2} \quad MeV \quad and \quad v_i \approx 0.1384112 \cdot \sqrt{T_i} \quad cm \cdot sec^{-1} \quad ,$$

where $i \neq j = (1,2)$. In the case of Eq.(1) we shall also use the symbols t and α to designate the tritium and ${}^4He$ nuclei, respectively. For the first reaction, Eq.(1), this gives $v_t \approx 0.228872 \cdot 10^{10}$ $cm \cdot sec^{-1}$ and $v_\alpha \approx 0.198209 \cdot 10^{10}$ $cm \cdot sec^{-1}$. These values are approximately 3 - 10 times larger, than the corresponding electron velocities in the incident lithium atom. On the other hand, the v_α and v_t velocities are in 10 - 15 times smaller than the speed of light c. In general, for the $(n;t)$−reactions with fast neutrons ($E_n \geq 2$ MeV), the kinetic energies and velocities of the charged nuclear fragments gradually increase with the kinetic energy of incident neutron. In fact, our present approach can be generalized to describe radiation emitted in the $(n;t)$−reactions with fast neutrons. However, in this study we do not want to discuss the $(n;t)$−reactions with fast neutrons. In particular, this related to the fact that the cross-sections of all $(n;t)$−reactions rapidly decrease with the neutron energy. For instance, for thermonuclear neutrons ($E_n \approx 14.059$ MeV) the cross-section of $(n,{}^6Li;t,{}^4He)$−reaction is approximately 0.026 $\cdot 10^{-24}$ cm^2. This value is significantly smaller than the cross-sections for any other channel in the $(n,{}^6Li)$−nuclear reaction, including the elastic and non-elastic neutron scattering.

In this study we shall use the atomic units. In these units $m_e = 1, \hbar = 1$ and $e = 1$, where m_e is the electron mass, e is the proton charge and $\hbar$ is Planck's constant divided by 2π. The fine structure constant $\alpha = \frac{e^2}{\hbar c} \approx 7.297352533 \cdot 10^{-3}$ [4] is dimensionless. The unit of length is the Bohr radius $a_0 = \frac{\hbar^2}{m_e e^2} \approx 0.5291772083 \cdot 10^{-10}$ cm. The unit of time (one atomic second) is $\tau_a = \frac{\hbar^3}{m_e e^4} = \frac{\hbar^2}{m_e e^2} \cdot \frac{\hbar c}{e^2} \cdot \frac{1}{c} = \frac{a_0}{\alpha c} \approx 2.418884331 \cdot 10^{-17}$ sec and c is the speed of light. In fact, in atomic units $c = \alpha^{-1}$, but we shall use the explicit notation c in all formulas below. The velocity, $v = \alpha c \approx 2.187691249 \cdot 10^8$ $cm \cdot sec^{-1}$ is the atomic electron velocity, which is exactly coincides with the electron velocity for the first Bohr orbit in the hydrogen atom.

These numerical values allows us to understand the natural limits of our present consideration. For instance, in terms of the classical approach we cannot consider the times t which are significantly shorter than τ_a (i.e. one atomic second). In fact, the predictions made for times $t \approx \tau_a$ cannot be strictly correct, but they are still useful for a qualitative understanding of the process. The corresponding classical limit for the frequency is $\omega \approx 4.13413732 \cdot 10^8$ GHz. However, if non-relativistic quantum mechanics is used, then the corresponding upper limits for the time and frequency ω are $\tau_c \approx \alpha\tau_a$ (i.e. $\approx 1/137$ of one atomic second) and $\omega_c \approx 3.016825743 \cdot 10^{10}$ GHz. For quantum electrodynamics there are no fundamental limits, but in this study we shall apply only the lowest order QED approximation (upon α) which has essentially the same limits as non-relativistic quantum theory [5].

Note that in the present case, i.e. in Eqs.(1) - (2), nuclear motion can always be considered in terms of classical mechanics. Indeed, the kinetic energies of nuclear fragments in the considered $(n;t)-$reactions are bounded between ≈ 0.3 MeV and ≈ 2 MeV. The total energies of the electron-nuclear Coulomb interaction in the case of Eqs.(1) - (2) are less than 150 eV. Moreover, for both reactions Eqs.(1) - (2), the velocities of daughter nuclei are larger than or at least comparable with the characteristic velocities of the electrons in their incident atomic orbits. The corresponding nuclear masses are significantly larger than the electron mass m_e (e.g. $M_t = 5496.92158$ m_e, $M_\alpha = 7294.2996$ m_e, etc). This means that the momenta of daughter nuclei in Eqs.(1) - (2) are significantly larger (since $M/m_e \gg 1$) than the corresponding electron momenta. In other words, for the $(n;t)-$reactions Eqs.(1)-(2) in atoms and ions we can neglect the momentum transfer from the electron to both daughter nuclei. This means that after nuclear $(n;t)-$reactions both daughter nuclei will move with constant velocities v_1 and v_2, along strait lines, in exactly opposite directions. For both daughter nuclei their maximal deflection from this strait line can be neglected for our purposes, i.e. the nuclear motion in the $(n;t)-$reactions can always be considered classically (see also [6]).

In contrast with the nuclear fragments, for electrons one can easily find a noticeable difference between classical and quantum results. However, for atomic electrons all essential quantum-mechanical modifications can be incorporated in the classical picture. For instance, the radiation spectrum and other properties obtained with the use of the classical approach coincides quite well with the values of such properties calculated using of the more rigorous and accurate quantum-mechanical approach. The remaining problems here are related to the electron-electron repulsion and Pauli principle for identical particles (electrons).

An important question for the considered $(n;t)-$reactions in atoms and ions is related to the interaction between atomic electrons and the incident neutron. It is clear, however, that when the incident neutron moves through the electron shells, it does not produce any noticeable interaction with atomic electrons [7]. A detailed discussion of this problem can be found in, e.g., [5]. In fact, there is a very small $\sim \alpha^2(\frac{m_e}{M_n}) \approx 5 \cdot 10^{-8}$ spin-spin electron-neutron interaction. However, the overall contribution from such an interaction is significantly smaller (by $\approx 5 \cdot 10^{-8}$ times) than other essential contributions to the problem. In other words, such a contribution is beyond the level of our present accuracy. Therefore, the electron-neutron interaction will be ignored in the present study.

1.3 Classical approach

Let us consider the $(n;t)$−reaction, Eq.(1), in the one-electron $^6Li^{2+}$−ion. As mentioned above, our analysis in the classical case is restricted to the times t for which $t \geq \tau_a \gg \frac{a}{c}$. But the last condition also means that $ct(= \lambda) \gg a$, i.e. the dimension of the considered system a is small in comparison to the radiated wavelength λ. Obviously, this corresponds to the case of dipole radiation [8]. The emitted radiation in this cases is determined by the second derivative of dipole moment of the considered system. Note that the considered lithium ion (Li^{2+}) has the total positive charge which differs from zero. Therefore, its dipole moment depends upon the origin of the coordinates. Below, the origin of the coordinates is always chosen at the center of mass of the considered system. In the present case after $(n;t)$−reaction, Eq.(1), in the Li^{2+} ion, we can write $M_1 + M_2 + m_e \approx M_1 + M_2$, since $m_e \ll M_i$ $(i = 1,2)$. From here, one easily finds

$$\mathbf{R}_c = \frac{\sum_i m_i \mathbf{r}_i}{\sum_i m_i} \approx \frac{M_1}{M_1 + M_2}\mathbf{r}_1 + \frac{M_2}{M_1 + M_2}\mathbf{r}_2 \quad , \tag{3}$$

where $\mathbf{r}_i$ $(i = 1,2)$ are the position vectors of the two nuclear fragments with masses M_1 and M_2, respectively. Analogously, $\mathbf{r}_3$ is the position vector of the electron $(m_3 = m_e = 1)$. The vectors $\mathbf{R}_i$ $(i = 1,2,3)$ are the position vectors of the same particles in respect to the center of mass. Now, we can write

$$q_1\mathbf{R}_1 = q_1(\mathbf{r}_1 - \mathbf{R}_c) = \frac{M_2}{M_1 + M_2}\mathbf{r}_{12} = \mu_2 q_1 \mathbf{r}_{21} \quad , \tag{4}$$

$$q_2\mathbf{R}_2 = q_2(\mathbf{r}_2 - \mathbf{R}_c) = \frac{M_1}{M_1 + M_2}\mathbf{r}_{12} = \mu_1 q_2 \mathbf{r}_{21} \quad , \tag{5}$$

$$q_3\mathbf{R}_3 = q_3(\mathbf{r}_3 - \mathbf{R}_c) = \mu_1 q_3 \mathbf{r}_{31} + \mu_2 q_3 \mathbf{r}_{32} \quad , \tag{6}$$

where $\mu_i = \frac{M_i}{M_1+M_2+m_e} \approx \frac{M_i}{M_1+M_2}$ $(i = 1,2)$ are the reduced (nuclear) masses, while q_1, q_2 and q_3 are the nuclear (q_1, q_2) and electron charges $(q_3 = q_e = -1)$. Also, in these formulas $\mathbf{r}_{ij} = \mathbf{r}_i - \mathbf{r}_j$, where $i \neq j = (1,2,3)$ and $\mu_3 = \mu_e = \frac{m_e}{M_1+M_2+m_e} \approx 0$. Finally, for the dipole moment $\mathbf{d}$ one finds

$$\mathbf{d} = q_1\mathbf{R}_1 + q_2\mathbf{R}_2 + q_3\mathbf{R}_3 = \mu_2 q_1 \mathbf{r}_{12} + \mu_1 q_2 \mathbf{r}_{21} - \mu_1 q_3 \mathbf{r}_{31} - \mu_2 q_3 \mathbf{r}_{32} \quad . \tag{7}$$

The second time-derivative of the dipole moment $\mathbf{d}$ takes the form

$$\ddot{\mathbf{d}} = -q_3(\mu_1 + \mu_2)\mathbf{r}_3 = -q_3\ddot{\mathbf{r}}_3 \quad , \tag{8}$$

since in our present case $\mu_1 + \mu_2 = 1$ and $\dot{\mathbf{r}}_1 = \mathbf{v}_1$ and $\dot{\mathbf{r}}_2 = \mathbf{v}_2$ are the constant (upon t) velocities. In atomic units $m_3 = m_e = 1$ and the equation of electron motion takes the form

$$\ddot{\mathbf{r}}_3 = -\nabla_3(\frac{q_3 q_1}{r_{31}}) - \nabla_3(\frac{q_3 q_2}{r_{32}}) = q_3 q_1 \frac{\mathbf{r}_{31}}{r_{31}^3} + q_3 q_2 \frac{\mathbf{r}_{32}}{r_{32}^3} \quad . \tag{9}$$

Moreover, in atomic units for the electron charge we have $q_3^2 = q_e^2 = 1$, and therefore,

$$\ddot{\mathbf{d}} = q_1 \frac{\mathbf{r}_{31}}{r_{31}^3} + q_2 \frac{\mathbf{r}_{32}}{r_{32}^3} \quad . \tag{10}$$

Let us introduce the vector $\mathbf{R}_0$ from the center of mass of the considered system to the point, where we determine the field. The corresponding unit vector is $\mathbf{n} = \frac{\mathbf{R}_0}{R_0}$. Below, we consider the electromagnetic fields produced by a system of moving charges at distances R large compared to the dimensions of the system a, i.e. for $R \gg a$ [8]. This corresponds to the case of the wave zone (or wave zone approximation). In the wave zone the scalar distance between the observation point and elementary charge i (where $i = 1, 2, 3 (= e)$) can be written in the form $\mid \mathbf{R}_0 - \mathbf{R}_i \mid \approx R_0 - \mathbf{R}_i \cdot \mathbf{n} = R_0(1 - \frac{\mathbf{v}_i}{c} \cdot \mathbf{n})$. In the present case, $\frac{v_i}{c} \ll 1$ $(i = 1, 2, 3)$, and therefore, the dimensionless value $(\frac{\mathbf{v}_i}{c} \cdot \mathbf{n})$ is small in comparison to unity. This means that in the wave zone one can replace the distances $\mid \mathbf{R}_0 - \mathbf{R}_i \mid$ $(i = 1, 2, 3)$ in the denominators of all formulas below by the distance R_0. However, such an replacement cannot be made in the numerators, since the electric charge distribution in the radiated system changes very rapidly. Finally, we can write the following expressions for the electric $\mathbf{E}$ and magnetic $\mathbf{H}$ field strengths in the wave zone (i.e. for $R \gg a$)

$$\mathbf{E} = \frac{1}{c^2 R_0}\Big[q_1 \mathbf{n}\frac{(\mathbf{n}\cdot\mathbf{r}_{31})}{r_{31}^3} - q_1\frac{\mathbf{r}_{31}}{r_{31}^3} + q_2\mathbf{n}\frac{(\mathbf{n}\cdot\mathbf{r}_{32})}{r_{32}^3} - q_2\frac{\mathbf{r}_{32}}{r_{32}^3}\Big] \quad , \tag{11}$$

$$\mathbf{H} = \frac{1}{c^2 R_0}\Big[q_1\frac{(\mathbf{n}\times\mathbf{r}_{31})}{r_{31}^3} + q_2\frac{(\mathbf{n}\times\mathbf{r}_{32})}{r_{32}^3}\Big] \quad . \tag{12}$$

These formulas can be used to calculate the energy density and energy flux for the emitted radiation. Furthermore, by using these formulas one can construct all sixteen components of the energy-momentum tensor and, therefore, determine all other properties of the radiation field. Presently, however, we do not want to discuss this problem and restrict ourselves to the consideration of total radiation I and radiation spectrum $\frac{dI(\omega)}{d\omega}$.

The intensity dI of dipole radiation emitted into the spherical angle $d\Omega$ is

$$dI = \frac{1}{4\pi c^3}\Big[q_1^2\frac{1}{r_{31}^6}(\mathbf{r}_{31}\times\mathbf{n})^2 + q_2^2\frac{1}{r_{32}^6}(\mathbf{r}_{32}\times\mathbf{n})^2 + 2q_1q_2\frac{1}{r_{31}^3 r_{32}^3}(\mathbf{r}_{31}\times\mathbf{n})(\mathbf{r}_{32}\times\mathbf{n})\Big]d\Omega \quad , \tag{13}$$

where $d\Omega = \sin\theta d\theta d\phi$. From this expression one finds the total radiation $I(t)$

$$I(t) = \frac{2}{c^3}\Big[q_1^2\frac{1}{r_{31}^4} + q_2^2\frac{1}{r_{32}^4} + 2q_1q_2\frac{1}{r_{31}^3 r_{32}^3}(\mathbf{r}_{31}\cdot\mathbf{r}_{32})\Big] \quad . \tag{14}$$

Now, it is easy to obtain the asymptotic form $I(t)$ in the considered case. Indeed, it is shown below that $r_{3i} = \sqrt{a^2 + (v_i t - b)^2}$, where $i = 1, 2$. Therefore, from the last formula one easily finds that $I(t) \sim t^{-4}$, when $t \to +\infty$. This means that the total radiation $I(t)$ is a decreasing function of t and tends to zero as $t \to +\infty$.

The energy radiated per unit frequency interval is

$$dI(\omega) = \frac{4}{3c^3} \mid \ddot{\mathbf{d}}_\omega \mid^2 \frac{d\omega}{2\pi} = \frac{2\omega^4}{3\pi c^3} \mid \mathbf{d}_\omega \mid^2 d\omega \quad , \tag{15}$$

where $\mathbf{d}_\omega$ is the Fourier-transform of the dipole momentum

$$\mathbf{d}_\omega = \int_0^{+\infty} \mathbf{d}(t)\exp(\imath\omega t)dt \quad . \tag{16}$$

The value $\frac{dI(\omega)}{d\omega}$ is called the spectral density (below, spectrum, for short) of the emitted (dipole) radiation. To obtain the explicit formula for the $\mathbf{d}_\omega$ vector we note that in the present case the two nuclear fragments move away with the constant velocities (v_1 and v_2) along strait lines in exactly opposite directions. Let as assume that the initial $^6Li-$nucleus is located at origin $x = 0, y = 0, z = 0$. The two daughter nuclei move along the positive and negative directions of OX, respectively. The electron (i.e. point charge e^-) is located at the $\mathbf{r}_3 = (x, y, 0)$ point relative to the origin. The equations of motion presented above, Eq.(9), mean that the electron motion proceeds in the $XY-$plane. Now, the vectors $\mathbf{r}_{31}$ and $\mathbf{r}_{32}$ in Eqs.(7)-(15) take the form

$$\mathbf{r}_{31} = (x + v_1 t, y, 0) \quad , \quad \mathbf{r}_{32} = (x - v_2 t, y, 0) \quad . \tag{17}$$

The scalar inter-particle (electron-nuclear) distances r_{31} and r_{32} are

$$r_{31}(t) = v_1 \sqrt{(t + \frac{x}{v_1})^2 + (\frac{y}{v_1})^2} \quad , \quad and \quad r_{32}(t) = v_2 \sqrt{(t - \frac{x}{v_2})^2 + (\frac{y}{v_2})^2} \quad , \tag{18}$$

respectively. The vector $\mathbf{d}(t)$ can be represent in the form $\mathbf{d}_+(t) + \mathbf{d}_-(t)$, where $\mathbf{d}_+(-t) = \mathbf{d}_+(t)$ and $\mathbf{d}_-(-t) = -\mathbf{d}_-(t)$. It can easily be shown that the odd vector $\mathbf{d}_-(t)$ is small ($\approx \mid \frac{a_0}{v_2} - \frac{a_0}{v_1} \mid$) and can be neglected. This simplifies the computation of the Fourier integral, Eq.(16).

Finally, the $\mathbf{d}_\omega$ vector is represented in the following two-component form $\mathbf{d}_\omega = (d_{\omega,x}, d_{\omega,y}, 0)$, where

$$d_{\omega,x} = q_1 \frac{\omega}{v_1^2} \Big[K_0(\frac{y\omega}{v_1}) + \frac{x}{y} K_1(\frac{y\omega}{v_1}) \Big] \exp(\imath \frac{x\omega}{v_1}) + q_2 \frac{\omega}{v_2^2} \Big[K_0(\frac{y\omega}{v_2}) + \frac{x}{y} K_1(\frac{y\omega}{v_2}) \Big] \exp(\imath \frac{x\omega}{v_2}) \tag{19}$$

is the $x-$component, while

$$d_{\omega,y} = q_1 \frac{\omega}{v_1^2} K_1(\frac{y\omega}{v_1}) \exp(\imath \frac{x\omega}{v_1}) + q_2 \frac{\omega}{v_2^2} K_1(\frac{y\omega}{v_2}) \exp(\imath \frac{x\omega}{v_2}) \tag{20}$$

is the $y-$component. In these equations $K_0(z)$ and $K_1(z)$ are the modified Bessel functions of orders 0 and 1, respectively. The spectrum takes the form

$$\begin{aligned} \frac{dI(\omega)}{d\omega} &= \frac{2\omega^6}{3\pi c^3} \Big\{ \frac{q_1^2}{v_1^4} \Big[\Big(K_0(\frac{y\omega}{v_1}) + \frac{x}{y} K_1(\frac{y\omega}{v_1}) \Big)^2 + K_1^2(\frac{y\omega}{v_1}) \Big] + \frac{q_2^2}{v_2^4} \Big[\Big(K_0(\frac{y\omega}{v_2}) + \frac{x}{y} K_1(\frac{y\omega}{v_2}) \Big)^2 \\ &+ K_1^2(\frac{y\omega}{v_2}) \Big] + \frac{2q_1 q_2}{v_1^2 v_2^2} \Big[K_0(\frac{y\omega}{v_1}) + \frac{x}{y} K_1(\frac{y\omega}{v_1}) \Big] \Big[K_0(\frac{y\omega}{v_1}) + \frac{x}{y} K_1(\frac{y\omega}{v_1}) \Big] \times \\ & \cos(\frac{x\omega}{v_1} - \frac{x\omega}{v_2}) + \frac{2q_1 q_2}{v_1^2 v_2^2} K_1(\frac{y\omega}{v_1}) K_1(\frac{y\omega}{v_2}) \cos(\frac{x\omega}{v_1} - \frac{x\omega}{v_2}) \Big\} \quad . \end{aligned} \tag{21}$$

These formulas contain the incident (Cartesian) $x-$ and $y-$coordinates of the electron which is bound inside of the atom. In fact, to obtain these formulas we assumed that the electron does not move appreciably during the $(n;t)-$reaction, i.e. its $x-$ and $y-$coordinates do not change noticeably during the process. Obviously, this assumption corresponds to the beginning of the process, i.e. to relatively short times t (i.e. $t \approx 0$), or in other words, to very large frequencies ω. For large t the observed spectrum can be different from the spectrum $\frac{dI(\omega)}{d\omega}$ predicted above. However, the case of large t corresponds to the low-frequency tail of energy spectrum which is not a specific interest for the present study. It is also shown below that the intensity of emitted radiation rapidly vanishes for $\omega < \omega_0$, where ω_0 is some characteristic atomic frequency. In general, the t dependence of the $x-$ and $y-$coordinates in Eqs.(19) - (21) cannot be neglected even for short times $t \approx 0$. This significantly complicates the final formulas for the $\mathbf{d}_\omega$ vector and for the spectral density, Eq.(21), presented above.

The explicit formulas for the $x(t)-$ and $y(t)-$dependencies one can derive, e.g., by using some realistic atomic models. In fact, in classical mechanics the one-electron atom is usually represented as a system, having a massive positively charged nucleus at the origin of coordinates and an electron, which is moving in a circular atomic orbit of radius a with a constant angular velocity ω_0, i.e. $x = a\cos(\omega_0 t + \phi_0)$ and $y = a\sin(\omega_0 t + \phi_0)$ [9]. The orbit radius a is related to the Bohr radius a_0 by the relation $a = a_0/Q$, where Q is the charge of the incident nucleus in Eqs.(1)-(2). Presently, we assume that this orbit lies in the same XY plane (see above). Such a model atom must be a stable system, i.e. it does not emit any radiation under normal conditions. Therefore, one can expect that if $\omega \rightarrow \omega_0$, then radiation emitted by the considered system will vanish. In fact, some (relatively weak) radiation can be detected even for $\omega \approx \omega_0$, but this case is required a separate consideration. Note only that the corresponding spectrum for $\omega \approx \omega_0$ can be obtained by using the same equations Eq.(16) - (18) presented above. Only now the incident electron $x-$ and $y-$coordinates in the expression for dipole momentum also depend on t and the final spectrum $\frac{dI(\omega)}{d\omega}$ takes a very complicated form. In the case, when $\omega \ll \omega_0$ radiation is not emitted. Indeed, for $\omega \ll \omega_0$ the fields created by moving nuclear fragments can be considered as a slow (adiabatic) perturbation. In this case the electron density distribution has enough time to adjust the corresponding changes in the nuclear density. Obviously, such a process cannot produce any noticeable emission of radiation.

The limiting forms of the energy spectrum for small and very large frequencies can be obtained from Eq.(21). The high-frequency part takes the form $\frac{dI(\omega)}{d\omega} \sim \omega^5 \exp(-\beta\omega)$. In fact, the exact high-frequency spectrum is represented as the sum of a few such terms with different β's. The low-frequency tail can formally be represented in the form $a_1\omega^5 + a_2\omega^6 \ln\omega$, where a_1 and a_2 are two positive constants. However, as we mentioned above, the low-frequency limit is not an actual limit, since our present method cannot be used, in principle, in the cases when $\omega \ll \omega_0$, where ω_0 is some characteristic atomic frequency.

Note that the spectral density $\frac{dI(\omega)}{d\omega}$, Eq.(21), contains the oscillating factors $\simeq \cos(\frac{x\omega}{v_1} - \frac{x\omega}{v_2})$. These terms correspond to the interference of radiation emitted by the same electron in the two different processes. In the case of Eq.(1), the first process is the electron acceleration which is produced by the fast tritium nucleus. In the second process analogous electron acceleration is produced by the fast helium-4 nucleus (or $\alpha-$particle). Since the radiation is emitted by the same electron, then it must be an interference between these two radiation fields. Finally, the dependence of spectral density $\frac{dI(\omega)}{d\omega}$ upon ω does take an oscillating shape (for $\omega > \omega_0$). In general, the amplitudes of such oscillations depend upon ω. In fact, the oscillating shape for the spectral density $\frac{dI(\omega)}{d\omega}$ can be found in the case of fission-type nuclear reaction in arbitrary atomic system, in which two (or more) rapidly moving and positively charged nuclear fragments are formed in the final stage. The velocities of such nuclear fragments must be significantly larger than the incident velocities of electrons in their atomic orbits.

In fact, our conclusion about the oscillating shape of spectral density $\frac{dI(\omega)}{d\omega}$ is restricted to the case, when radiation is emitted in the $(n;t)-$nuclear reaction in one atomic (ionic) system. For many-electron systems the observed radiation spectrum will be different from the spectrum predicted above. In particular, the amplitudes of oscillations in such a spectrum rapidly decrease when the number of electrons grow. Formally, in many-electron systems Eqs.(19) - (21) must be averaged over all possible directions and absolute values of the electron coordinates. This procedure essentially gives zero phase factors, i.e. the electron accelerations produced by different nuclear fragments can be considered as truly independent. The radiation fields which are created in the result of such accelerations are also completely independent (i.e. no interference). This explains why the emission of thermal bremsstrahlung inside of thermonuclear explosive devices and stars can always be computed quite accurately by using the single-nucleus Coulomb model, which ignores all contributions from other nuclear fragments. This conclusion is remarkable, since the densities in actual thermonuclear systems can be extremely high (e.g. $\rho \geq 10^4 \; g \cdot cm^{-3}$ and even higher, see, e.g., [2]). For such high densities the nuclear-nuclear distances ℓ_{NN} can be very short, e.g., they can be comparable be comparable or even smaller than the corresponding coherence length ℓ_{coh} [10], [11] for non-relativistic bremsstrahlung. Formally, if $\ell_{NN} \approx \ell_{coh}$, then the presence of neighboring nuclei cannot be ignored in the consideration. However, in thermonuclear applications the observed radiation is simultaneously emitted by very large number of electrons N_e, where N_e is comparable with the Avogadro number $N_A \approx 6.02214199 \cdot 10^{23}$. The resulting phase factor essentially vanishes, since $N_e \gg 1$. This means that the single-nucleus Coulomb model [12] is quite accurate for such systems.

1.4 Quantum-mechanical approach

The classical theory developed above allows one to explain the basic properties of radiation emitted in the $(n;t)-$nuclear reactions in various atoms and ions. However, a full understanding of the considered process requires a rigorous quantum-mechanical treatment. In quantum mechanics to analyze the free-free radiation emission one needs to compute the probability of dipole radiation, which is written in the form (in atomic units) [10]

$$dP = \frac{\omega^3}{(2\pi)^4 c^3} \mid (\mathbf{e} \cdot \mathbf{d}_{if}) \mid^2 do_{\mathbf{k}} d^3\mathbf{p}_f = \frac{\alpha^3 \omega}{(2\pi)^4} \mid (\mathbf{e} \cdot \ddot{\mathbf{d}}_{if}) \mid^2 do_{\mathbf{k}} d^3\mathbf{p}_f \quad , \tag{22}$$

where $\mathbf{e}$ is the polarization vector of the emitted photon, $\mathbf{k}$ is the momentum of this photon (i.e. $\mathbf{k} \perp \mathbf{e}$ and $\mathbf{k}^2 = \omega^2$) and $\mathbf{d}$ is the dipole moment of the system, i.e.

$$\mathbf{d}_{if} = \langle \mathbf{d}_{if} \rangle = \int_0^{+\infty} \oint \psi_i^*(r_3, \mathbf{n}_3) \mathbf{d}(r_3, \mathbf{n}_3) \psi_f(r_3, \mathbf{n}_3) r_3^2 dr_3 d\mathbf{n}_3 \quad ,$$

where ψ_i and ψ_f are the initial and final electron wave functions. Also in these equations and below, the product $(\mathbf{a} \cdot \mathbf{b}) = \mathbf{a}^* \cdot \mathbf{b}$ is the scalar product of the two vectors $\mathbf{a}$ and $\mathbf{b}$ and $\alpha = c^{-1}$ is the fine structure constant. The expression Eq.(22) determines the probability dP to emit a photon $\mathbf{k}$ into a solid angle $do_{\mathbf{k}}$ and produce the final (scattered) electron into the range of states $d^3\mathbf{p}_f$. By using the classical results obtained above (see, Eqs.(7) - (10)) and applying the correspondence principle one can reduce the last formula to the following form

$$\begin{aligned} dP &= \frac{\alpha^3 \omega q_3^2}{(2\pi)^4} \mid (\mathbf{e}^* \cdot (\ddot{\mathbf{r}}_3)_{if}) \mid^2 do_{\mathbf{k}} d^3\mathbf{p}_f = \frac{\alpha^3 \omega}{(2\pi)^4} \mid q_1 \langle \frac{1}{r_{31}} \mathbf{e} \cdot \nabla_3 (\psi_i \psi_f) \rangle \\ &+ q_2 \langle \frac{1}{r_{32}} \mathbf{e} \cdot \nabla_3 (\psi_i \psi_f) \rangle \mid^2 do_{\mathbf{k}} d^3\mathbf{p}_f \quad , \end{aligned} \tag{23}$$

where in atomic units $q_3^2 = q_e^2 = 1$ and ψ_i and ψ_f are the incident and final wave functions of the considered system. To produce this expressions we also performed the integration by parts. Note also, that in these equations we have $r_{ij} = \mid \mathbf{r}_i - \mathbf{r}_j \mid = R_{ij} = \mid \mathbf{R}_i - \mathbf{R}_j \mid$ $(i \neq j = 1, 2, 3)$ and $\nabla_3 = \frac{\partial}{\partial \mathbf{r}_3} = \frac{\partial}{\partial \mathbf{R}_3}$. Here the notations from previous section are used. The origin of the coordinates is also chosen at the center of mass of the considered system, i.e. $\mathbf{R}_i = \mathbf{r}_i$, where $i = 1, 2, 3$. This simplifies significantly all formulas presented below.

By integrating over $do_{\mathbf{k}}$ and performing summation over the polarizations of the emitted photon one finds the following expression

$$dP = \frac{\alpha^3\omega\sqrt{Q^2 - 2n^2\omega}}{6\pi^3 n}\Big[q_1^2\mathbf{a}_{if}^2 + 2q_1q_2(\mathbf{a}_{if}\cdot\mathbf{b}_{if}) + q_2^2\mathbf{b}_{if}^2\Big]d\omega d\mathbf{n}_f \quad , \tag{24}$$

where $Q = q_1 + q_2$, n is the principal quantum number of the initial hydrogen-like system, $\mathbf{n}_f = \mathbf{p}_f/p_f$ is the unit vector corresponding to the $\mathbf{p}_f$ vector, while the vectors $\mathbf{a}_{if}$ and $\mathbf{b}_{if}$ are

$$\mathbf{a}_{if} = \langle\frac{1}{r_{31}}\nabla_3(\psi_i\psi_f)\rangle \quad and \quad \mathbf{b}_{if} = \langle\frac{1}{r_{32}}\nabla_3(\psi_i\psi_f)\rangle \quad . \tag{25}$$

These values can also be considered as the matrix elements computed with the ψ_i and ψ_f wave functions. The frequency $\omega_{up} = \frac{Q^2}{2n^2}$ determines the so-called upper limit of the radiation spectrum (in atomic units). This frequency corresponds to the case when the final (free) electron has zero velocity (or zero momentum). The radiation spectrum cuts off for $\omega \geq \omega_{up}$. This indicates clearly that we are dealing with the bound-free emission, rather than with bremsstrahlung (or free-free emission). In the last case, the observed spectrum does not have any upper boundaries.

Now, let us produce the explicit expressions for the $\mathbf{a}$ and $\mathbf{b}$ vectors from Eq.(25). The wave function of the incident hydrogen-like ion takes the form (in atomic units) [13]

$$\psi_i(r) = C\exp(-Qr) = \frac{Q\sqrt{Q}}{\sqrt{\pi}}\exp(-Qr) \quad , \tag{26}$$

where we assume that the incident ion was in its ground $S(L = 0)-$state. Note that ψ_i is a real function. The generalization of our present approach to the case of excited states in the incident atoms (and ions) is straightforward. Note that the incident wave function corresponds to the bound state spectrum. Rigorously speaking, bremsstrahlung is the radiative transition between two states in continuum in which photon is emitted. Formally, in the present case we consider the bound-free radiative transitions in atomic systems, rather than bremsstrahlung. On the other hand, all systems which arise immediately after the $(n;t)-$reaction in light atoms and ions (see, e.g., Eqs.(1) - (2)) are unbound. However, the bound state wave function can be used for such systems, since: (1) the two nuclear fragments move significantly faster than atomic electrons, and (2) the electron density distribution cannot change instantly at $t = 0$.

The wave function of the final electron can be written as a plane wave, i.e. $\psi_f = \exp(\imath\mathbf{p}_f\mathbf{r}_3)$, where $\mathbf{p}_f$ is the momentum of the final (i.e. free) electron. Now, we can write

$$\nabla_3(\psi_i\psi_f) = \frac{Q\sqrt{Q}}{\sqrt{\pi}}\Big(-Q\mathbf{n}_3 + \imath\mathbf{p}_f\Big)\exp(-Qr_3 + \imath\mathbf{p}_f\cdot\mathbf{r}_3) \quad , \tag{27}$$

where $\mathbf{n}_3 = \frac{\mathbf{r}_3}{r_3}$ is the unit vector in the direction of $\mathbf{r}_3$. It is clear that the matrix elements of the $\mathbf{a}_{if}$ and $\mathbf{b}_{if}$ vectors can be expressed as the linear combinations of four following auxiliary vector-integrals $I_{1,i}$ and $I_{2,i}$

$$\begin{aligned} I_{1,i} &= \frac{Q^2\sqrt{Q}}{\sqrt{\pi}}\langle\frac{1}{r_{3i}}\mathbf{n}_3\cdot\exp(-Qr_3 + \imath\mathbf{p}_f\cdot\mathbf{r}_3)\rangle = \frac{Q^2\sqrt{Q}}{\sqrt{\pi}} \times \\ &\int_0^{+\infty}\oint\frac{\mathbf{n_3}}{r_{3i}}\exp(-Qr_3 + \imath\mathbf{p}_f\cdot\mathbf{r}_3)r_3^2dr_3d\mathbf{n}_3 \quad , \end{aligned} \tag{28}$$

$$I_{2,i} = \imath \mathbf{p}_f \frac{Q\sqrt{Q}}{\sqrt{\pi}} \langle \frac{1}{r_{3i}} \cdot \exp(-Q r_3 + \imath \mathbf{p}_f \cdot \mathbf{r}_3) \rangle = \imath \mathbf{p}_f \frac{Q\sqrt{Q}}{\sqrt{\pi}} \times \int_0^{+\infty} \oint \frac{1}{r_{3i}} \exp(-Q r_3 + \imath \mathbf{p}_f \cdot \mathbf{r}_3) r_3^2 dr_3 d\mathbf{n}_3 \quad , \tag{29}$$

where $i = 1, 2$. Our next goal is to obtain the explicit formulas for these auxiliary integrals.

In actual calculations the plane wave is represented in terms of the spherical Bessel functions (see, e.g. [5])

$$\exp(\imath \mathbf{p}_f \cdot \mathbf{r}_3) = 4\pi \sum_{\ell=0}^{\infty} \imath^\ell j_\ell(p_f r_3) \sum_{m=-\ell}^{\ell} Y^*_{\ell m}(\mathbf{n}_f) Y_{\ell m}(\mathbf{n}_3) \quad , \tag{30}$$

where $j_\ell(y)$ are the spherical Bessel functions (i.e. $j_\ell(y) = \sqrt{\pi/(2y)} J_{\ell+\frac{1}{2}}(y)$) and $Y_{\ell m}(\mathbf{n})$ are the spherical harmonics. Also, in this formula $p_f =\mid \mathbf{p}_f \mid, r_3 =\mid \mathbf{r}_3 \mid$, while $\mathbf{n}_f = \mathbf{p}_f / p_f$ and $\mathbf{n}_3 = \mathbf{r}_3 / r_3$ are two unit vectors. The $\frac{1}{r_{31}}$ function is represented in the form

$$\frac{1}{r_{31}} = \frac{4\pi}{r_>} \sum_{\ell_1=0}^{\infty} \frac{1}{2\ell_1 + 1} \Bigl(\frac{r_<}{r_>} \Bigr)^{\ell_1} \sum_{m_1=-\ell_1}^{\ell_1} Y^*_{\ell_1 m_1}(\mathbf{n}_3) Y_{\ell_1 m_1}(\mathbf{n}_1) \quad , \tag{31}$$

where $r_> = max(r_1, r_3)$ and $r_< = min(r_1, r_3)$. Analogously, for the $\frac{1}{r_{32}}$ function we have

$$\frac{1}{r_{32}} = \frac{4\pi}{r_>} \sum_{\ell_1=0}^{\infty} \frac{(-1)^{\ell_1}}{2\ell_1 + 1} \Bigl(\frac{r_<}{r_>} \Bigr)^{\ell_1} \sum_{m_1=-\ell_1}^{\ell_1} Y^*_{\ell_1 m_1}(\mathbf{n}_3) Y_{\ell_1 m_1}(\mathbf{n}_2) \quad , \tag{32}$$

where $r_> = max(r_2, r_3)$ and $r_< = min(r_2, r_3)$. Here, the relation $\mathbf{n}_2 = -\mathbf{n}_1$ is used (this relation is obeyed in the present case). Note also, another useful relation

$$\oint Y^*_{\ell_1 m_1}(\mathbf{n}_3) \mathbf{n}_3 Y_{\ell m}(\mathbf{n}_3) d\mathbf{n}_3 = \sqrt{\frac{3(2\ell+1)}{4\pi(2\ell_1+1)}} C^{\ell_1 0}_{10;\ell 0} \sum_{\mu=-1}^{1} C^{\ell_1, m_1}_{1,\mu;\ell,m} \mathbf{e}_\mu \quad , \tag{33}$$

where $\mu = -1, 0, 1$ and $\mathbf{e}_{-1} = -\frac{1}{\sqrt{2}}(\mathbf{i} + \imath \mathbf{j}), \mathbf{e}_0 = \mathbf{k}, \mathbf{e}_{+1} = \frac{1}{\sqrt{2}}(\mathbf{i} - \imath \mathbf{j})$ are the spherical basis vectors [5]. Here, $\mathbf{i}, \mathbf{j}$ and $\mathbf{k}$ are the corresponding (Cartesian) unit vectors. The coefficients $C^{\ell_1, m_1}_{1,\mu;\ell,m}$ are the Clebsch-Gordan coefficients [14] (or $CG-$coefficients, for short). Note that the CG-coefficient $C^{\ell_1 0}_{10;\ell 0}$ in this expression differs from zero only if $\ell + \ell_1 + 1$ is an even number, i.e. $\ell_1 = \ell \pm 1$. The last formula allows one to obtain the following explicit expression for the $I_{1,i}$ auxiliary integrals

$$I_{1,i} = 8\pi Q^2 \sqrt{3Q} \cdot \sum_{\ell=0}^{\infty} \imath^\ell j_\ell(p_f r_3) \frac{\exp(-Q r_3)}{r_>} \Bigl[\sqrt{\frac{2\ell+1}{(2\ell+3)^3}} \Bigl(\frac{r_<}{r_>} \Bigr)^{\ell+1} C^{\ell+1,0}_{1,0;\ell,0} \times \sum_{\mu=-1}^{+1} \sum_{m=-\ell}^{\ell} C^{\ell+1, m+\mu}_{1,\mu;\ell,m} Y^*_{\ell m}(\mathbf{n}_f) Y_{\ell+1, m+\mu}(\mathbf{n}_i) \mathbf{e}_\mu + \sqrt{\frac{2\ell+1}{(2\ell-1)^3}} \Bigl(\frac{r_<}{r_>} \Bigr)^{\ell-1} C^{\ell-1,0}_{1,0;\ell,0} \times \sum_{\mu=-1}^{+1} \sum_{m=-\ell}^{\ell} C^{\ell-1, m+\mu}_{1,\mu;\ell,m} Y^*_{\ell m}(\mathbf{n}_f) Y_{\ell-1, m+\mu}(\mathbf{n}_i) \mathbf{e}_\mu \Bigr] \quad , \tag{34}$$

where $i = 1, 2$, $r_> = max(r_i, r_3)$ and $r_< = min(r_i, r_3)$.

Analogous computations for the $I_{2,i}$ auxiliary integral is even simpler, since the spherical harmonics are pairwise orthogonal to each other, i.e. $\oint Y^*_{\ell_1 m_1}(\mathbf{n}_3)Y_{\ell_2 m_2}(\mathbf{n}_3)d\mathbf{n}_3 = \delta_{\ell_1\ell_2}\delta_{m_1 m_2}$. The explicit expression for the $I_{2,i}$ integral is

$$I_{2,i} = 4Q\sqrt{\pi Q}\cdot \mathbf{p}_f \cdot \frac{\exp(-Qr_3)}{r_>}\sum_{\ell=0}^{\infty} \imath^{\ell+1}\Big(\frac{r_<}{r_>}\Big)^{\ell} j_\ell(p_f r_3)P_\ell(\mathbf{n}_i\cdot\mathbf{n}_f) \quad , \tag{35}$$

where $i = 1, 2$, $r_> = max(r_i, r_3)$, $r_< = min(r_i, r_3)$ and $P_n(x)$ is the corresponding Legendre polynomial of order ℓ. The following integration over the radial r_3 variable is straightforward in all integrals $I_{1,i}$ and $I_{2,i}$ $(i = 1, 2)$. Note that each of two integrals $I_{2,i}$ differs from zero, if $\ell + \ell_1 + 1 (= 2\ell + 1)$ is an odd number. The integer $\ell + \ell_1 + 1$ cannot be even and odd simultaneously. This means that all matrix elements which contain the products of $\mathbf{n}$ and $\mathbf{p}_f$ vectors equal zero identically. This corresponds to the product of $I_{1,i}$ and $I_{2,j}$ auxiliary integrals. The identity $I_{1,i}I_{2,j} = 0$ (where $i = 1, 2$ and $j = 1, 2$) simplifies computation of the corresponding terms $\mathbf{a}^2_{if}$, $\mathbf{b}^2_{if}$ and $\mathbf{a}_{if}\cdot\mathbf{b}_{if}$ in Eq.(24).

The actual computation of these terms includes an integration over $\mathbf{n}_f$ (see, Eq.(24)). This procedure is also straightforward. Indeed, the original expression for the I_{2i} integral Eq.(29) can be re-written in the form

$$I_{2,i}(\mathbf{n}_f) = \mathbf{p}_f \cdot \sum_{\ell=0}^{\infty} d_\ell(i)P_\ell(\mathbf{n}_i\cdot\mathbf{n}_f) \quad , \tag{36}$$

where $d_\ell(i)$ are some scalar coefficients which are easily determined from Eq.(35). From this equation one finds the following expression for the $I_{2,i}(\mathbf{n}_f)I_{2,j}(\mathbf{n}_f)$ product averaged over the $\mathbf{n}_f$ directions

$$\oint I_{2,i}(\mathbf{n}_f)I_{2,j}(\mathbf{n}_f)d\mathbf{n}_f = 4\pi p_f^2\sum_{\ell=0}^{\infty}[\delta_{ij} + (-1)^\ell(1-\delta_{ij})]\frac{d_\ell(i)d_\ell(j)}{2\ell+1} \quad , \tag{37}$$

where $p_f^2 = \frac{Q^2}{n^2} - 2\omega$ and $\delta_{ij} = 1$, if $i = j$ and zero otherwise. This relation $p_f = p_f(\omega)$ follows from the energy conservation. Note, however, that in the present case there is no conservation of momentum for the electron-photon system, since the atomic nuclei are heavy compared with the electron (see discussion in [13]).

Analogous procedure for the product of $I_{1,i}$ and $I_{1,j}$ auxiliary integrals is slightly more complicated. Formally, the original expression for these integrals, Eq.(34), can be written in the form

$$\begin{aligned} I_{1,i}(\mathbf{n}_f) \;&=\; \sum_{\ell=0}^{\infty} a_\ell(i)\sum_{\mu=-1}^{1}\sum_{m=-\ell}^{\ell} C^{\ell+1,m+\mu}_{1,\mu;\ell,m} Y^*_{\ell m}(\mathbf{n}_f)Y_{\ell+1,m+\mu}(\mathbf{n}_i)\mathbf{e}_\mu + \\ &\sum_{\ell=0}^{\infty} b_\ell(i)\sum_{\mu=-1}^{1}\sum_{m=-\ell}^{\ell} C^{\ell-1,m+\mu}_{1,\mu;\ell,m} Y^*_{\ell m}(\mathbf{n}_f)Y_{\ell-1,m+\mu}(\mathbf{n}_i)\mathbf{e}_\mu \quad , \end{aligned} \tag{38}$$

where $a_\ell(i)$ and $b_\ell(i)$ $(i = 1, 2)$ are some scalar coefficients which can easily be determined from Eq.(34). In general, the product of auxiliary integrals $I_{1,i}(\mathbf{n}_f)$ and $I_{1,j}(\mathbf{n}_f)$ averaged over the $\mathbf{n}_f$ directions can be expressed as a sum of four terms T_1, T_2, T_3 and T_4, where the first term T_1 has the form

$$T_1 = \sum_{\ell=0}^{\infty} a_\ell(i)a_\ell(j)\sum_{\mu=-1}^{+1}\sum_{m=-\ell}^{\ell}\Big(C^{\ell+1,m+\mu}_{1,\mu;\ell,m}\Big)^2 Y^*_{\ell+1,m+\mu}(\mathbf{n}_i)Y_{\ell+1,m+\mu}(\mathbf{n}_j) \quad . \tag{39}$$

Three remaining terms T_2, T_3 and T_4 in the integral $\oint I_{1,i}(\mathbf{n}_f)I_{1,j}(\mathbf{n}_f)d\mathbf{n}_f$ can be represented (and then treated) in analogous manner. Now, we note that if $j \neq i$, then $\mathbf{n}_j = -\mathbf{n}_i$ and $Y_{\ell,m}(\mathbf{n}_j) = Y_{\ell,m}(-\mathbf{n}_i) = (-1)^{\ell} Y_{\ell,m}(\mathbf{n}_i)$. Also, the formula

$$Y^{*}_{\ell_a,m_a}(\mathbf{n}_i)Y_{\ell_b,m_b}(\mathbf{n}_i) = (-1)^{m_a} \sum_{\ell_c} \sqrt{\frac{(2\ell_a+1)(2\ell_b+1)}{4\pi(2\ell_c+1)}} C^{\ell_c,0}_{\ell_a,0;\ell_b,0} C^{\ell_c,m_c}_{\ell_a,-m_a;\ell_b,m_b} Y_{\ell_c,m_c}(\mathbf{n}_i) \quad (40)$$

is very useful in the present case. Finally, the expression for the T_1 term is given by

$$T_1 = \sum_{\ell=0}^{\infty} \frac{(2\ell+3)}{\sqrt{4\pi}} a_{\ell}(i) a_{\ell}(j) \sum_{L=0}^{2\ell+2} \frac{C^{L,0}_{\ell+1,0;\ell+1,0}}{\sqrt{(2L+1)}} \sum_{\mu=-1}^{+1} \sum_{m=-\ell}^{\ell} \left(C^{\ell+1,m_1}_{1,\mu;\ell,m} \right)^2 \times \quad (41)$$
$$C^{L,M}_{\ell+1,-m_1;\ell+1,m_1} Y_{LM}(\mathbf{n}_i) \quad ,$$

where $m_1 = m + \mu$. In fact, the last expression can be re-written in some different forms, which are more convenient for numerical calculations. However, for our purposes it is more important to note that the sum of T_i terms (i = 1, 2, 3, 4) determines the explicit dependence of the total probability of dipole radiation P upon the direction of nuclear motion $\mathbf{n}_i$. In general, the probability $P(\mathbf{n}_i)$ includes contributions from partial components with different L. The isotropic component (or $L = 0-$component) is usually larger (and even significantly larger) than the corresponding non-isotropic (or $\mathbf{n}-$dependent) components with $L \geq 2$. It can easily be predicted, since the isotropic components in all T_1, T_2, T_3 and T_4 terms are larger than non-isotropic. Moreover, the contribution from the term, which is represented by Eq.(37), also contains only the isotropic part. The non-isotropic contributions are represented by the terms $\sim Y_{LM}(\mathbf{n}_i)$, where $L = 2, 4, ..., 2k+2, ...$. In general, such contributions rapidly decrease with L ($\sim (2L+1)^{-3}$).

Thus, by using formulas obtained above (see Eqs.(23) - (25) and Eqs.(37) - (41)) one can analyze the emission of bremsstrahlung in the $(n;t)-$reactions in light atoms and ions. Note, however, that our analysis is restricted to the consideration of one-electron systems only. The generalization of this approach to the case of many-electron atoms and ions is straightforward. In fact, in the first approximation, for many-electron systems the number of free electrons N_f can be included in the corresponding formulas as a common factor. Indeed, the system remaining after the $(n;t)-$reaction contains only a few electrons. As is well known (see, e.g., [15]) such a system cannot emit radiation in the dipole approximation. This essentially means that in the first approximation the results obtained above can directly be used for many-electron systems. In the next approximation, however, there are two additional problems for many-electron systems. First, the electron-electron repulsion cannot be ignored in the incident and final systems. The incident and final wave functions in many-electron systems take very complex forms. Such functions depend upon a large number of variables and can be determined only approximately (i.e. numerically). Moreover, in many-electron systems the corresponding equations of motion also include the electron-electron repulsions. Finally, such equations take a very complex form. Analytical solution of these equations does not exists, but even their numerical solution is extremely complicated. Second, any many-electron wave function, according to the Pauli principle, must be antisymmetric function of all electron coordinates, i.e. $\Psi(..., i, ..., j, ...) = -\Psi(..., j, ..., i, ...)$, where i and j designate the spin and spatial coordinates of $i-$th and $j-$th electrons. This also complicates the problem significantly. Note, however, that there are some approximate methods which can be used to analyze the $(n;t)-$reactions in many-electron atoms and ions. In fact, the consideration of such reactions in many-electron systems is the central problem for our future research.

1.5 Radiation emitted in the neutron-tritium reaction in-flight

In this section we consider the $(n;t)-$reactions in light atoms in-flight. The $(n;t)-$reactions in-flight are of specific interest in numerous applications, since they can be used to produce a very intense beams of visible and UV-radiation. Moreover, the study of radiation emitted in the $(n;t)-$reactions in-flight allows one to investigate the electron-electron position correlations in incident atomic systems. In general, bremsstrahlung emitted during the $(n;t)-$reactions in the fast atom (ion) can be obtained from our formulas presented above. Indeed, the field strengths $\mathbf{E}$ and $\mathbf{H}$ exhibited in Eqs.(11) - (12) must now be transformed according to the formulas (see, e.g., [9])

$$\mathbf{E}' = \gamma(\mathbf{E} + \frac{\mathbf{v}}{c} \times \mathbf{H}) - \frac{\gamma^2}{\gamma+1}\frac{\mathbf{v}}{c}(\frac{\mathbf{v}}{c} \cdot \mathbf{E}) \tag{42}$$

$$\mathbf{H}' = \gamma(\mathbf{H} - \frac{\mathbf{v}}{c} \times \mathbf{E}) - \frac{\gamma^2}{\gamma+1}\frac{\mathbf{v}}{c}(\frac{\mathbf{v}}{c} \cdot \mathbf{H}) \tag{43}$$

to produce the actual field strengths $\mathbf{E}'$ and $\mathbf{H}'$. In the last equations $\mathbf{v}$ is the velocity of incident atom (ion) in Eqs.(1) - (2), while $\gamma = \frac{1}{\sqrt{1-\frac{v^2}{c^2}}}$ is the Lorentz $\gamma-$factor. By using these field strengths $\mathbf{E}'$ and $\mathbf{H}'$, one can easily determine the canonical and symmetric stress tensors $T^{\alpha,\beta}$ and $\Theta^{\alpha,\beta}$, respectively [8], [9] (here $\alpha = 0,1,2,3$ and $\beta = 0,1,2,3$). As we mentioned above these tensors allows us to determine, in principle, an arbitrary property of the considered electromagnetic fields. In general, the energy and momentum densities (i.e. the (0,0) and (0,1), (0,2), (0,3) components of the symmetric stress tensor) of the emitted field increase approximately quadratically with γ. In fact, in actual applications it is very difficult to produce one-electron lithium ions with $\gamma \geq 3$. For instance, the ions with $\gamma \approx 2$ correspond to the kinetic energies ≈ 1 GeV per nucleon in the incident nucleus. A few-electron ions with such energies cannot be easily created in modern experiments. On the other hand, the corresponding relativistic amplification of radiation emitted in the $(n;t)-$reaction is relatively small for $\gamma \leq 3$.

Note, however, that the considered $(n;t)-$reaction in light atoms and ions produces a number of free electrons. This is also true for the nuclear fission in heavy atoms with $A \geq 232$ and $Q \geq 90$. For our purposes the difference between the $(n;t)-$reactions in light atoms and nuclear fission in heavy elements can be found only for the number of free electrons in the final state N_f. For the light atoms $N_f \approx 1-3$, while after the nuclear fission in heavy atoms one finds $N_f \approx 6-10$ [7]. It is assumed below that the incident atom (ion) moves with a constant velocity $\mathbf{v}$ which is significantly larger than the appropriate velocities of atomic electrons in light atoms. This means that after $(n;t)-$reaction (or nuclear fission) an observer (in the laboratory frames) will see a sudden creation of the group of fast electrons. In general, such a sudden creation of a few (N_f) electrons will be accompanied by the emission of radiation. Let us discuss the basic properties of such radiation.

The intensity distribution in frequency (ω) and angle (Ω) is written in the form

$$\frac{d^2 I(\omega,\Omega)}{d\omega d\Omega} = \frac{e^2\omega^2}{4\pi^2 c} \mid \sum_{j=1}^{N_f} \int_0^{+\infty} [\mathbf{b}_j - \mathbf{n}(\mathbf{b}_j \cdot \mathbf{n})] \exp(\imath\omega(t - \frac{\mathbf{r}_j(t)\cdot\mathbf{n}}{c}))dt \mid^2 \quad , \tag{44}$$

where t is the charge's own time, $\mathbf{b}_j = \frac{\mathbf{v}_j}{c}$ $(1 \leq j \leq N_f)$ and $\mathbf{n}$ is the unit vector in the direction from the point where radiation has been emitted $(\mathbf{r}(t))$ to the observation point $(\mathbf{x})$, i.e. $\mathbf{n} = (\mathbf{x} - \mathbf{r}(t))/ \mid \mathbf{x} - \mathbf{r}(t) \mid$. In our present case, all electrons are moving with the same (constant) velocity $\mathbf{v}$, i.e. $\mathbf{b}_j \approx \mathbf{b}$ (where $\mathbf{b} = \frac{\mathbf{v}}{c}$) and $\mathbf{r}_j(t) \approx \mathbf{r}_j(0) + c\mathbf{b}t$ ($1 \le j \le N$). Therefore, the last integral can be transformed to the following form:

$$\frac{d^2I(\omega,\Omega)}{d\omega d\Omega} = \frac{e^2\omega^2}{4\pi^2 c} \mid \int_0^{+\infty} \Bigl[\mathbf{b} - \mathbf{n}(\mathbf{b}\cdot\mathbf{n})\Bigr] \exp(\imath\omega t(1-\mathbf{b}\cdot\mathbf{n}))dt \mid^2 \times \Bigl[N_f + \sum_{i\neq j=1}^{N_f} \exp(-\imath\frac{\omega}{c}\mathbf{n}\cdot(\mathbf{r}_i(0)-\mathbf{r}_j(0)))\Bigr] \quad . \tag{45}$$

Or, in other words

$$\frac{d^2I(\omega,\Omega)}{d\omega d\Omega} = \frac{e^2N_f}{4\pi^2 c}\frac{\mathbf{b}^2-(\mathbf{b}\cdot\mathbf{n})^2}{(1-\mathbf{b}\cdot\mathbf{n})^2}\Bigl[1 + \frac{2}{N_f}\sum_{i>j=1}^{N_f} \cos(\frac{\omega}{c}\mathbf{n}\cdot(\mathbf{r}_i(0)-\mathbf{r}_j(0)))\Bigr] \quad , \tag{46}$$

where N_f is the number of the electrons from the incident atom (ion) which became free after the $(n;t)-$reaction. Here $\mathbf{r}_i(0)$ ($1 \le i \le N_f$) are the vectors which designate the incident positions of electrons in the incident atom (ion). The time $t = 0$ corresponds to the moment when the nuclear $(n;t)-$reaction occurs. The difference between the two $\mathbf{r}_i(0)$ and $\mathbf{r}_j(0)$ vectors is designated below by $\mathbf{R}_{ij}(0)$.

In our case, the numerical value of $\beta = \mid \mathbf{b} \mid \approx \frac{1}{2} < 1$. Therefore, the radiated energy $I(\omega,\Omega)$ can be represented by its Taylor series expansion $I(\omega,\Omega) = I^{(0)}(\omega,\Omega) + \beta^2 I^{(2)}(\omega,\Omega) + \beta^4 I^{(4)}(\omega,\Omega) + \ldots$, where β^2 is a small parameter. Now, from the last equation one easily finds the corresponding equations for the $I^{(0)}(\omega,\Omega)$ and $I^{(2)}(\omega,\Omega)$ functions. In fact, we have

$$\frac{d^2I^{(0)}(\omega,\Omega)}{d\omega d\Omega} = \frac{e^2N_f^2}{4\pi^2 c}\cdot\frac{\mathbf{b}^2-(\mathbf{b}\cdot\mathbf{n})^2}{(1-\mathbf{b}\cdot\mathbf{n})^2} \tag{47}$$

and

$$\frac{d^2I^{(2)}(\omega,\Omega)}{d\omega d\Omega} = \frac{e^2\omega^2}{4\pi^2 c^3}\cdot\frac{\mathbf{b}^2-(\mathbf{b}\cdot\mathbf{n})^2}{(1-\mathbf{b}\cdot\mathbf{n})^2}\cdot\sum_{i>j=1}^{N_f}(\mathbf{n}\cdot\mathbf{R}_{ij}(0))^2 \quad , \tag{48}$$

respectively. After integration over the angular variables Ω we obtain the two following equations

$$\frac{dI^{(0)}(\omega)}{d\omega} = \frac{e^2N_f^2}{\pi c}\Bigl[\frac{1}{\beta}\ln(\frac{1+\beta}{1-\beta}) - 2\Bigr] \tag{49}$$

and

$$\frac{dI^{(2)}(\omega)}{d\omega} = \frac{2e^2\omega^2}{15\pi c^3}\Bigl[2\beta^2(\sum_{i>j=1}^{N_f} R_{ij}^2(0)) - \sum_{i>j=1}^{N_f}(\mathbf{b}\cdot\mathbf{R}_{ij}(0))^2\Bigr] \quad . \tag{50}$$

The first term $dI^{(0)}(\omega)d\omega$ represents a typical bremsstrahlung spectrum emitted by a group of few (N_f) free electrons. The factor N_f^2 shows that we are dealing with coherent bremsstrahlung. When $N_f = 1$ the result Eq.(49) coincides exactly with the known answer for radiation emitted during β^--decay. The second term ($\sim I^{(2)}$) contains $R_{ij}^2(0)$ and $\mathbf{R}_{ij}(0)$ which represents the electron-electron (ij)-position correlation in the incident atom. All highest order terms $I^{(n)}(\omega,\Omega)$ (where $n \ge 4$) also contain $\mathbf{R}_{ij}(0)$ ($= \mathbf{r}_i(0) - \mathbf{r}_j(0)$) and $R_{ij}^2(0)$. This means that by measuring the bremsstrahlung spectrum experimentally, one can study the electron-electron position correlations in incident atom and ions.

1.5.1 Circular motion of the incident lithium ion

The case considered above corresponds to the motion of incident atom (ion) with a constant velocity v ($v \gg v_e$). After the $(n;t)-$reaction in such an atom a group of N_f electrons is suddenly created and this process is obviously accompanied by the emission of radiation. In general, the considered $(n;t)-$reaction can also be observed in the presence of magnetic (or electric) fields. If the field strength, e.g. **H**, is sufficient to accelerate the incident heavy ions, then the same field will produce a significant acceleration of the final (i.e. free) electrons. This means that the radiation emission will rapidly increase with the field strength. In general, by varying the field strength one can increase the intensity of such radiation to arbitrarily high values.

Below, we shall consider the presence of magnetic field only. Let us assume that the ^6Li ions have been accelerated to the energies 1 - 3 GeV per nucleon. Then such ions begin to move in some strong magnetic field with field strength **H**. For simplicity, the vector **H** is assumed to be perpendicular to its velocity **v**. If at this moment the nuclear $(n;t)-$reaction occurs, then the electrons, which remain free after this reaction, will be accelerated by this magnetic field. The electron acceleration means the emission of magnetic bremsstrahlung or synchrotron radiation [8]. But in contrast with radiation emitted in the electron synchrotrons, such a radiation has some specific properties. The main difference follows from the fact, that in the present case the energy losses for moving electrons are not compensating, e.g. by some electric field (in contrast with electron synchrotrons). This means that these electrons will lose energy (due to radiation), slow down and finally come to a complete stop. In other words, the electron trajectory will be changed in the process. In this section we want to discuss the properties of emitted radiation.

For simplicity, we shall assume that the magnetic field strength **H** is a constant upon time and spatial coordinates. In such a constant magnetic field the electron with initial velocity **v** will move along the circular trajectory. Its radius of curvature is given by

$$r = (\gamma - 1)\frac{mcv}{eH} = \beta\frac{\mathcal{E}}{eHc} \quad , \tag{51}$$

where $H = \mid \mathbf{H} \mid$, while the basic frequency ω_H takes the form:

$$\omega_H = \frac{eH}{(\gamma - 1)mc} = \frac{eHc}{\mathcal{E}} \tag{52}$$

In these equations $m = m_e$ is the electron mass, v is its velocity, $\beta = \mid \mathbf{b} \mid = \frac{v}{c}$, $\gamma = \frac{1}{\sqrt{1-\beta^2}}$ and $\mathcal{E} = (\gamma - 1)mc^2$ is the kinetic energy of the particle. If $\gamma = 1$, then from these equations one finds $\mathcal{E} = 0$ and $r = 0$ (the case of ω_H is considered below).

The total radiated power $P(t)$ can be found from the relation:

$$P(t) = \gamma^2\beta^2\frac{2e^4H^2}{3m^2c^3} \tag{53}$$

where (and below) t is the charge's own time. The energy balance is given by the equation:

$$\frac{d\mathcal{E}}{dt} = -P(t) = -\gamma^2\beta^2\frac{2e^4H^2}{3m^2c^3} = -(\gamma^2 - 1)\frac{2e^4H^2}{3m^2c^3} \tag{54}$$

or, in other words:

$$\frac{d\gamma}{dt} = -(\gamma^2 - 1)\frac{2e^4H^2}{3m^3c^5} = -(\gamma^2 - 1)b \tag{55}$$

where $b = \frac{2e^4H^2}{3m^3c^5}$ is a constant for the constant field strength **H**. The last equation can easily be solved

$$\gamma = \frac{\gamma_0 \cosh(bt) + \sinh(bt)}{\gamma_0 \sinh(bt) + \cosh(bt)} \quad , \tag{56}$$

where the initial condition is obvious: $\gamma(t=0) = \gamma_0$. For the velocity $v(t)$ one finds:

$$v(t) = \frac{cv_0}{c\cosh(bt) + \sqrt{c^2 - v_0^2}\sinh(bt)} \quad , \tag{57}$$

and for the radius $r(t)$

$$r(t) = \frac{mc\gamma_0 v_0}{eH\Big[\gamma_0 \sinh(bt) + \cosh(bt)\Big]} \quad , \tag{58}$$

where the velocity v_0 is the initial ion (or atom) velocity before the $(n;t)-$reaction. From this formula one immediately finds that at large t the total radiated power $P(t)$ decreases exponentially, i.e. $P(t) \simeq \exp(-2bt)$. Note also that in the considered case the relation between the differentials $d\omega_H$ and $d\gamma$ is given by the following equation

$$\frac{d\omega_H(t)}{\omega_H(t)} = -\frac{d\gamma(t)}{\gamma(t) - 1} \quad . \tag{59}$$

From this equation one finds that at large t

$$\omega_H(t) = \omega_H(0)(\gamma_0 - 1)\exp(-bt) \quad , \tag{60}$$

i.e. the basic frequency also decreases exponentially (at large t). The computation of the radiation spectrum in this case is straightforward. Some other properties of the considered 'decaying' synchrotron radiation can be found in [6].

In fact, the formula for the total radiated power presented above corresponds to the one-electron case, i.e. $N_f = 1$. In the case of a few electron systems one has to multiply this expression by the factor $N_f\Big[1 + \frac{2}{N_f}\sum_{i>j=1}^{N_f}\cos(\frac{\omega}{c}\mathbf{n}\cdot(\mathbf{r}_i(0) - \mathbf{r}_j(0)))\Big]$ mentioned above. This factor includes the differences between electron position vectors $\mathbf{r}_i(0) - \mathbf{r}_j(0)$ $(i \neq j = 1, 2, \ldots, N_f)$. Therefore, by studying the emitted radiation one can obtain some information about the electron-electron position correlations in incident ions.

1.6 Conclusion

In this study an attempt is made to develop the theory of bremsstrahlung emitted during the $(n;t)-$nuclear reactions in light atoms and ions. A few realistic models have been developed for describing the properties of the emitted radiation. In particular, it is shown that radiation emitted during such reactions in atoms and ions has some unique properties which cannot be detected in ordinary, single-nucleus Coulomb bremsstrahlung. The spectrum of the emitted radiation and its asymptotics at large and small ω are discussed in detail. The use of quantum-mechanical approach allows us to predict many important properties of the emitted radiation which cannot be analyzed in terms of the classical approach. In particular, the angular distribution of dipole radiation probability $P(\mathbf{n}_i)$ has been obtained and discussed quite thoroughly. A detail investigation of properties of emitted bremsstrahlung opens a new avenue for studying of atomic excitations in the $(n;t)-$reactions in incident atoms and ions.

In fact, the emitted radiation can significantly be amplified, if the $(n;t)-$reaction proceeds in rapidly moving atoms and ions. A sudden creation of rapidly moving, free electrons is a source of radiation. In general, the intensity of bremsstrahlung gradually increases with the velocity of incident atomic system. Moreover, the total radiation output can be increased even further, if the $(n;t)-$reaction proceeds in the presence of strong EM-fields. In particular, the case of a strong magnetic field is discussed in details. The radiation loss ($\sim \mathbf{H}^2$) slows down the electrons, which finally come to a complete stop. By studying the properties of emitted radiation one can obtain a useful information about the electron-electron position correlations in incident atoms (ions).

As mentioned above the $(n;t)-$reactions are used in a large number of thermonuclear applications. In addition to this, the $(n;t)-$reactions can be used in many other applications, including the cancer treatment. Currently, the radiation cancer treatment is based on the use of high-energy $X-$rays ($E_f \approx 10 - 15$ MeV) and beams of protons ($E_p \approx 100 - 250$ MeV). However, when such high-energy beams pass though the human body, they interact with tissues and deposit radiation dose along the way [16]. This causes a lethal damage to a significant number of normal cells. In the new procedure, a solution containing the ^6Li atoms is proposed to be injected directly into the tumor. At the second stage a few beams of slow neutrons must be used. The reactions of the lithium-6 nuclei with slow neutrons inside of the cancer cells will kill such cells instantly. In general, the low intensity beam of slow neutrons $E_n \leq 3$ eV is almost harmless to human body [17]. However, a principal problem is related to the tritium nuclei which will remain in the dead cells after $(n;t)-$reaction. In this sense, the exothermic $(n;p)-$ and $(n;\alpha)-$reactions have the advantage of being free of tritium at the final stage [18], [19]. Briefly, we can say that the new approach seems to be quite promising in medical applications, but there are some questions about its overall safety. In any case, this approach warrants further theoretical study. This example illustrates the importance of the $(n;t)-$reactions in some applications which are not directly related to the nuclear fusion (or tritium production).

1.7 Acknowledgements

It is pleasure to thank the Natural Sciences and Engineering Research Council of Canada for financial support.

References

[1] Highly compressed ^6LiD ($\rho \geq 150\ g \cdot cm^{-3}$) can be ignited, e.g. by a strong shock wave in the presence of intense neutron and thermal fluxes (see, e.g., [20]). Thermal energy released during the complete thermonuclear combustion of 1 kg of ^6LiD is equivalent to the explosion of $\approx$ 23.9 kt (i.e. 23,900,000 kg) of TNT.

[2] G.S. Fraley, E.J. Linnebur, R.J. Mason and R.L. Morse, *Phys. Fluids*, **17**, 474 (1974).

[3] By incident atomic state we mean a state in the original atomic (or ionic) system which forms immediately after the completion of nuclear reaction.

[4] E.R. Cohen and B.N. Taylor, *Physics Today*, **53**(8), 9 (2000).

[5] A.I. Akhiezer and V.B. Berestetskii, *Quantum Electrodynamics*, (Interscience, New York, (1965)).

[6] A.M. Frolov, Nuclear Instruments and Methods in Physics Research B: *Beam Interactions with Materials and Atoms*, **155**, 238 (1999).

[7] A.B. Migdal and V. Krainov, *Approximation Methods in Quantum Mechanics*, (Benjamin, New York, 1969).

[8] L.D. Landau and E.M. Lifshitz, The Classical Theory of Fields, 4th edn, (Pergamon Press, New York, (1975)).

[9] J.D. Jackson, *Classical Electrodynamics*, 2nd edn, (J.Wiley and Sons, New York, (1975)).

[10] V.B. Berestetskii, E.M. Lifshitz and L.P. Pitaevskii, *Quantum Electrodynamics*, (Interscience, New York, (1965)).

[11] R. Loudon, *Quantum Theory of Light*, (Clarendon Press, Oxford, (1983)).

[12] H.A. Bethe and L. Maximon, *Phys. Rev.* **93**, 768 (1954).

[13] H.A. Bethe and E.E. Salpeter, *Quantum Mechanics of One- and Two-Electron Atoms*, (Springer-Verlag, Berlin, (1957)).

[14] A.R. Edmonds, *Angular Momentum in Quantum Mechanics*, (Princeton University Press, Princeton, NJ (1974)).

[15] B.M. Barker and R.F. O'Connel, *Can. J. Phys.* **58**, 1659 (1980).

[16] A.L. Boyer, M. Goitein, A.J. Lomax and E.S. Pedroni, *Phys. Today* **55**, 38 (2002).

[17] In the case of slow neutrons the total radiation dose obtained by a human body can be represented as a sum of three terms, which correspond to the two $(n, \gamma)-$reactions (with the $^1H-$ and $^{14}N-$nuclei, respectively) and $(n,p)-$reaction (with the $^{14}N-$nuclei). All other reactions, including the neutron scattering contribute less than 15 %. The situation changes drastically in the case of fast neutrons.

[18] G. Locher, *Am. J. Roent. Radium Ter.* **36**, 1 (1936).

[19] B.F. Bayanov et al., *Nucl. Instr. and Meth. in Phys. Res.* A **413**, 397 (1998).

[20] A.M. Frolov, V.H. Smith, Jr. and G.T. Smith, *Can. J. Phys.* **80**, 43 (2002).

In: Frontiers in Quantum Physics Research ISBN 1-59454-002-2
Editor: F. Columbus and V. Krasnoholovets, pp. 185-203

Absorptive Photon Switching by Quantum Interference

Min Yan, Edward G. Rickey and Yifu Zhu

Department of Physics

Florida International University

Miami, Florida 33199

Abstract

Quantum coherence and interference manifested by electromagnetically induced transparency suppress the linear susceptibilities and enhance the third-order susceptibilities in a four-level atomic system. Under appropriate conditions, the atomic system absorbs two photons, but not one photon. This feature may be used to realize an optical switch in which a laser pulse controls absorption of another laser pulse at different frequencies. In the ideal limit, the switch may operate at single photon levels.

1.0 Introduction

An absorbing medium may be rendered transparent by a properly coupled laser field. This phenomenon has been referred to as electromagnetically induced transparency (EIT) [1-2]. Linear and nonlinear optical properties of an atomic medium may be modified and controlled by manipulating coherence and quantum interference in a variety of ways based on electromagnetically induced transparency [1-9]. In particular, because the vanishing linear absorption and large enhancement of nonlinear polarizability in an EIT medium, nonlinear optics may be studied at low light levels [10]. Recently, Harris and Yamamoto described a four-level EIT atomic system that exhibits greatly enhanced third-order susceptibility but has vanishing linear susceptibility [11]. In dispersive response, the EIT system exhibits giant Kerr nonlinearity that may be used to obtain a large phase shift by the cross-phase modulation at very weak pump intensities [12]. In absorptive response, the EIT system can be made to absorb two photons, but not one photon. This feature may be used to realize an optical switch operating down to single photon levels in which a laser pulse controls absorption of another laser field at different frequency [10]. Here we briefly discuss some of the theoretical calculations for the linear and nonlinear susceptibilities of the four-level EIT system and present the experimental studies of large enhancement of nonlinear absorption and absorptive photon switching in the four-level EIT system realized with cold rubidium atoms confined in a magneto-optical trap [13-14].

The four-level EIT system is depicted in Fig. 1. A coupling field drives the transition |2>-|3> and a probe field driving the transition |1>-|3>, which creates the standard Λ-type configuration for EIT [1]. A switching field drives the transition |2>-|4> with Rabi frequency 2Ω. The Hamiltonian describing the four-level system is given by (set $\hbar = 1$)

$$\wp = \sum_{j=1}^{4} \varepsilon_j |j\rangle\langle j| - \left(\Omega_c e^{-i\theta_c} |3\rangle\langle 2| + h.c.\right) - \left(\Omega_p e^{-i\theta_p} |3\rangle\langle 1| + h.c.\right)$$
$$- \left(\Omega e^{-i\theta} |4\rangle\langle 2| + h.c.\right) \qquad (1)$$

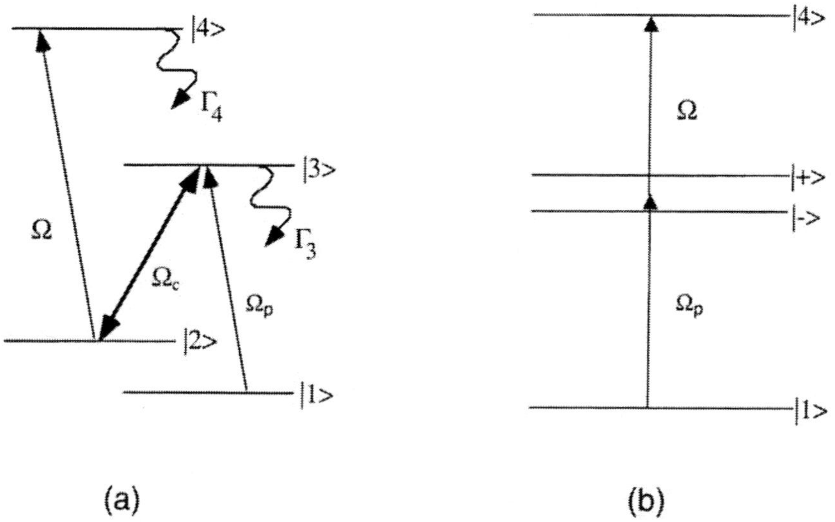

Fig. 1 (a) Four-level atomic system and laser coupling scheme. Γ_3 (5.3 MHz) and Γ_4 (5.9 MHz) are the spontaneous decay rates. (b) Dressed state picture showing the quantum interference between the two excitation paths for the two-photon excitation.

where $\theta_j = \omega_j t - \vec{k}_j \cdot \vec{r}$, and $\Omega_j (j = 1,2,c,p,m)$ are one-half of the Rabi frequencies for the respective transitions ($\Omega_c = D_{23}E(\omega_c)/(2\hbar)$, $\Omega_p = D_{13}E(\omega_p)/(2\hbar)$, $\Omega = D_{24}E(\omega)/(2\hbar)$, with D_{ln} denoting the dipole moment for the transition between levels $|\ell\rangle$ and $|n\rangle$), and ε_j is the energy of the atomic state $|j\rangle$ (taking $\varepsilon_1 \equiv 0$ for the ground state |1>). In the interaction picture, the Hamiltonian can be written as

$$H = -\Delta_c|2\rangle\langle 2| - \Delta_p|3\rangle\langle 3| - \Delta|4\rangle\langle 4| - (\Omega_c e^{i\vec{k}_c\cdot\vec{r}}|3\rangle\langle 2| + \Omega_p e^{i\vec{k}_p\cdot\vec{r}}|3\rangle\langle 1| + \Omega e^{i\vec{k}\cdot\vec{r}}|4\rangle\langle 2| + h.c.) \tag{2}$$

where $\Delta_p = \omega_p - \varepsilon_2$ is the single-photon detuning, $\Delta_c = (\omega_p - \omega_c) - \varepsilon_2$ is the Raman two-photon detuning, and $\Delta = \omega_p - \omega_c + \omega - \varepsilon_4$ is the three-photon detuning respectively.

From Schrödinger equation in the interaction picture, $i\frac{\partial}{\partial t}|\Psi\rangle = H|\Psi\rangle$, and define the atomic state as $|\Psi\rangle = A_1|1\rangle + A_2 e^{i(\vec{k}_p-\vec{k}_c)\cdot\vec{r}}|2\rangle + A_3 e^{i\vec{k}_p\cdot\vec{r}}|3\rangle + A_4 e^{i(\vec{k}_p-\vec{k}_c+\vec{k})\cdot\vec{r}}|4\rangle$, one readily obtains [12]

$$\frac{\partial}{\partial t}A_1 = i\Omega_p^* A_{34} \tag{3a}$$

$$\left(\frac{\partial}{\partial t}+\gamma_2 - i\Delta_c\right)A_2 = i\Omega_c^* A_3 + i\Omega^* A_4 \tag{3b}$$

$$\left(\frac{\partial}{\partial t}+\gamma_3 - i\Delta_p\right)A_3 = i\Omega_p A_1 + i\Omega_c A_2, \tag{3c}$$

$$\left(\frac{\partial}{\partial t}+\gamma_4 - i\Delta\right)A_4 = i\Omega A_2, \tag{3d}$$

where γ_j (j=2-4) is the decay rate. The induced polarization at the probe frequency can be written as $\mathrm{P}(\omega_p) = \varepsilon_0\chi(\omega_p)\mathrm{E}(\omega_p)$ where the susceptibilities $\chi(\omega_p) = \chi^{(1)} + \chi^{(3)}\,|E(\omega)|^2 + \ldots$ $\chi^{(1)}$ represents the linear susceptibility and $\chi^{(3)}$ represents the third-order nonlinear susceptibility. Considering all fields are monochromatic and assuming the usual EIT condition $\Omega_c >> \Omega_p$, which leads to $|A_1| \approx 1$, the steady-state solution to the Schrodinger equation can be readily derived. In the steady state, when both the coupling field and the probe field are on resonance with their respective transitions, EIT suppresses the linear susceptibility $\chi^{(1)}$ and greatly enhances the third-order susceptibility $\chi^{(3)}$ as shown in [11-12]. The susceptibility at the probe frequency is given by

$$\chi(\omega_p) = \frac{-i\mathrm{K}(|\Omega^2| + (\gamma_2 - i\Delta_p)(\gamma_4 - i\Delta_p))}{(\gamma_3 - i\Delta_p)(|\Omega^2| + (\gamma_2 - i\Delta_p)(\gamma_4 - i\Delta_p)) + |\Omega_c^2|(\gamma_4 - i\Delta_p)}. \tag{4}$$

At $\Delta_p=0$, $\chi(\omega_p) = \frac{-i\mathrm{K}(|\Omega^2| + \gamma_2\gamma_4)}{\gamma_3|\Omega^2| + |\Omega_c^2|\gamma_4 + \gamma_2\gamma_4}$, i.e., $\chi(\omega_p)$ is imaginary and purely absorptive. The term proportional to γ_2 represents the residual absorption due to decay of the ground state coherence. The term proportional to $|\Omega|^2$ contributes to the third-order nonlinear absorption. In alkaline atoms, the conditions $\Omega_c<\gamma_3$, $\gamma_{3(4)}>>|\Omega|$ and $|\Omega|^2>>\gamma_2\gamma_4$ can be met because of a very small γ_2, and the EIT system then exhibits the greatly enhanced nonlinear absorption with vanishing linear absorption loss. When all laser fields are on resonance with their respective transitions, we derive the population probability P_3 (P_4) in the excited state ||3> (|4>) as follows

$$P_3 = \frac{|\Omega|_p^2 (|\Omega|^2 + \gamma_2\gamma_4)^2}{(\gamma_3 |\Omega^2| + |\Omega_c^2| \gamma_4 + \gamma_2\gamma_3\gamma_4)^2}, \tag{5}$$

$$P_4 = \frac{|\Omega|_p^2 |\Omega_c^2| |\Omega|^2}{(\gamma_3 |\Omega^2| + |\Omega_c^2| \gamma_4 + \gamma_2\gamma_3\gamma_4)^2}. \tag{6}$$

Eqs. (4) – (6) show that when γ_2 is negligible, the probe field experiences absorption only when the switching field is present. In alkaline atoms, γ_2 is usually negligibly small and can be neglected. Under this condition, the absorption process in the four-level EIT system is intrinsically nonlinear. However, there is population in the excited state |3>, indicating that there is no one to one correlation between absorption of the probe photons and the switching photons. In this regard, P_3 may be viewed as the measure of the linear absorption while P_4 is the measure of the pure three-photon nonlinear absorption [15]. The population ratio, $P_3 / P_4 = (|\Omega^2| + \gamma_2\gamma_4)/|\Omega_c^2|$. If $|\Omega_c|^2>>|\Omega^2|+\gamma_2\gamma_4$, the ratio approaches zero, and the four-level EIT system exhibits the enhanced nonlinear absorption with near perfect correlation of the absorption of the probe photons and the switching photons [11,15]. Under such ideal conditions, however, the nonlinear absorption rate is small and it requires a dense EIT medium in order to achieve the extinction ratio of a practical optical switch. Thus, for certain practical applications that requires a large absorption rate, the four-level EIT may be operated in a regime that maximizes the probe absorption amplitude but lack the ideal correlation between the probe laser and the switching laser.

Fig. 1 plots Im ($\chi(\omega_p)$) versus Δ_p and Ω_c (Fig. 1a), and versus Δ_p and Ω (Fig. 1b) for the four-level EIT system. The calculations are obtained by numerically solving the density matrix equations of the four-level system and are consistent with the analytical results of Eq. (4). Fig. 1 shows that for the weak pump and probe fields, the probe absorption is maximized at a weak coupling field. In particular, if $\gamma_3 \approx \gamma_4$ and $|\Omega_c| \approx |\Omega|$ $\gg \sqrt{\gamma_2 \gamma_4}$, then $\mathrm{Im}(\chi(\omega_p))=-K/(2\gamma_3)$. This value should be compared with the resonant linear absorption in a simple two-level system. In a simple two-level system (states |1> and |3> only) coupled by a weak field with a Rabi frequency Ω_p, the linear absorption amplitude at the resonance is given by Im ($\chi(\omega_p)$)$=-K/\gamma_3$. Therefore, the amplitude of the probe absorption in the four-level EIT system may approach 50% of the amplitude of the linear absorption in an isolated two-level system [10]. Thus, the four-level system exhibits unusually large, controllable absorption amplitude and provides an ideal system to demonstrate that EIT enhancement leads to the possibility of studying nonlinear optical phenomena at low light intensities.

Since the nonlinear absorption coefficient, $\alpha = \dfrac{2\pi}{\lambda}\mathrm{Im}(\chi)$, is proportional to $|\Omega|^2$, the probe absorption can be turned on and off by the switching field, thus, realizing absorptive photon switching. In the adiabatic limit, assuming the intensity of the input probe field is $I_{in}(t)$, then the intensity of the output probe field is given by $I_{out}(t) = I_{in}(t)e^{\alpha(t)l}$ (l is the absorption length). The switching time is determined by the reciprocal of the EIT width, which is equal to $\Gamma = \sqrt{\gamma_3^2 + 4\Omega_c^2} - \gamma_3$, and therefore, can be actively controlled by the coupling field [10]. In the ideal case, it can be shown that switching of medium from transparent to opaque only requires the pump energy density of a single photon focused to a spot size of a half wavelength [11].

Our experiments are done in a vapor cell MOT produced in the center of a 10-ports, 4-1/2" diameter, stainless-steel vacuum chamber pumped down to a pressure ~ 10^{-9}

torr. The MOT is produced with the standard six-beam configuration [16]. The rubidium vapor pressure of ~ 10^{-8} torr is maintained with three rubidium getters connected in series and placed close to the center of the vacuum chamber. An extended-cavity diode laser with output power of ~40 mW is used as the cooling and trapping laser that supplies six σ^+ and σ^- polarized beams in three perpendicular spatial directions and its frequency is locked to ~ $3\Gamma_4$ below the D_2 F=2→F'=3 transition. Another extended cavity diode laser with output power of ~15 mW is used as the repump laser and its frequency is locked to the D_2 F=1→F'=2 transition. The diameter of the trapping laser beams and the repumping laser beam is ~ 1.5 cm. The trapped ^{87}Rb atom cloud is ~ 2 mm in diameter and contains ~ 10^8 atoms. The simplified experimental scheme is depicted in Fig. 3. The weak probe field driving the D_1 F=1-F'=1 transition at 795 nm is provided by a third extended-cavity diode laser with a beam diameter ~ 0.8 mm and power attenuated to ~ 1 μW. The coupling field driving the D_1 F=2-F'=1 transition at 795 nm is provided by a fourth extended-cavity diode laser with a beam diameter ~ 2 mm and output power ~ 20 mW. A Ti:Sapphire laser (Coherent 899-21) with a beam diameter ~ 3 mm is used as the switching laser and couples the D_2 F=2-F'=3 transition at 780 nm. The probe laser and the switching laser are linearly polarized parallel with each other and perpendicular to the linearly polarized coupling laser. The induced transitions among the magnetic sub-levels by the three lasers can be grouped together according to the selection rules and form a manifold of four-level systems. To a good approximation, the coupled Rb system can be viewed as equivalent to a generic four-level system treated in Ref. [11]. The validity of such a simplification has been supported by several previous EIT-type studies in alkaline atoms [17-19]. The laser intensities are varied by neutral density filters. The linewidth of all lasers is ~ 1 MHz. The probe laser, the coupling laser, and the switching laser are overlapped with the trapped Rb cloud and the transmitted light of the probe laser is recorded by a fast photodiode. The propagating directions of the three lasers are mutually perpendicular among each other. This non-collinear excitation scheme ensures that the effect of the nonlinear wave mixing is minimized since the phase matching conditions are not met. For example, for the four-wave mixing process discussed in ref. [10], the energy and the phase matching conditions are $\omega_S = \omega_p + \omega - \omega_c$ and $\kappa_S = \kappa_p + \kappa - \kappa_c$ respectively. It is easy to show that for our non-collinear excitation scheme, the two conditions cannot be met simultaneously.

The experiment is run in a sequential mode with a repetition rate of 10 Hz and the time sequence is shown in Fig. 2. All lasers, except the weak probe laser, are turned on and off by Acousto-Optic Modulators (AOM) according to the time sequence described below. For each period of t_5-t_1=100 ms, t_5-t_4=99.4 ms is used for cooling and trapping of the Rb atoms during which the trap laser and the repump laser are turned on by two separate AOMs and the coupling laser and the switching laser are off. Although the weak probe laser is continuously on, it does not disturb the MOT. The time for the probe transmission measurements lasts ~ t_4-t_1=0.6 ms during which the trap laser and the repump laser are turned off and the coupling laser with its frequency fixed on the D_1 F=2→F'=1 transition is turned on by a third AOM. For the measurement of the probe absorption spectrum, the switching laser with its frequency fixed on the D_2 F=2→F'=3 transition is turned on by a fourth AOM at the same time t_1 as the coupling laser, and the pulse durations of the coupling laser and the switching laser are equal. After a 0.1 ms delay, which ensures that most of the Rb atoms are optically pumped to the F=1 ground hyperfine state, the probe laser frequency is scanned across the D_1 F=1→F'=1 transition

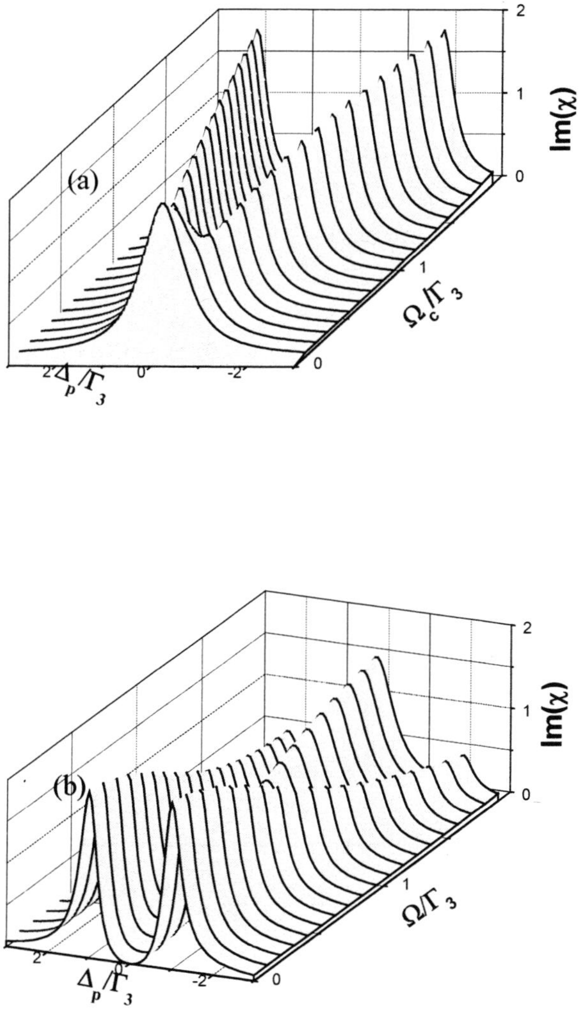

Fig. 2 (a) Calculated imaginary part of the susceptibility, Im(χ), versus the probe detuning Δ/Γ_3 and the coupling laser Rabi frequency Ω_c/Γ_3. The numerical calculations are obtained with Γ_4=1.1Γ_4, and Ω_p=0.1Γ_3 and Ω=0.5Γ_3 respectively. (b) Calculated imaginary part of the susceptibility, Im(χ), versus the probe detuning Δ/Γ_3 and the switching laser Rabi frequency Ω/Γ_3. The numerical calculations are obtained with Γ_4=1.1Γ_4, and g=0.1Γ_3 and Ω_c=Γ_3 respectively.

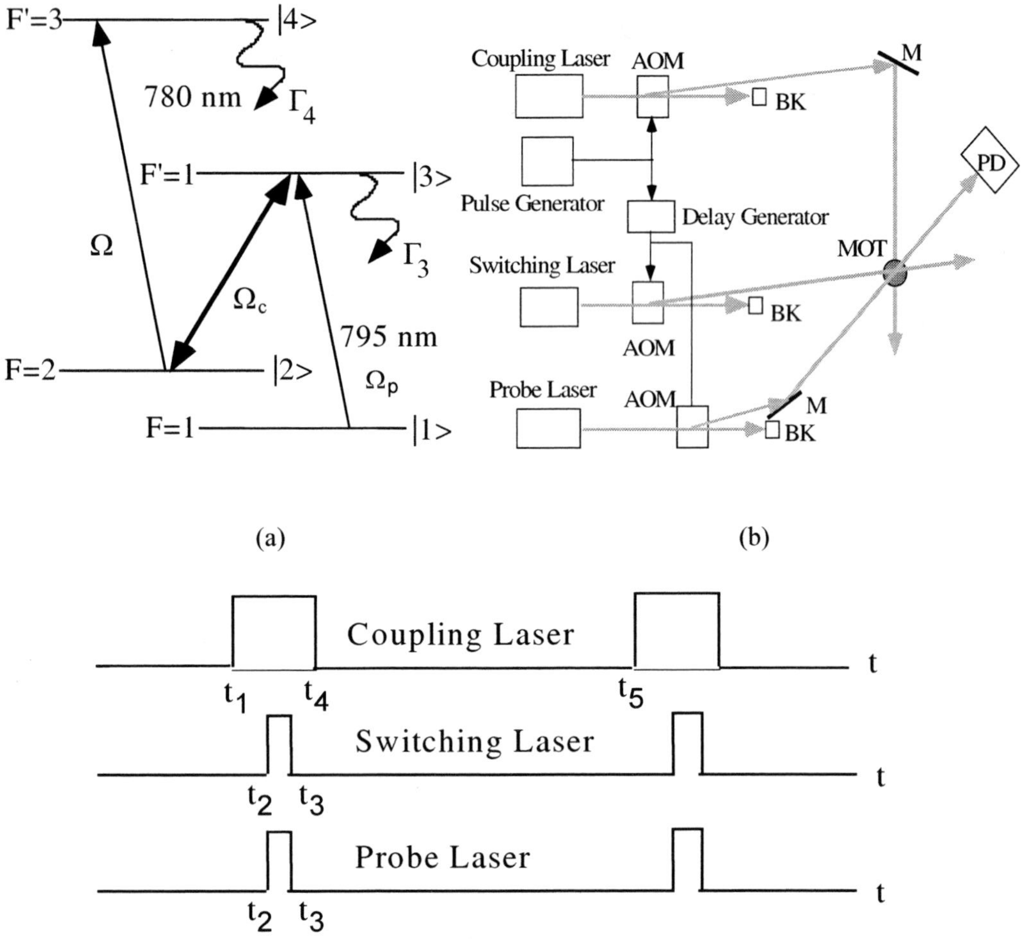

(a) (b)

(c)

Fig. 3 (a) Laser coupling scheme of the ^{87}Rb system. (b) Schematic diagram of the experimental set up. (c) Experimental time sequence. The experiment is running at a repetition rate of 10 Hz. For each period, the two-photon switching experiment runs for t_4-t_1=0.6 ms, and the cooling and trapping lasts for t_5-t_4=99.4 ms.

in ~0.1 ms and the probe transmission versus its frequency is recorded. For the photon switching demonstration, the switching laser is turned on 0.1 ms after the coupling laser pulse (see Fig. 3) and its pulse duration t_3-t_2 can be varied from 1 μs to 200 μs. The frequency of the weak probe laser is fixed on the D_1 F=1→F'=1 transition and the probe transmission versus time is recorded.

Fig. 4 shows the measured transmission spectrum of the weak probe laser versus its frequency detuning Δ. The experimental data are plotted in solid lines while the dotted lines are numerical calculations of the four-level system depicted in Fig. 1. Fig. 4a shows the measured probe spectrum when the switching laser is on. The data are recorded under the condition of $\Delta'=\Delta_c=0$. The central peak (at $\Delta=0$) in Fig. 4a corresponds to the EIT enhanced probe absorption while the two side peaks at $\Delta\sim\pm\Omega_c/2$ represent the linear absorption of the dressed states. For comparison, Fig. 4b shows the measured probe spectrum under identical experimental conditions but without the switching laser. The absorption peak at $\Delta=0$ disappears. It represents the EIT spectrum observed in the Λ-type Rb system. The coherence dephasing rate γ_{21}, due to the atom collisions and the off-resonant laser excitation, is estimated to be ~ 10^4 s^{-1}, which is negligibly small under our experimental conditions. Small residual absorption (~2%) at $\Delta=0$ in the EIT spectrum is likely due to the finite laser linewidths. From the EIT spectrum, the Rabi frequency of the coupling laser is deduced to be $\Omega_c\approx9$ MHz The Rabi frequencies of the switching laser and probe laser are estimated from the laser powers and beam diameters, which are $\Omega\approx15$ MHz, $\Omega_p\approx1.2$ MHz respectively. These parameters are then used in the numerical calculations shown by the dotted lines in Fig. 4. Since the laser pulse durations are much greater than the atomic decay times $1/\Gamma_3$ (30 ns) and $1/\Gamma_4$ (27 ns), the measurements are carried out essentially in the steady-state regime and the steady-state analysis is validated. Fig. 4a shows that the observed single-pass, probe absorption at $\Delta=0$ is ~ 30%. From the

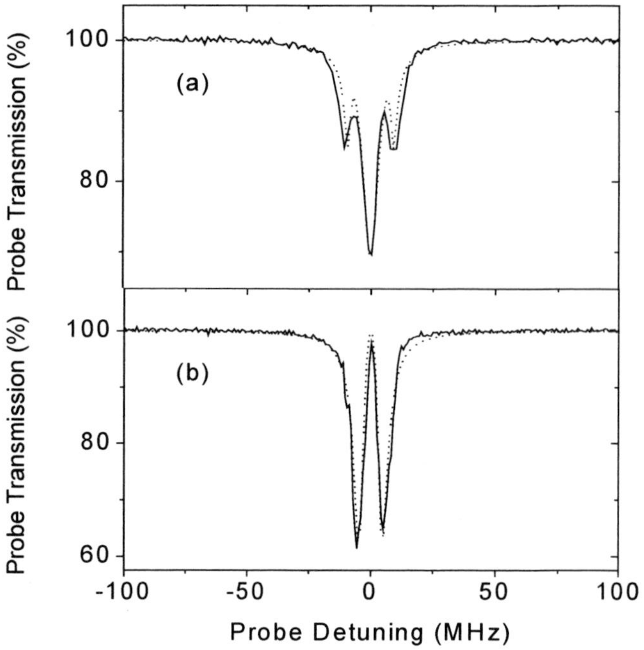

Fig. 4. (a) Measured probe transmission versus the probe frequency detuning Δ while the coupling laser and the switching laser are on resonance, i.e. $\Delta_C = \Delta' = 0$. The central peak corresponds to the EIT enhanced absorption and the two side peaks represent the Autler-Townes' doublet transitions. For comparison, the measured EIT spectrum of the probe transmission versus the probe detuning Δ is plotted in (b), which was taken without the switching laser. The solid lines represent the experimental data while the dotted lines are the theoretical calculations.

measured percentage absorption of the two peaks in the EIT spectrum of Fig. 4b, we deduce that the linear probe absorption at the center of the D_1 F=1→F'=1 transition without the coupling laser is ~ 70%. So the EIT enhanced nonlinear absorption is ~ 42% of the linear absorption in a simple two-level system, which agrees with the theoretical calculation. Under our experimental conditions, the calculation shows that the 50% maximum value is reached at $\Omega_c \approx 6$ MHz. However, due to the finite laser linewidths, the residual absorption at the line center becomes appreciable when the Rabi frequency $\Omega_c \leq$ 6 MHz.

The absorptive photon switching is done first with a pulsed switching laser while the probe laser is continuous. The probe absorption is turned on and off by the switching pulse and leads to a pulsed, stepwise probe transmission profile. The bandwidth of the probe absorption at the line center is determined by the EIT width, $\Gamma = \sqrt{\gamma_3^2 + 4\Omega_c^2} - \gamma_3$. Therefore, the switching speed is set by the EIT width [11]. Although a small EIT width is necessary for obtaining the maximum probe absorption, it may compromise the optical switching time. Under our experimental conditions, a near complete EIT requires Ω_c to be ≥ 9 MHz. Therefore, the optimal Rabi frequency under our experimental conditions that does not limit the switching time, yet still leads to a large probe absorption, is $\Omega_c \sim 2\Gamma_3$. As Ω_c decreases, the decay (fall) time of the probe absorption increases rapidly after the switching laser is turned off. The numerical calculation shows that for $\Omega_c = 2\Gamma_3 = 10.6$ MHz, the decay time associated with the EIT width is $\sim 1/(2\pi\Gamma_3) \approx 30$ ns; for $\Omega_c = \Gamma_3 = 5.3$ MHz, the decay time increases to $\sim 2.5/(2\pi\Gamma_3) \approx 70$ ns. However, the rise and fall time of the switching laser pulse in our experiments is ~ 150 ns (limited by the turn-on time of the AOM). Thus, the switching time of the probe absorption essentially follows the rise and fall time of the switching pulse. Fig. 5a plots the calculated probe absorption (Im (ρ_{13})) versus time for a pair of Ω_c

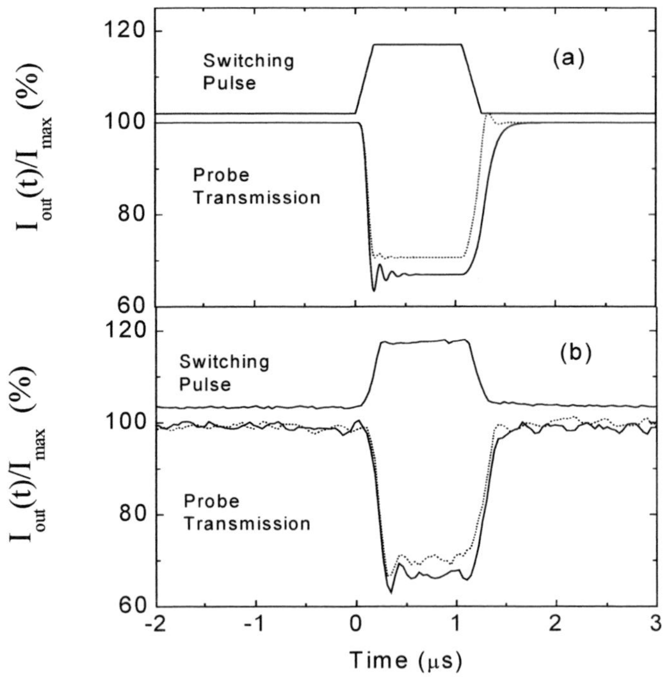

Fig. 5. (a) Calculated $I_{out}(t)/I_{max}$ of a cw probe laser versus time when a switching pulse simulated from the experiment (top curve) is turned on at $t=t_2=0$. I_{max} is the amplitude of the input probe laser. The rise time of the switching pulse is 150 ns. The relevant parameters are $\Omega=3\Gamma_3$, $\Omega_g=0.2\Gamma_3$, and $\Omega_c=\Gamma_3$ (solid line) and $2\Gamma_3$ (dotted line) respectively. (b) The measured switching pulse (top curve in arbitrary units) and the probe transmission (lower curves) versus time. $\Omega_c\approx 10$ (5) MHz for the dotted (solid) curve.

values with a simulated switching pulse derived from the experiment. When $\Omega_c=2\Gamma_3$ (dashed curve), the switching time of the probe absorption follows the rise and fall time of the switching pulse; when $\Omega_c=\Gamma_3$ (solid curve), the decay time of the probe absorption becomes larger due to the smaller EIT width. The experimental data for the photon switching demonstration are plotted in Fig. 5b. The switching pulse with 1 μs duration (the upper curve) is turned on at t=0 (0.1 ms after the coupling laser is turned on). The probe transmission versus time (the two lower curves) shows that the probe absorption is switched on and off by the switching pulse and the peak absorption is about 30%. The observed switching time of the probe absorption in Fig. 5b follows the rise and fall time of the switching laser pulse for $\Omega_c\approx10$ MHz (dashed line). At $\Omega_c\approx5$ MHz (solid line), the decay time of the probe absorption becomes longer due to the smaller EIT width, in agreement with the theoretical calculations in Fig. 5a. We have varied the pulse duration of the switching laser from 1 μs to 200 μs and observed that the time duration of the probe absorption follows exactly that of the switching laser pulse.

Fig. 6 shows the absorptive switching of a pulsed probe laser by a pulsed switching laser. The two pulses have the same duration (~1 μs) and overlap with each other in time as shown in Fig. 2(c). Fig. 6(a) plots the theoretical calculations of the normalized output intensity of the probe pulse $I_{out}(t)$. The rise (fall) time of the switching pulse and the input probe pulse are taken to be 150 ns to match the experimental conditions. The upper dashed line is the output probe pulse when the switching pulse is absent. The two lower traces are the output probe pulses when the switching pulse is on. The Rabi frequency of the coupling laser is $\Omega_c\approx10$ MHz (solid line) and 5 MHz (dot-dashed line). The output pulse of the probe laser shows a transient peak near the switching time t=0 in a time interval about equal to the lifetime of the excited state |3> and then decays to a quasi-steady-state value. The decay time depends on the EIT width and is longer for smaller Ω_c

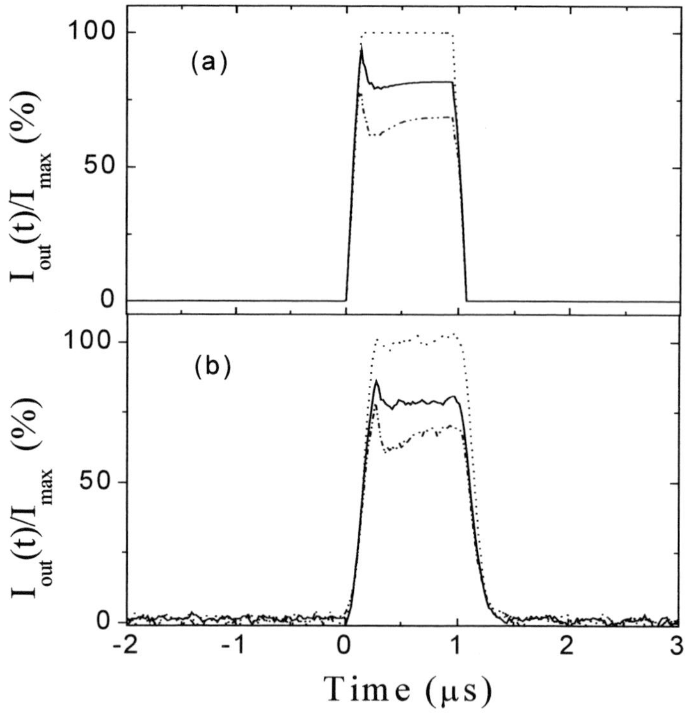

Fig. 6. (a) Calculated $I_{out}(t)/I_{max}$ of a pulsed probe laser versus time. The top dotted curve, $I_{out}(t)= I_{in}(t)$ (no switching pulse). I_{max} is the amplitude of the input probe pulse. The two lower curves show the output probe pulse $I_{out}(t)$ when the switching pulse is turned on at $t=t_2=0$. The relevant parameters are $\Omega=3\Gamma_3$, $\Omega_p=0.2\Gamma_3$, and $\Omega_c=2\Gamma_3$ (solid line) and Γ_3 (dot-dashed line) respectively. (b) Measured output probe pulses, $I_{out}(t)$, versus time under the conditions given in (a).

values. When the switching pulse and the input probe pulse are turned off, the output probe pulse is switched off smoothly and follows the fall time of the input pulse. Fig. 6b presents the measured output probe pulse without the switching laser pulse (the upper dashed line) and with the switching laser pulse (the two lower traces, corresponding to the Ω_c values in Fig. 4a). With the switching pulse on, the measured output pulses show the quick transient rise at the switching time and then exhibit the noticeable slower decay at the smaller Ω_c value due to the reduced EIT width, in agreement with the theoretical calculations. The intensity attenuation of the output probe pulse is about 30% and 25% for the two Ω_c values respectively. We have also varied the pulse duration of the switching laser and the probe laser from 1 μs to ~100 μs and observed that the transmitted probe pulse shows the transient rise and fall at the turn on time and then switches off smoothly with a lineshape similar to that shown in Fig. 4(b).

In the ideal case discussed in Ref. [11], the two-photon switch may operate at an invariant switching energy per area equal to that of a single photon per square wavelength. A single photon in a 1 μs pulse duration and focused to a spot size of a half wavelength (at λ=780 nm) will have an intensity ~ 0.17 mW/cm^2. Our calculation shows that in this regime, the optimized probe absorption corresponds to a coupling Rabi frequency Ω_c~0.2 $\Gamma_3 \approx$1.1 MHz. At such low Ω_c values, the line-center linear absorption is dominant in our experiment due primarily to the limitation of the finite laser linewidths (~ 1 to 2 MHz). In order to suppress the linear absorption and obtain a deep EIT depth, we used large coupling Rabi frequencies ($\Omega_c \geq$5 MHz), under which the probe absorption becomes very small if the switching laser intensity is below the saturation value. For the optimized probe absorption, we then used a moderate switching laser with an intensity ~ 13 mW/cm^2 for the measurements of the probe absorption spectrum and also the absorptive photon switching. Our experiment is done essentially in the quasi-steady state regime and the probe laser intensity, although below the saturation limit, is far above the single photon level. Nevertheless, with a moderate switching laser (mW) and a weak probe laser (μW), the experiment provides an example of resonant nonlinear optics manifested by quantum interference under low light intensities. The novel idea of the photon switching controlled by EIT in a four-level atomic system has been demonstrated.

In conclusion, we have observed the EIT controlled probe absorption and demonstrated the EIT absorptive photon switching in cold ^{87}Rb atoms. The experiment has been carried out in a regime in which the EIT controlled probe absorption is comparable in amplitude to the single-photon absorption in a two-level system, but the lacks the ideal one-to-one correlation between absorptions of the probe photons and the switching photons. The experimental results agree with the theoretical calculations based on the four-level EIT system. Refinement of the experimental set up and tight focus of the probe and switching laser beams may render it possible to study the four-level EIT system and related photon switching near single photon levels. The EIT enhanced third-order susceptibilities should be useful in a variety of applications in quantum optics, quantum measurements, and nonlinear spectroscopy [20-23].

This work is supported by the National Science Foundation, Office of Naval Research, and the U. S. Army Research Office.

References

1. S. E. Harris, *Phys. Today* 50, 36 (1997).
2. E. Arimondo, in *Progress in Optics*, edited by E. Wolf (Elsevier Science, Amsterdam, 1996), p. 257.
3. K. Hakuta, L. Marmet, and B. P. Soicheff, *Phys. Rev. Lett.* **66**, 596 (1991)
4. M. D. Lukin, P. R. Hemmer, M. Loffler, and M. O. Scully, *Phys. Rev. Lett.* **81**, 2675

(1998).

5. G. S. Agarwal and W. Harshawardhan, *Phys. Rev. Lett.* **77**, 1039 (1996).
6. Y. Li and M. Xiao, *Opt. Lett.* **21**, 1064 (1996).
7. M. Jain, et al, *Phys. Rev. Lett.* **75**, 4385 (1995).
8. L. V. Hau, S. E. Harris, Z. Dutton, and C. H. Behroozi, *Nature* (London) **397**, 594 (1997).
7. M. M. Kash et al, *Phys. Rev. Lett.* **82**, 5229 (1999).
9. D. Budker, D. F. Kimball, S. M. Rochester, and V. V. Yashchuk, *Phys. Rev. Lett.* **83**, 1767 (1999)
10. S. E. Harris and L. V. Hau, *Phys. Rev. Lett.* **82**, 4611 (1999)
11. S. E. Harris and Y. Yamamoto, *Phys. Rev. Lett.* **81**, 3611 (1998).
12. H. Schmidt and A. *Imamoglu, Opt. Lett.* **21**, 1936 (1996).
13. M. Yan, E. Rickey, and Y. Zhu, *Opt. Lett.* **26**, 548-550 (2001);
14. M. Yan, E. Rickey, and Y. Zhu, *Phys. Rev.* A **64**, 041801 (2001).
15. W. Xu and J. Gao, *Phys. Rev.* A 67, 033816(2003).
16. K. Lindquist, M. Stephens, and C. Wieman, *Phys. Rev.* A **46**, 4082 (1992).
17. Y. Q. Li and M. Xiao, *Phys. Rev.* A **51**, R2703 (1995).
18. G. G. Padmabandu, et al; *Phys. Rev. Lett.* **76**, 2053 (1996).
19. J. Kitching, and L. Hollberg, *Phys. Rev.* A **59**, 4685 (1999).
20. D. Vitali, M. Fortunato, P. Tombesi, *Phys. Rev. Lett.* **85**, 445 (2000).
21. M. D. Lukin and A. Imamoglu, *Nature* **413**, 273 (2001).
22. J. Clausen, L. Knoll L, D. G. Welsch, J. *Opt. B-Quantum S Opt.* **4** (2), 155(2002).
23. A. Andre and M. D. Lukin, *Phys. Rev. Lett.* **89**, 143602 (2002).

INDEX

N

O

P

Q

R

S

T

U

V